Biology and Ideology from Descartes to Dawkins

Biology and Ideology from Descartes to Dawkins

EDITED BY DENIS R. ALEXANDER
AND RONALD L. NUMBERS

THE UNIVERSITY OF CHICAGO PRESS   CHICAGO AND LONDON

SUPPORTED BY A GRANT FROM THE TEMPLETON PUBLISHING SUBSIDY PROGRAM.

Denis R. Alexander is director of The Faraday Institute for Science and Religion, St. Edmund's College, Cambridge, and has worked in the biological research community for the past forty years. Ronald L. Numbers is Hilldale Professor of History of Science and Medicine at the University of Wisconsin–Madison and coeditor of *When Science and Christianity Meet*, also published by the University of Chicago Press.

The University of Chicago Press, Chicago 60637
The University of Chicago Press, Ltd., London
© 2010 by The University of Chicago
All rights reserved. Published 2010
Printed in the United States of America
19 18 17 16 15 14 13 12 11 10   1 2 3 4 5

ISBN-13: 978-0-226-60840-2 (cloth)
ISBN-13: 978-0-226-60841-9 (paper)
ISBN-10: 0-226-60840-9 (cloth)
ISBN-10: 0-226-60841-7 (paper)

Library of Congress Cataloging-in-Publication Data

Biology and ideology from Descartes to Dawkins / edited by Denis R. Alexander and Ronald L. Numbers.
    p. cm.
  Includes bibliographical references and index.
  ISBN-13: 978-0-226-60840-2 (cloth : alk. paper)
  ISBN-10: 0-226-60840-9 (cloth : alk. paper)
  ISBN-13: 978-0-226-60841-9 (pbk. : alk. paper)
  ISBN-10: 0-226-60841-7 (pbk. : alk. paper) 1. Biology—Philosophy. 2. Biology—Religious aspects. 3. Evolution (Biology)—Philosophy. 4. Genetics—Philosophy.
5. Eugenics—Philosophy. I. Alexander, Denis R. II. Numbers, Ronald L.
  QH331.B477 2010
  570.1—dc22

                                                2009034186

♾ The paper used in this publication meets the minimum requirements of the American National Standard for Information Sciences—Permanence of Paper for Printed Library Materials, ANSI Z39.48-1992.

Contents

Introduction 1
    Denis R. Alexander and Ronald L. Numbers

CHAPTER 1. The cultural authority of natural history in early modern Europe 11
    Peter Harrison

CHAPTER 2. Biology, atheism, and politics in eighteenth-century France 36
    Shirley A. Roe

CHAPTER 3. Eighteenth-century uses of vitalism in constructing the human sciences 61
    Peter Hanns Reill

CHAPTER 4. Biology in the service of natural theology: Paley, Darwin, and the *Bridgewater Treatises* 88
    Jonathan R. Topham

CHAPTER 5. Race, empire, and biology before Darwinism 114
    Sujit Sivasundaram

CHAPTER 6. Darwin's choice 139
    Nicolaas Rupke

CHAPTER 7. Biology and the emergence of the Anglo-American eugenics movement 165
Edward J. Larson

CHAPTER 8. Genetics, eugenics, and the Holocaust 192
Paul Weindling

CHAPTER 9. Darwinism, Marxism, and genetics in the Soviet Union 215
Nikolai Krementsov

CHAPTER 10. Evolution and the idea of social Progress 247
Michael Ruse

CHAPTER 11. Beauty and the beast? Conceptualizing sex in evolutionary narratives 276
Erika Lorraine Milam

CHAPTER 12. Creationism, intelligent design, and modern biology 302
Ronald L. Numbers

CHAPTER 13. The ideological uses of evolutionary biology in recent atheist apologetics 329
Alister E. McGrath

Acknowledgments 353
Notes 355
Contributors 429
Index 435

Introduction
Denis R. Alexander and Ronald L. Numbers

Why should we be concerned about biology and ideology? One good reason is that the use of biology for non-biological ends has been the cause of immense human suffering. Biology has been used to justify eugenic programs, enforced sterilization, experimentation on living humans, death camps, and political ambitions based on notions of racial superiority, to name but a few examples. We should also be concerned because biological ideas continue to be used, if not in these specific ways, then in other ways that lie well beyond science. Investigating the past should help us to be more reflective about the science of our own day, hopefully more equipped to discern the ideological abuse of science when it occurs.

Not so many decades ago science represented the antithesis of ideology. Indeed, science rested securely on a pedestal, enshrined as the very "norm of truth." According to the founding father of the history of science, George Sarton (1884–1956), the "main purpose" of science, pursued by disinterested scholars, was "the discovery of truth." Convinced that science was the only human activity that "is obviously and undoubtedly cumulative and progressive," he described the history of science as "the story of a protracted struggle, which will never end, against the inertia of superstition and ignorance, against the liars and hypocrites, and the deceivers and the self-deceived, against all the forces of darkness and nonsense."[1]

By the late nineteenth century, practicing scientists, as well as science educators and popularizers, were increasingly attributing the success of science to something called "the scientific method," a slippery but rhetorically powerful slogan. In the words of the distinguished American astronomer Simon Newcomb, who devoted considerable thought to scientific methodology, "the most marked characteristic of the science of the present day . . . is its entire rejection of all speculation on propositions which do not admit of being brought to the test of experience."[2]

To such devotees, science was not only true but edifying, totally unlike the "grubby worlds" of business and politics. As Harvard president Charles W. Eliot, an erstwhile chemist, declared at the opening of the American Museum of Natural History in 1878, science produced a "searching, open, humble mind . . . having no other end than to learn, prizing above all things accuracy, thoroughness, and candor." Many of its practitioners, asserts the historian David A. Hollinger, saw science "as a religious calling," "a moral enterprise." Those who used science for ideological purposes often found themselves denounced as charlatans and pseudo-scientists.[3]

Until well into the twentieth century neither scientists themselves nor the scholars who studied science linked science with ideology, a term coined in the late eighteenth century and typically employed pejoratively to designate ideas in the use of particular interests. Among the first to connect ideology and science were Karl Marx and his followers, who identified "ideologies" as ideas that served the social interests of the bourgeoisie. Western historians of science first encountered the linkage between science and ideology at the Second International Congress of the History of Science and Technology, held in London in 1931, when a delegation from the Soviet Union contrasted "the relations between science, technology, and economics" under the capitalist and socialist systems. The Russian physicist Boris Hessen, under intense political pressure at home to prove his Marxist orthodoxy, delivered an iconoclastic paper on "The Socio-Economic Roots of Newton's *Principia*," which described Newtonian science in the service of the ideological (that is, industrial and commercial) needs of the rising bourgeoisie. Despite his bravura effort, he died in a Soviet prison five years later, falsely convicted of terrorism.[4]

Such "vulgar Marxism" exerted little influence on the writing of the history of science outside the Soviet Union. It was not until the 1960s that Marxism penetrated Anglo-American historiography, largely through the efforts of Robert M. (Bob) Young, an expatriate Texan working in Cambridge, England. In 1970, at a conference on "The Social Impact of Modern

Biology," he delivered a paper on "Evolutionary Biology and Ideology," in which he "treated science *as* ideology." He acknowledged that the term "ideology" traditionally had derogatory and political connotations that were connected with its popularization by Marx, who concentrated his use of it as a term of abuse for ideas that served as weapons for social interests. But Marxists were soon subjected to their own critique, and this led to Young's general definition of ideology:

> When a particular definition of reality comes to be attached to a concrete power interest, it may be called an ideology. . . . In its early manifestations the concept of ideology conveyed a sense of more or less conscious distortion bordering on deliberate lies. I do not mean to imply this. . . . [T]he effort to absorb the ideological point of view into positive science only illustrates the ubiquitousness of ideology in intellectual life. . . . We need to see that ideology is an inescapable level of discourse.

In contrast to earlier Marxists, who had damned ideology as inimical to good science, Young argued that all facts are theory-laden and that no science is value-free. The late historian Roy Porter described the efforts of Young and his fellow New Marxists as concentrating on "exposing the dazzling conjuring trick whereby science had acquired and legitimated authority precisely while claiming to be value-neutral." Their goal was to liberate humanity from the thrall of science by demoting it from its privileged intellectual position and relocating it on the same level as other belief systems. Thus, at a time when some observers were declaring "the end of ideology," a small group of historians of science was rushing to embrace it.[5]

Meanwhile, scholars of a less radical persuasion were also undermining the notion of science as a value-neutral enterprise. In 1958 the philosopher Norwood Russell Hanson, who would soon found the Indiana University program in the history and philosophy of science, published *Patterns of Discovery*, which described all observations as "theory-laden." Influenced in part by Hanson, the historian of science Thomas Kuhn published his best-selling *The Structure of Scientific Revolutions* (1962), by far the most influential book ever written about the history of science and one of the most important books on any topic published in the twentieth century. In his slight monograph, Kuhn challenged Sarton's cherished notion that science was cumulative, arguing instead that scientific paradigms are incommensurable and therefore that science does not progressively approach a truthful description of nature. Although he insisted that "there is no standard higher than the assent of the relevant community" in determining the boundaries of good science, he shied away from equating science and ideology. In fact, he used the latter term only to dismiss a commitment to the cumulative nature of science as "the ideology of the scientific profession." Some critics denounced Kuhn's work for promoting "irrationality and relativism"—and many postmodernists and other denigrators of science drew inspiration from it in their attempts to undermine the privileged status of science—but Kuhn never joined the revolutionaries. He took pride in the description of *The Structure of Scientific Revolutions* as "a profoundly conservative book."[6]

Outside of radical circles, in the 1970s and 1980s few historians of science paid much attention to ideology. In 1971 Ronald C. Tobey published a study of attempts to generate popular support for science entitled *The American Ideology of National Science*, but for him "ideology" implied little more than a commitment to "traditional social values." Similarly, the British historians Jack Morrell and Arnold Thackray in their history of the early years of the British Association for the Advancement of Science, founded in 1831, credited the "Gentlemen of Science," who founded the association, with articulating "a particular ideology of science," by which they hoped "to consolidate the role of science as the dominant mode of cognition of industrial society":

> The deliberate creation of boundaries between natural and religious or political knowledge, the conceptualization of science as a sharply edged and value-neutral domain of knowledge, the subordination of the biological and social to the physical science, the harnessing of a rhetoric of science, technology, and progress—these were some of the ways in which an ideology of science was constructed.

The most influential blow to the traditional separation between science and ideology came in the 1970s and 1980s from a group of scholars in the Edinburgh University Science Studies Unit dedicated to creating a thoroughgoing sociology of scientific knowledge. Unlike such pioneers in the sociology of science as Robert K. Merton, who explored the impact of social factors on the growth of scientific institutions but left scientific knowledge untainted by ideologies, the Edinburgh scholars advocated a "strong programme" that treated science like any body of knowledge, vulnerable to psychological, social, and cultural factors. These "constructivists" insisted on treating "true" and "false" scientific claims identically and on exploring the role played by "biasing and distorting factors" in both cases, not just for unsuccessful or pseudo-science. Contrary to the claims of some of their critics, they never asserted that science was "purely social" or "that knowledge depended *exclusively* on social variables such as interests." "The strong programme says that the social component is always present and always constitutive of knowledge," explained David Bloor, one of the founders of the Science Studies Unit. "It does not say that it is the *only* component, or that it is the component that must necessarily be located as the trigger of any and every change."[7]

As Bloor later discovered, the "strong programme" in the sociology of knowledge had been anticipated nearly a century and a half earlier by the church historian Ferdinand Christian Baur and his colleagues at the University of Tübingen. In the early nineteenth century, historians of Christian doctrine had distinguished between truth (which they explained supernaturally) and heresy (explained naturally—or diabolically). Baur insisted on using the same historical methods to explain the development of both Christian dogma and heresy, bringing the former within the realm of historical inquiry. The parallels for Bloor were obvious: "In the place of the historical unfolding of divine inspiration we have the unfolding of rational enquiry, the 'internal' history of science. In the place of heresy we have irrationality and the socio-psychologically caused deviations from the true scientific method, the 'external' history of science. Doctrinal error in theology has given way to ideological bias in science."[8]

In the early 1980s a young historian of science at Edinburgh, Steven Shapin, collaborated with Simon Schaffer on a landmark book that dramatically illustrated the applicability of the "strong programme" to the history of science. In *Leviathan and the Air Pump: Hobbes, Boyle, and the Experimental Life*, which the authors described as "an exercise in the sociology of scientific knowledge," Shapin and Schaffer sought to identify the role played by ideology in establishing trust in the experimental way of producing knowledge about the workings of nature. As good constructivists, they treated the views of Thomas Hobbes (the loser) symmetrically with the opinions of Robert Boyle (the winner). In the end they concluded that "scientific activity, the scientist's role, and the scientific community have always been dependent: they exist, are valued, and supported insofar as the state or its various agencies see point in them."[9]

By the 1990s the sometimes acrimonious debate over ideology and science was dying down. Although a few historians of science held out for value-free science, the great majority, it seems, had come to accept a moderate form of constructivism—not so much for ideological reasons but because the evidence supported it. While rejecting the radical claim that science was *merely* social, they readily granted the propriety, indeed the necessity, of exploring the constitutive role of ideologies in the making of science. Ideologies had morphed from antiscience to the heart of the scientific enterprise.

This scholarly consensus did not, however, bring the study of ideology and science to an end or make it less important, most strikingly in the context of biology.[10] Perhaps more than other disciplines, biology has been particularly susceptible to ideological manipulation and application, a trend that shows no sign of abating. The essays in this volume illustrate the many and varied ways in which biology has been utilized for a wide range of political, religious, and social purposes from 1600 to the present day. The purposes may be beneficial, benign, or harmful in their outcomes, but all are "ideological" in the broadest sense of not being intrinsic to biology itself.

But the flow has gone both ways, not only "outwards" from biology into the worlds of politics, philosophy, or social structures, but also "inwards," with whole scientific programs being shaped by ideological concerns, as Chapter 6 in particular suggests. At other times there is more of an iterative process of "co-evolution," as occurred in theories about "racial hygiene" (Chapters 7 and 8), whereby the ideology shaped the biology, which in turn was used to prop up the ideology.

It is not therefore the goal of this volume to attempt a resolution of the vexed question raised in our historiographic discussion as to the precise meaning of the term "ideology" in its relationship to biology, but rather to illustrate the sheer diversity of that relationship over the centuries, with different authors highlighting its various facets, often explicitly, sometimes implicitly. Their essays do suggest that ideology provides an interpretative framework that serves a social purpose, motivated by ethical, religious, or political convictions. The history of biology does certainly evince ideologies as either motivating or as being justified by certain kinds of scientific research and declaration, and most of the contributors investigate episodes in the history of biology in which biological science has become thoroughly entangled with social causes.

The newly emergent cultural authority of biology's precursor, natural history, is described in Chapter 1 by Peter Harrison, who shows how the first systematic investigations of the natural world in the early modern period attracted prestige by their support for natural theology and for the moral order. Even Descartes' idea of animals as machines without souls, invoking thereby a sharp demarcation between human and animal, was employed as part of the argument for design. Yet nature did not always speak with a clear moral voice, and in the eighteenth century the vigorous debate surrounding the spontaneous generation of life was shaped by concerns about atheism and its perceived threat to morality, as Shirley Roe discusses in Chapter 2. Biological ideas connecting life and matter played a central role in the materialistic arguments of the French *philosophes*, which in turn were employed in the subversion of the social order. Yet, as Peter Hanns Reill reviews in Chapter 3, the eighteenth century also saw something of a reaction against the mechanistic analogies that had proven so influential in the natural philosophy of the preceding century, reformulating an "Enlightenment vitalism" that sought to revive ideas of nature as a dynamic system. This renewed emphasis on the internal driving forces and systematic organization of living things was used to generate a new science of humanity, which in turn was deployed to argue for particular economic and political structures. From the structure of organisms to the structure of societies has often been a short step in the history of biology.

One of the striking insights highlighted by this collection of essays is the way in which the ideological application of biological concepts is shaped by place as well as time. In some cases the same biological ideas have been used during the same period for quite opposite ideological purposes in different countries. The biology that in France was utilized by the *philosophes* to subvert the social order was in Britain used as a key resource for natural theology, whereas in Germany it was being used politically as an analogy for the structure of nation states.

Sometimes a synergistic relationship between the biology and its nonbiological extrapolation is particularly apparent. As Jonathan Topham discusses in Chapter 4, natural theology in the first half of the nineteenth century drew heavily upon the latest biology to make its case; in turn the progressive and law-like view of the history of the creation that natural theology promoted paved the way for Charles Darwin's *Origin of Species*. In Chapter 5, Sujit Sivasundaram shows how concepts of race before Darwin were based on contemporary biological understandings of anatomy and physiology, and in turn the racial construction of societies defined new programs of biological investigation, lending itself conveniently to the requirements of colonial subjugation.

It should not be surprising that in this bicentenary of Darwin's birth (1809) and sesquicentenary of *Origin of Species* (1859), this volume should include several essays on the ideological uses and abuses of Darwinism. Would evolutionary theory have developed in the particular way that it did without Darwin's own casting of his theory as a stark choice between natural selection and the non-scientific, religious doctrine of special creation? Nicolaas Rupke thinks not, arguing in Chapter 6 that "if Darwin had not boarded *HMS Beagle* and the *Origin of Species* had never been written, today's evolutionary biology could well have looked different, possibly fundamentally so." The point here is that Darwin ignored a third and well-established possible theory, one held by many of the great biological minds of the previous half-century, namely autogenesis—the spontaneous origin of species.

Evolutionary biology became increasingly intertwined with eugenics, at first in a relatively benign way as a result of the interactions between Francis Galton, founder of the eugenics movement, and his first cousin Charles Darwin. In Chapter 7, Edward Larson describes these early roots of the eugenics movement in the late nineteenth century, and the way in which genetic arguments were then used to justify the more aggressive eugenics programs of the first half of the twentieth century, leading in the USA to segregation and enforced sterilization of those supposed to be the bearers of "defective" genes. American eugenics became a model for other countries, not least for Hitler's Third Reich in Germany, where the Law for the Prevention of Genetically Diseased Progeny, passed in 1933, was modeled in part on California's compulsory-sterilization law. In Chapter 8, Paul Weindling focuses on the twelve brief years of National Socialism in Germany, commenting that "biology lay at the core of the Nazi racial culture. Both in terms of ideology and application, biology fueled the racial war to exterminate, kill, and resettle total populations." Yet, as Weindling points out, the relationship between Darwinism and Nazi ideology is complex, and, in particular, Nazi values of race purity and psychological types sit uneasily with Darwinism. Whereas biology in general played a powerful role in Nazi rhetoric, it was not as dependent upon evolutionary biology as is sometimes thought.

In the Soviet Union, however, Darwinism was central to the competing claims of rival Marxist factions, but as Nikolai Krementsov explains in Chapter 9, the ramifications of biology were far wider and more complex than the well-known affair involving Lysenko's attempt to wed a discredited

Lamarckian notion of inheritance to socialist doctrine. Ironically, Western observers believed Lysenko's own version of events, that his biological views were Marxist and Darwinist, while portraying genetics as anti-Marxist and anti-Darwinist. But meanwhile Lysenko's opponents within the Soviet Union believed precisely the opposite: that it was Lysenko's views that were incompatible with both Marxism and Darwinism. In reality there was fierce competition over which of the many interest groups would set the terms, forms, and norms of the meaning of Marxist-Darwinism, and this characterized much of the development of Soviet biology.

A powerful idea providing an important backdrop for the deployment of Darwinism in so many different contexts was the close linkage between evolution and the notion of progress, as Michael Ruse discusses in Chapter 10. Whereas in formal scientific literature the linkage between the two declined as evolution became more established as an empirical science, in biologists' more popular writings the idea continued to flourish and continues to remain implicit in much popular evolutionary writing.

Less progressive was the role of biology in shaping notions of sex and gender, as Erika Lorraine Milam recounts in Chapter 11. Whereas early evolutionary theory promoted femaleness as being less evolved than, or a degeneration of, maleness, eugenic rhetoric ironically "secured for women an evolutionary status equal to that of men, as it was recognized that they bore the racial future of Anglo-American society." Other images nurtured by evolutionary discourse included man the hunter and woman the sexually available mother. "Nature" became, and to a large extent continues to be, the key arbiter of difference between women and men. The ideological application of biological arguments is in no topic more striking than in the context of sex difference.

The final two essays in this volume both serve to bring the central theme right into the present. It is precisely the ideological transformation of Darwinism into a perceived threat to the moral, social, and religious order that helps us to understand the strident creationist opposition to Darwinism in the USA that gathered such pace and energy from the 1960s onward. No longer was Darwinism "merely" a theory to explain the origins of biological diversity, but it was now invested with a naturalistic antireligious agenda. No wonder this became a target for those seeking to preserve young minds from contamination in U.S. schools, and Ronald Numbers brings the story further into the twentieth century in Chapter 12 with an overview of the Intelligent Design movement and the battle for its access to the U.S. educational system.

Should the combined anti-Darwinian efforts of proponents of creationism or intelligent design ever lack suitable ideological ammunition to justify their cause, this is amply provided by Richard Dawkins, until recently the Professor of the Public Understanding of Science at Oxford, who, as Alister McGrath describes in Chapter 13, has made a concerted effort to invest evolutionary biology with an atheistic agenda. Here "universal Darwinism" is no longer just a highly successful biological theory but, to use Daniel Dennett's evocative phrase, a "universal acid" destined to eat away at those values and aspirations that many people find make life worth living. Ideologically transformed from the status of biology to a universal worldview, it is hardly surprising that anti-Darwinian movements continue to thrive and proliferate, feeding off such rich evolutionary rhetoric.

So today there is no sign of a let-up in the continuing traffic between biology and ideology. The flow both ways continues as strongly as ever. If this collection of essays has any message, it is that biological ideas promoted in good faith within the academy can find eventual non-biological application in ways remote from the original goals of the scientific investigator. Biologists would therefore do well to be cautious before proclaiming on the broader applications of their findings, recognizing that they themselves are embedded within particular places and cultures, and therefore only too prone to impress their own presuppositions upon their data. A wiser stance would be to emphasize the limited scope of biological theories, generated to explain particular scientific questions, and to resist the ideological hijacking of their scientific ideas to support all kinds of non-biological agendas.

Hopefully this volume will also encourage critical attitudes more generally amongst non-scientists towards the utilization of biological ideas in the service of political, racial, social, religious or antireligious ideologies. This is not merely because such extrapolations from biology often have little to do with the biology itself, thereby subverting the proper public understanding of science, but because such extrapolations can actually be dangerous, as several of the essays that follow most powerfully illustrate.

CHAPTER ONE

The cultural authority of natural history in early modern Europe

Peter Harrison

In the twenty-first century, the social utility of biology, and of the natural sciences more generally, is taken for granted. Most Western societies devote considerable resources to the funding of scientific activities: science faculties occupy pride of place in the major research universities, and natural scientists themselves, while perhaps not enjoying the unchallenged prestige of their immediate predecessors, still wield considerable cultural authority. By way of contrast, the various disciplines of the humanities are viewed with mixed feelings, and the contributions of those who labor in Faculties of Arts, while welcome, are not generally thought to contribute in tangible ways to progress and prosperity. During the European Renaissance a rather different set of priorities was in place. The Italian scholar and poet Francesco Petrarch (1304–74) posed the pointed question: "What is the use—I beseech you—of knowing the nature of quadrupeds, fowls, fishes and serpents and not knowing or even neglecting man's nature, the purpose for which we are born, and whence and whereto we travel?"[1] The question identified by Petrarch as the most pressing of all—the nature of the human person—is often now thought to be best answered by those with training in the sciences. Indeed, in certain circles, biology and its controversial offspring evolutionary psychology are regarded as providing the only reliable account of human nature, in contrast to the vague and ungrounded pronouncements of theologians and philosophers. Petrarch, however, clearly sought the answers elsewhere. Fittingly for the father of Renaissance humanism, he advocated a turn away from the study of nature, and indeed his dismissive remarks provide a fairly typical assessment of the lowly place occupied by natural history in the intellectual world of the Renaissance. For Renaissance thinkers, the study of nature was a distraction from the intellectual activities that really mattered—the pursuit of those disciplines believed to contribute in direct ways to the education of well-rounded citizens and to the general welfare of society. This animus against natural history persisted well into the seventeenth century, when the Polish naturalist and physician John Johnston (1603–75) could complain that natural history "now a dayes is so despised, that it is of no esteem at all."[2]

How, in the period that separates the Renaissance from our own time, did such a reversal of fortunes come about? Here, I offer a general account of what might be termed the first phase in the transformation of natural history—one that took place in the seventeenth and eighteenth centuries. This period witnessed significant changes to both the practice and the status of natural history. Three aspects of this change are of particular significance. First was the manner in which, during the seventeenth century, all of the new sciences sought to establish their social legitimacy in the face of competing Aristotelian approaches to the natural world and humanist priorities, both of which devalued the study of nature. Apologists for the new sciences advanced the argument that their activities could promote the moral and religious goals associated with the traditional humanist education, and do so more effectively than the Aristotelian sciences. Second, and in some tension with the first element, was the attempt to provide additional justifications for scientific endeavors. These justifications sought to expand the traditional goals of scholarship and learning, focusing more on the provision of material benefits to society. These new utilitarian goals were still justified in terms of prevailing religious values, however, and again the new sciences were argued to be superior to Aristotelian approaches in achieving them. Third was the manner in which natural history itself was repositioned in the hierarchy of the disciplines concerned with the study of nature, moving from the margins of intellectual life and coming to be regarded as a serious systematic and scientific activity, and fundamental to the whole scientific enterprise.

Natural history at the margins

William A. Wallace has observed of the Renaissance university curriculum that "little or no attention" was paid to natural history.[3] There is no lack of contemporary witnesses to corroborate Wallace's assessment. According to Gilbert Jacchaeus (1577–1628), a well-traveled Scot who became professor of philosophy and medicine at the University of Leiden, "teachers do not lecture on the works of natural history."[4] John Johnston, who, as we have seen, judged natural history to have been "despised" by his contemporaries, claimed that natural history was not "entirely professed in any University or School (except Bolognia, where Aldrovandus was)."[5] By the end of the seventeenth century, natural history enjoyed a somewhat higher profile, but John Ray (1627–1705), a leading naturalist and pioneer of modern taxonomy, could still complain about a continuing curricular bias toward the humanities. We rest content with the study of the languages, philosophy, and history, he grumbled, but neglect what is far more important, "Natural History and the Works of the Creation."[6] As a consequence, histories of animals were "in a great measure neglected by English men."[7]

The lowly position of natural history in the sixteenth and seventeenth centuries is not difficult to account for. The humanist priorities outlined by Petrarch in the fourteenth century were still widely influential. Siennese scholar Alessandro Piccolomini (1508–79), in spite of his keen interest in astronomy, wrote that the Aristotelian natural sciences were inferior to "the more honorable sciences which teach the art of living." It is these, Piccolomini claimed, that promote virtue and happiness.[8] These same values were affirmed by the Elizabethan poet Sir Philip Sidney (1554–86) in his classic *Defence of Poesy*. Sidney repeated the formula articulated by Petrarch and Piccolomini, stating that the ultimate aim of learning is "the knowledge of a mans selfe, in the Ethike and Politique consideration, with the end of well doing, and not of well knowing onely."[9] The most obvious candidates for this task, in Sidney's view, were moral philosophy, history, and poetry (poesy). At first sight, moral philosophy may seem to be the most promising prospect. However, in a telling passage from which we learn that little has changed in the discipline of philosophy over the centuries, Sidney points out that philosophers are rather good with distinctions, divisions, and definitions, but operate at so abstract a level that no one really understands them. Compounding their incomprehensibility is an utter failure to inspire or motivate anyone to the pursuit of virtue. So much for philosophy. History fared little better. Sidney suggests that there is a fundamental problem with attempting to derive general moral lessons from the haphazard events of history. History is bound "not to what should be, but to what is, to the particular truth of things." Sidney proceeds to the unsurprising conclusion that the study of literature, and poetry in particular, is the best way to achieve the goals of humanist education.[10]

Sidney's treatment of the relative merits of philosophy and history, and their relegation to a station beneath that of literature, is relevant because the two major contemporary approaches to the study of nature were historical and philosophical. The closest counterparts to our modern natural sciences in the early modern period (roughly 1500–1800) were the disciplines of natural history and natural philosophy. Natural history, as its name suggests, was a historical discipline (along with civil and ecclesiastical history). It was historical because, although its subject matter was the natural world rather than human affairs, it was a descriptive enterprise that dealt with discrete facts—"the particular truth of things" as Sidney put it.[11] Natural philosophy was regarded as the more scientific approach to nature because it went beyond mere description to offer causal explanations. Natural philosophy was also regarded as a proper science in a way that natural history was not, for its conclusions were offered in the form of logical demonstrations.[12] Thus, even if the formal study of nature had enjoyed a more privileged status in the Renaissance curriculum, natural history would still have been regarded as inferior to natural philosophy. Gilbert Jacchaeus therefore proposed that one of the reasons that natural history was undervalued—along with the fact that it could not easily be accommodated within the two- or three-year philosophy course and was in any case regarded as a "soft option"—was that it did not involve demonstrative science.[13] Sentiments akin to these are evident in physicist Ernest Rutherford's more recent observation that "all science is either physics or stamp collecting."[14]

By the end of the seventeenth century, and in spite of John Ray's pessimistic pronouncements on the topic, the fortunes of natural history were on the rise. Apologists for the new sciences mounted effective arguments suggesting that both natural history and natural philosophy could make a major contribution to the traditional moral goals of formal education. Their argument drew upon a range of traditional views about the moral relevance of the study of nature. But beyond the mere rehearsal of traditional justifications for this study were arguments that pointed to the particular relevance of the recent advances made in the realm of science that were claimed to enhance greatly the moral and theological relevance of natural history. Natural historians were now assisted with new magnifying instruments and had at their disposal an ever-increasing mass of data that far surpassed anything to which the ancients or their medieval successors had had access.

Closely associated with the moral role of the new natural history was the claim that natural history could provide important foundations for natural theology, and in particular the argument from "design" (see "Designs on nature," below in this chapter). In its strongest form, this argument amounted to the claim that natural history was actually a form of religious activity, and that its practitioners had been called to a theological vocation. In addition to these elaborations of some of the standard justifications for the pursuit of natural history, the discipline was also brought into a closer relationship with natural philosophy. While it remained primarily a "historical" or descriptive discipline rather than an explanatory one, it was given a new role as the activity that provided the factual basis for the more scientific conclusions of natural philosophy. The success, in the eighteenth century, of new classificatory schemes also lent natural history a new authority, as it was recognized as the only enterprise that could imbue the chaotic world of living things with a systematic order.

Natural history and moral edification

The seventeenth-century biography of the eminent French naturalist Nicolas-Claude Fabri de Peiresc (1580–1637) recounts what to modern readers is a rather curious story about an encounter between a louse (Figure 1.1) and a flea. Peiresc had been observing the behavior of these two creatures through one of the newly invented microscopes and, according to his own recollection of the event, experienced a kind of moral epiphany. As his biographer attests:

> And he was wont to say, that nothing did ever so much prevail with him to rule his passion, as a fight which he happened to see in an Augmenting-glasse, or Microscope. For, having inclosed therein a Lowse and a Flea; he observed, that the Lowse, setting himself to wrastle with the Flea, was so incensed, that his blood ran up and down from head to foot, and from foot to head again. Whence, he gathered, how great a Commotion of Humors and Spirits, and what a disturbance of all the faculties, anger must needs make; and what harm that man avoids, who quits that passion.[15]

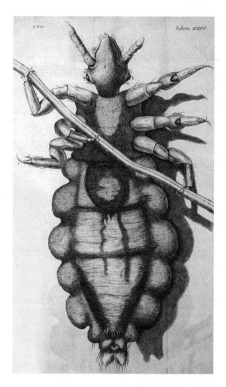

FIGURE 1.1 A louse. Microscopic examination could sometimes show internal structures and the fluxes of bodily fluids in small creatures such as the louse. The motions of these fluids, and of the humors in particular, were associated with emotional states, and hence were thought to have relevance for an understanding of human moral behavior. (From Robert Hooke, *Micrographia*, 1665.)

This account is informative about the status of seventeenth-century natural history and its practitioners at a number of levels. The biography itself calls to mind the classic "lives of the philosophers" genre of late antiquity.[16] Because philosophy—which in ancient and medieval times included the formal study of nature—was ultimately about "the good life," recounting biographical detail from the lives of eminent philosophers was thought to have a salutary moral affect on the reader. More specifically, narratives of this kind reinforced the message that the study of nature could teach important moral lessons—in this instance the general principle that the passions (or emotions, as we call them) must be brought under the control of reason. In the conduct of life, we should emulate the flea, and not the louse. Natural history, in other words, could contribute to the moral formation that was the principal purpose of education. Finally, this episode makes a subtle appeal to the authority of novel approaches to natural history—ones that relied on new instruments and experimental methods. The fact that the microscope enabled Peiresc to see directly the motions of the spirits and humors in at least one of these insects made this a uniquely edifying experience.

The idea that nature can serve as our moral tutor was not new. The ancient Stoics had taught that one should live in accordance with nature. If we are to know how to act appropriately in the cosmos, they argued, it is important to understand something of its rational order.[17] Significantly, for our purposes, Stoic philosophy had undergone a revival during the sixteenth and seventeenth centuries, and its lessons about bringing the passions under control and living in harmony with the principles that governed the natural world can be encountered in numerous contemporary moral writings.[18] Stoicism was not the sole source for the idea that the natural world could teach important moral lessons. The Judeo-Christian tradition also provided support for this view. In Proverbs (6: 6) we encounter the familiar injunction "Go to the ant, thou sluggard."[19] From ancient times, the idea that the industry of ants was a silent indictment of human sloth was a common moral trope. In late antiquity and into the Middle Ages, such passages justified the drawing of moral lessons from animal behaviors. The patristic Ambrose of Milan (339–397) wrote, against the background of both Stoic thought and the Hebrew Bible, that nature was "a guide and teacher for what is actual and true," and for this reason the animals can serve as "examples for our own way of life."[20]

The Christian tradition also taught that animals, plants, and stones symbolized important theological truths. Because Scripture relied upon a range of natural metaphors—Christ as "lion of Judah," for example—its interpretation required knowledge of the natures of various creatures. Nature also mirrored particular theological doctrines. It was commonly assumed, for instance, that the metamorphosis of insects foreshadowed the post-mortem resurrection of the body.[21] For reasons such as these, Augustine of Hippo (354–430) cautiously endorsed the value of natural history as an exegetical tool: "An ignorance of things makes figurative expressions obscure when we are ignorant of the natures of animals, or stones, or plants, or other things which are often used in the Scriptures for purposes of constructing similitudes."[22] Subsequently, the English Friar Bartholomaeus Anglicus (fl. 1250) was to write in his encyclopedic *De proprietatibus rerum* that his work on natural history would prove useful for those who sought "to understand the obscurities or darkness of the holy Scriptures, which are given to us under figures, parables, and semblances of natural and artificial things."[23]

Of course, natural histories also provided important information on the practical uses of animals and plants—for cures, for food, as labor-saving devices, and for the provision of other beneficial commodities. Because of the pharmacological applications of living things, natural history, in spite of its relatively lowly status, was a component of the formal training of physicians in a number of medieval and early modern universities.[24] This practice preserved the priorities of the first-century physician Theophrastus of Eresos, author of the enormously influential *De materia medica* [On medical material], who had taught that one of the major reasons for studying nature was to extend the range of available natural remedies.[25] Indeed, until the radical medical innovations of Paracelsus (1593–1441), who first introduced chemical cures, all pharmacological preparations were based on plant or, less frequently, animal material. While these medical uses of natural history were of great importance, it should be remembered that at the time they were understood to promote the health of the body—a lower priority than the edification of the soul.

Renaissance natural histories typically invoked these justifications when attempting to establish their own relevance. Conrad Gesner's five-volume *Historiae animalium* [On the history of animals], published between 1551 and 1587, was perhaps the most impressive digest of natural history produced during the sixteenth century. In the preface he observes that, in addition to their practical uses and medical applications, animals and plants teach us important moral lessons, serve as emblems of divine truths, and point to the wisdom of the Deity.[26] Most similar works of the period, many of them based on Gesner's encyclopedic productions, offered similar justifications for the pursuit of natural history.

In his *Historie of Four-footed Beastes*, a popular digest of natural history drawn largely from Gesner, Edward Topsell (1572–1625), an Anglican cleric, repeated the idea that the similitudes of Scripture could not be fully understood without a knowledge of animals and plants. Passages of Scripture, he asserted, would remain obscure without some acquaintance with natural history. Renaissance natural histories would typically list the biblical references to particular plants and animals under their respective entries. The *Herbal for the Bible* by the Dutch physician Levinus Lemnius therefore took as its subject the "Similitudes, Parables, and Metaphors, both in the olde Testament and the Newe, as are borrowed and taken from Herbs, Plants, Trees, Fruits, and Simples."[27] Francis Bacon (see below) also made reference to the importance of natural history for biblical exegesis, arguing that the "book of nature" could be of great use in the interpretation of "the book of scripture."[28]

The idea of nature as a moral authority and the conviction that studying nature would yield significant moral dividends became increasingly important in the early modern period, partly, as already mentioned, on account of the revival of Stoicism. The French essayist Michel de Montaigne (1533–92) wrote that "we cannot erre in following nature and ... the soveraigne document is, for a man to conforme himself to her." For this reason, he concluded, "our wisedome should learne of the beasts, the most profitable documents."[29] In his enormously popular *Of Wisdom* (1601), Montaigne's disciple, the French moral philosopher Pierre Charron, also enunciated the Stoic principle that we should live in accordance with nature: "God intending to Shame us into Vertue, sends us to School in Scripture, and bids us grow wiser by the Example of these Creatures."[30] While it was a venerable notion, the idea that we might learn important moral and religious lessons from nature gained traction from the particular circumstances of the sixteenth and seventeenth centuries. The religious controversies engendered by the Reformation raised in an acute manner the question of the problem of moral and religious authority. In this context, the Stoic counsel to follow the precepts of nature, rather than fallible and contestable human authorities, seemed to some at least to make good sense. Natural religion and natural theology were thus the beneficiaries of religious controversy. It could be claimed, as the German mathematician and astronomer Johannes Kepler once put it, that in nature we encounter the "clearer voice" of God.[31] Moreover, while the book of Scripture was contested territory (and was in any case available to only a select few) the book of nature was universal and, it was claimed, unambiguous.[32] Abbé Pluche (1688–1761) declared in his voluminous work of natural theology that "Nature is the most learned and complete of all Books, since she comprehends at once the Objects of every Science, and never confines her Instructions to any particular Language or People."[33]

Also distinguishing the newer from the older justifications of natural history was the great expansion of natural knowledge that followed the explorers and their voyages of discovery, the invention of new magnifying instruments, and the development of more systematic and coordinated approaches to the observation of the natural world. A reformed and augmented natural history, it was claimed, would yield more valid moral insights than the limited and erroneous records passed down from antiquity. Edward Topsell wrote in his *Historie of Foure-Footed Beastes* of the importance of following truth and not "deceivable Fables" about animals (although, admittedly, he did not always follow his own advice). Robert Boyle (1627–91) also distinguished between the fanciful inventions of such writers as Aesop and the painstaking and systematic observations from which genuine moral conclusions could be drawn. Proper scrutiny of the creatures will show them to be "not only teachers of ethics, but ... doctors of divinity." René Réaumur (1683–1757) complained similarly that naturalists had in the past represented animals "otherwise than nature has made them" in order to draw moral lessons. However, this did not obviate the principle that nature could serve as our moral tutor, but rather established the importance of accurately observing it.[34]

All of this meant that traditional moral defenses of the utility of natural history could be reinforced with the growing body of data drawn from a variety of new authorities. Nowhere was this more apparent than in the study of insects, where a number of ancient myths were exploded, to be replaced by equally startling information about the lives of these diminutive creatures. Following the advice of the author of Proverbs, Thomas Moffett (1553–1604), who had played a role in editing Gesner's work on insects, declared ants to be "exemplary for their great piety, prudence, justice, valour, temperance, modesty, charity, friendship, frugality, perseverance, industry, and art."[35] "Every particle of an insect," Thomas Blount wrote in his *Natural History* (1693), "carries with it an impress of Ethicks or Divinity."[36] Calvinist clergyman John Edwards (1637–1716) described the moral qualities of a range of creatures, including the insects, before concluding that a natural historian will almost invariably be "a Good Moralist."[37] Indeed, students of the much-despised insects labored to establish the moral usefulness of their discipline. These justifications for the study of animals persisted until well into the eighteenth century. René Réaumur wrote in his classic *Natural History of Bees* that these little creatures set out "useful instructions" and "examples fit for imitation."[38]

Close observation of insect behavior not only might reinforce conventional virtues but also had the potential to assist in the adjudication of such questions as whether monarchy or democracy was the more natural form of government. In *The Readie and Easie Way to Establish a Free Commonwealth* (1660), the poet John Milton pointed to the organization of the ant colony as providing a clear sanction for parliamentary democracy. The polity of the "pismire" (ant), he declared:

evidently shews us, that they who think the nation undon without a king, though they look grave or haughtie, have not so much true spirit and understanding in them as a pismire: neither are these diligent creatures hence concluded to live in lawless anarchie, or that commended, but are set the examples to imprudent and ungovernd men, of a frugal and self-governing democratie or Commonwealth; safer and more thriving in the joint providence and counsel of many industrious equals, then under the single domination of one imperious Lord.[39]

The lesson of the verse from Proverbs was, according to Milton, a political one. But if nature showed that a democratic commonwealth was possible, it also provided evidence for alternatives. Later in the century, John Edwards, a staunch Whig and supporter of William and Mary, pointed to the sagacity of bees and to the fact that the hive was ruled by a "Female Monarchy."[40] Neither would be the last to draw anthropological conclusions from the study of insect colonies.

These attempts to establish the social utility of natural history in the spheres of moral edification and biblical exegesis were still wedded, to a large degree, to the humanist understanding of the ends of education. On this view, whatever authority natural history might have had arose out its capacity to contribute to the goals of reforming one's own life and its role in the provision of instructive moral examples for others. While the importance of such justifications should not be underestimated—the role of the natural sciences in promoting morals persisted into the nineteenth century—they fell somewhat short of a robust defense of the study of nature. Some seventeenth-century thinkers, such as René Descartes, contended that the science of nature would provide a more scientific foundation for moral philosophy. Other apologists for natural history (and natural philosophy) argued that their activities could make a major contribution in the sphere of theology; they also sought to modify the conceptions of the purposes of education. This latter exercise involved moving the goal posts (as it were), so that the aims of study could incorporate the kinds of material benefits that it was claimed the sciences could provide.

Morals and mechanism

The most remarkable early modern theory about animals was the Cartesian doctrine of the beast-machine. René Descartes (1596–1650), although now best known for his philosophical writings, was during his lifetime a

FIGURE 1.2 René Descartes (1596–1650). (Engraving by W. Holl, after a painting by Franz Hals. Courtesy of Science Museum/SSPL.)

figure of major importance in the developing natural sciences (Figure 1.2). Notoriously, he suggested that animals were non-sentient machines, devoid of a soul and indeed bereft of anything comparable to the rich inner mental lives of human beings. Animals, in the words of one of his disciples, "eat without pleasure, cry without pain, grow without knowing it; they desire nothing, fear nothing, know nothing."[41] Then, as now, this was widely regarded as an outrageous idea. Yet, in its favor, it was consistent with the seventeenth-century tendency toward mechanistic explanation, it provided an elegant if counterintuitive solution to the perplexing question of whether animals had souls, and it absolved God of complicity in animal suffering. For these reasons, the idea of the beast-machine had some influential supporters.

It is sometimes asserted that the hypothesis of the beast-machine licensed the cruel treatment of animals and gave a special impetus to their scientific exploitation. According to this view, the beast-machine hypothesis signaled the beginning of a new exploitative attitude toward nature, an indifference to the animate world, and, more specifically, an increased willingness to subject animals to cruel and invasive experiments. There is little evidence for this view. The fact is that the Cartesian hypothesis was not widely accepted and, in any case, vivisectionists needed no dispensation from Descartes to sanction their activities. Many of those who subjected animals to unsavory procedures in vacuum chambers or on dissecting tables explicitly rejected the Cartesian view of animal insensitivity.[42]

More relevant to our present discussion, we might also expect that the beast-machine hypothesis would be inconsistent with moral applications for the study of nature. After all, the notion that the creatures possessed shadows of the virtues, and could act as moral exemplars, assumed the essential relatedness of all living things. This assumption was in turn usually premised upon the venerable concept of a graded "Chain of Being" according to which all living things ascended by degrees from the less perfect to the more perfect. Against this, Descartes argued for a radical rupture in the hierarchy of being, positing a great gulf between human beings and all the other earthly creatures—now imagined to be devoid of souls and sensations. He also rejected the idea that the behaviors of animals were analogous to our own and, indeed, attacked "reasoning by analogy" in the sciences more generally.[43] Not surprisingly, then, Descartes had little time for the idea that animals could be our moral teachers.

Yet Descartes did not abandon the idea that the study of animals had moral relevance, for it was his conviction that an understanding of the physical basis of animal behavior could yield vital insights into human behaviors and motivations. This was because, although human beings uniquely possessed an immaterial soul, that soul was nonetheless embedded in a material, and hence mechanical, body. Moreover—and this element of Descartes' thought is sometimes forgotten—body and soul in human beings were intimately conjoined. For this reason Descartes believed that investigation of the material and mechanical basis of animals' inner drives—anger, aggression, fear, hunger, thirst, and so on—would lead to the discovery of therapeutic strategies that would assist human beings in the moral task of exerting control over the impulses and drives of their own mechanical bodies. Although animals lack reason and perhaps thought, he wrote, "all the movements of the spirits and of the gland which produce passions in us are nevertheless present in them too, though in them they serve to maintain and strengthen only the movements of the levers and the muscles which usually accompany the passions. . . . "[44] In a sense, what Descartes sought in these ambitious studies of the mechanics of morals was a moral psychology that was grounded in comparative physiology. It is in this light that we are to understand his insistence that the three principal sciences are medicine, mechanics, and morals, and that these sciences are connected in important ways. Moral science, which for Descartes was the most important science of all, required a knowledge of all the other sciences.[45] Descartes, in short, still advocated morality as the ultimate goal of the sciences, and his reductionistic studies of animal behavior and animal physiology were directed toward that end.

Descartes' insistence that animals were mere machines had another implication, although it was not one that he himself pursued with much vigor. This was the idea that mechanism suggests a designer. While Descartes tended to be silent on this implication of his views, other advocates of the beast-machine developed it more enthusiastically. The English philosopher Kenelm Digby (1603–65) pointed out that "animals are but material instruments to performe without their knowledge or reflexion, a superior reasons counsels: even as in a clocke, that is composed of severall pieces and wheels, all the parts of it doe conspire to give notice of the . . . time, which the maker hath ordered it for." For Digby, the implication of animal mechanism was that instead of marveling at the intelligence of the creature, the diligent observer should "looke with reverence and duty upon the immensity of that immeasurable Architect out of whose hands these masterpieces issue."[46] Mechanism, for Digby and many of his contemporaries, pointed ineluctably to divine design.

Designs on nature

One of the remarkable features of the history of the argument from organismic design—the idea that the structure and function of living things provides evidence of the wisdom and benevolence of God—is how small a role it plays in theological discussions prior to the seventeenth century. Of course, it had long been assumed that the beauty and orderliness of nature pointed to the existence of God. The Psalmist had written that "the heavens declare the glory of God" (Psalms 19:1), while in the New Testament St. Paul suggested that the invisible things of God were evident in the created order (Romans 1:20). These sentiments were reinforced by arguments found in Plato and the Stoics (although these arguments tended to be based more on the regularities of the heavens than the structures of living things).⁴⁷ And while few in the Christian West doubted that the world was the product of divine wisdom, it must be said that the idea that this provided a justification for the formal study of nature was not particularly prevalent. Of Thomas Aquinas's famous "five ways," only the last may be said to be an argument from design, and even then it bears little resemblance to the kinds of design arguments that rose to prominence in the early modern period.⁴⁸

This was to change dramatically in the seventeenth century, when the argument from design came to dominate both natural theology and natural history. While, as we have seen, the moral and exegetical relevance of natural history was still emphasized, increasingly justifications for the pursuit of natural history stressed this element of divine design. Natural history, together with natural philosophy, was said to furnish unambiguous evidence of the wisdom and power of God. Cambridge Platonist Henry More (1614–87)—a keen advocate of Cartesian philosophy until he became disenchanted with Descartes' views about animals—was one of the first to make this argument explicit, and to enumerate many instances of it. The "exquisite contrivance" of the parts of animals, More pointed out in his *Antidote Against Atheisme*, "is an undeniable Demonstration that they are the effects of Wisdome."⁴⁹ Natural historians were subsequently to make much of this. John Ray declared that he had published his *Ornithology* for "the illustration of Gods glory, by exciting men to take notice of, and admire his infinite power and wisdom."⁵⁰ This emphasis on design in nature was present in many of the Boyle Lectures (see below), reached its acme in William Paley's *Natural Theology* (1802), and enjoyed its final hurrah in the Bridgewater Treatises (1833–40; and see Topham, Chapter 4, this volume).⁵¹

It is often assumed that this close association of the argument from design with the new natural history and natural philosophy was an apologetic strategy, adopted with a view to offering rational confirmation of religious belief through appeals to a "scientific" discipline that was more epistemically secure than the religious beliefs to which it lent support. Given the present status of the natural sciences this is an obvious conclusion to draw. References in many of these works of pious natural history to "the confounding of atheists and infidels" add further plausibility to this claim. That said, it is more likely that this close association was intended more to secure the social legitimacy of new approaches to the study of nature than to bolster the status of relatively secure religious beliefs. Consider, for example, the Boyle Lectures, endowed by Robert Boyle for the purpose of refuting "notorious Infidels, *viz.,* Atheists, Theists, Pagans, Jews and Mahometans."[52] The fact is that atheists, on any reckoning, were rather thin on the ground in the seventeenth century. In any case, atheists, pagans, Jews, and Muslims were unlikely to be present in the congregations of the London churches that served as the venues for these lectures. "Pretext" is too strong a word, but perhaps the specter of infidelity and atheism provided a convenient way of justifying new ways of studying nature in the context of a social and intellectual climate that was not always hospitable toward novel scientific practices. Consider, again, a standard juxtaposition of the religious aims of natural history and the new methods by which those aims were to be accomplished. The Dutch physician and naturalist Bernard Nieuwentijt wrote that his *Religious Philosopher: or the Right Use of Contemplating the Works of the Creator* was intended "to convince *Atheists* of the Wisdom, Power and Goodness of God" with examples that "are only taken from the *modern Observations,* and *probable Discoveries* in *Natural Philosophy.*"[53] Nieuwentijt's emphasis on the use of exclusively modern sciences is intended to highlight the religious utility of those sciences in an era in which novelty and modernity were still not self-evidently positive qualities.

It is partly for reasons such as these that such attention was paid to small and hitherto insignificant creatures. The discovery and publication of the microscopic perfections of insects lent an exclusive and unprecedented authority to those naturalists who had access to the microscope.[54] While for most of polite society such studies were at best irrelevant and at worst undignified and repulsive, apologists for microscopical studies highlighted both the moral and theological relevance of their work. Robert Hooke, whose beautifully produced *Micrographia* (1665) first introduced the reading public to the invisible microscopic world, wrote that the more objects are magnified with the microscope, "the more we discover the imperfections of our senses, and the Omnipotency and Infinite perfections of the great Creatour."[55] Such sentiments become commonplace in the late seventeenth century.[56] These new microscopic endeavors were contrasted with the other sciences and even with more traditional forms of learning. Abbé Pluche pointed out that natural history is the equal of astronomy in providing evidence of God, for "the Deity is as conspicuous in the structure of a Flys paw, as he is in the bright Globe of the Sun."[57] The Cartesian philosopher Nicolas Malebranche observed that "one insect is more in touch with Divine wisdom than the whole of Greek and Roman history."[58]

It was no great distance to move from pronouncements such as these, which showed how natural history could provide external rational support for theology, to the notion that natural history was itself a theological enterprise. John Johnston wrote that God "is comprehended under the title of natural history."[59] John Edwards expressed the proposition more cautiously, insisting rather that "our skill in Natural History must lead us to Theology."[60] Some naturalists also contended that the practice of natural history was a form of religious worship. Henry Power characterized natural history as a "homage due to [the] Creator."[61] Robert Boyle thought that the study of nature was "the first act of religion, and equally obliging in all religions." Accordingly, he and his peers were to be thought of as engaging in the "philosophical worship of God."[62] In his apologetic *History of the Royal Society*, Bishop Thomas Sprat went so far as to claim that the formal study of nature had been the original religion of Adam in Paradise. It followed that, in the present age, the religious pronouncements of the scientifically literate are more suitable to the divine nature than "the blind applauses of the ignorant."[63] John Ray regarded natural history as a "Preparative to Divinity."[64] For reasons such as these, the study of nature was considered an appropriate activity for the Lord's Day. Edward Topsell thought that reading works of natural history might enable the reader "to passe away the Sabbaths in heavenly meditations upon earthly creatures." Ray similarly believed the study of the natural world to be "part of the business of a Sabbath-day," for, as he goes on to explain, the Sabbath seems to have been instituted "for a commemoration of the Works of the Creation."[65]

With the new theological and religious status of natural history came a new status for its practitioners. The claims that natural historians and natural philosophers were best equipped to interpret the book of nature, and that their findings were of great theological import, bore the implication that natural history was a religious vocation. On occasions, the identification of the study of nature as a clerical activity was made explicitly. Johannes Kepler moved from the idea of the world as God's book to invest the interpreters of that book with a specific religious calling—"Priests of the most high God with respect to the book of nature."[66] Robert Boyle also thought that investigators of nature enjoyed a special religious calling, invoking a similar descriptor: "Priests of nature."[67] These phrases give some indication of the way in which natural historians sought to carve out for themselves a new vocational space from within which they could exercise their rightful authority.[68]

Not surprisingly, these more robust claims about the moral and theological authority of natural historians were met with some skepticism. The story of Peiresc and his life-changing encounter with the louse and flea was therefore repeated by Meric Casaubon (1599–1671), but it was now related with overtones of incredulity. Casaubon was an advocate of humanistic values, and a learned and outspoken critic of the new sciences. The anecdote of the louse and the flea was "a pretty story," which Casaubon suspected to have been concocted by French philosopher Pierre Gassendi. The Bible enjoins us to "Go to the ant" only in the event that we are so destitute of moral sense that we need to seek the example of an irrational creature. For Casaubon, the virtues of industry are well enough attested elsewhere.[69]

Many others shared Casaubon's views, and were unpersuaded by claims about the moral and religious utility of natural history. Critics of the moral authority of naturalists also found implausible the contention that the use of instruments and experiments conferred advantages on modern authors that the ancients had not shared. The apparent preoccupation of natural historians with such undignified creatures as fleas and lice also provided considerable ammunition for those who wished to deflate their lofty pretensions. Fellows of the Royal Society, sniffed one critic, admire nothing except fleas, lice, and themselves.[70] Another detractor observed that the proponents of the new sciences were good for only two things: "diminishing a Commonwealth, and multiplying a louse."[71] Such mockery was common currency in England amongst Restoration wits, and these attitudes persisted for some time.[72] Decades later, in *Gulliver's Travels* (1726), Jonathan Swift lampooned those naturalists who deferred to the moral authority of the insect world. When Gulliver visited the grand academy of Lagado (a thinly disguised simulacrum of the Royal Society), he encountered an ingenious architect who had contrived a new method for building houses, beginning at the roof and working down to the foundation. This method, we are told, was justified by appealing to the practice of "those two prudent insects, the bee and the spider."[73]

What these reactions suggest is that, while natural history had become far more prominent over the course of the seventeenth century, and while a strong case had been advanced for its importance, there were still those who remained to be convinced. One thing that could be argued for the theological virtues of natural history was that, in principle, its conclusions were ones that could be universally acknowledged, since they were not based in contested revealed traditions but rather on appeals to the universal criterion of nature.[74] But natural history was not yet at the point where it was widely recognized as making a significant contribution to agreed social and intellectual objectives. One way in which a stronger case could be made was to expand the goals of learning and set out new standards for knowledge.

The practical utility of natural history

If the prevailing attitude toward learning in the early seventeenth century looked to promote self-knowledge and moral formation, proponents of the new sciences proposed the inclusion of additional goals, extending self-mastery to the mastery of nature, and insisting on the importance of practical applications of knowledge. Perhaps no seventeenth-century figure better exemplifies these tendencies than Francis Bacon (1561–1626), although it is important to recognize that he was by no means alone in setting out these new priorities (Figure 1.3). Bacon insisted that "the improvement of man's mind and the improvement of his lot are one and the same thing," linking the accepted goal of learning—self improvement—with the broader goals that he had in mind for a reformed science of nature.[75]

Much of Bacon's case for a reorientation of the study of nature was based on his reading of the Genesis narratives of the Creation and the Fall. Once Adam had enjoyed dominion over the whole of nature, and his knowledge of the creation had been apparent in his capacity to name the creatures and bend them to his will. With the Fall, however, he had been dispossessed of both his original moral perfection and his knowledge of nature and dominion over it. While it was the goal of a reformed religion to help to ameliorate the loss of mankind's moral perfections, it was the objective of a reformed science of nature to assist in the recovery of Adam's lost dominion and knowledge. Bacon believed that a systematic knowledge of the natural world would bring about "a restitution and re-investing (in great part) of man to the sovereignty and power . . . which he had in his first state of creation."[76] Nature, he concluded, will be "at length and in some measure subdued to the supplying of man with bread; that is, to the uses of human life."[77]

This was a remarkable claim. In essence, it allocates a vital redemptive role to the formal study of nature and places it alongside the Christian religion. Yet this view is completely congruent with the subsequent judgments of Thomas Sprat and Robert Boyle that the study of nature was a

FIGURE 1.3 Francis Bacon (1561–1626). A bust of Charles II (1630–85), Patron of the Royal Society, is crowned by Fame. Francis Bacon, who provided the intellectual inspiration for the Society, is depicted on the right. To the left is the Society's first president, Lord Brouncker. (From Thomas Sprat, *History of the Royal Society*, 1667, frontispiece. Engraving by Wenceslaus Hollar. © The Royal Society.)

form of religious activity. Indeed, Sprat's claim that the original religion of Adam had been nothing other than the contemplation of nature is no doubt Baconian in inspiration. Bacon further reinforces his argument by suggesting that the ultimate goal of the study of nature is not knowledge, but Christian charity. For him there was only one legitimate purpose for studying nature, "the glory of the creator and the relief of man's estate."[78] It was also a relevant consideration for Bacon, as it was for a number of his contemporaries, that many prominent Old Testament figures had been able natural historians—most notable among them, Solomon and Job. It was widely believed that the canon of Scripture had once contained works of natural history by Solomon and Job, but that these had been lost to posterity.[79]

These positive arguments for a reorientation of knowledge were accompanied by criticisms of the previous inward-looking models of learning. The barrenness of the natural sciences owed much to the fact that they had been diverted away from their proper charitable and Christian ends. This distortion of the true aims of the sciences and their degeneration into a pedantic concern with texts and ancient authorities, metaphysical arguments, and pointless disputations, were the consequences of a defection away from an original biblical model. This corruption of the study of nature was attributed to the baleful influence of pagan philosophy, or monkish religion, or a fatal combination of both.[80]

Bacon's expanded conception of the goals of the study of nature was enormously influential. It was taken up in the middle decades of the seventeenth century by the "projectors"—individuals and groups who sought to apply new knowledge to a range of practical problems.[81] John Webster (1611–82), a radical critic of the university curriculum, urged two applications for the study of nature: that we might come to know the Creator through his creatures; and that we might gain knowledge of causes and effects in order "to make use of them for the general good and benefit of mankind, especially for the conservations and restauration of the health of man."[82] The utilitarian imperative also informed the ideology of the Royal Society, founded in the 1660s. Thomas Sprat, in an account of the activities of that august group, insisted that the formal study of nature could fulfill moral and religious goals while at the same time contribute to practical and technological advancement. Scientific investigations had advantages "above other *Studies*, to the wonted Courses of *Education*; to the Principles, and instruction of the minds of Men in general; to the

Christian Religion, to the Church of England; to all Manual Trades; to *Physic*." In instructing men's minds and improving their natures, moreover, the new philosophy was far superior to "the rigid praecepts of the Stoical, or the empty distinctions of the *Peripatetic Moralists*." But more than this, the new sciences were productive, and sought ways of knowing and mastering the created world. In imparting to us "the uses of all the creatures," Sprat argued, the society would promote the recapturing of a lost Adamic wisdom and dominion.[83] Cleric Joseph Glanvill was another who sought to defend the Royal Society against such critics as Meric Casaubon. He too insisted on the theological, moral, and practical advantages of the work of the Royal Society: "*The study of Nature and God's works, is very serviceable to Religion* . . . and the *Divine Glory* is written upon the *Creatures*." Such study, moreover, is morally edifying, for it will "assist and promote our *Vertue*, and our *Happiness*; and incline us to imploy our selves in living according to it." And, not least, in keeping with Bacon's "mighty design," the collation of the history of God's works leads to "the *Discovery* of *Causes*, and *Invention* of *Arts*, and *Helps* for the *benefit* of *Mankind*."[84]

The posited practical outcomes of natural history and natural philosophy seem ready-made for an economic argument to the effect that these activities would enhance the standing and wealth of the nation. Certainly such arguments were advanced. Glanvill, for example, spoke of "the accelerating and *bettering* of *Fruits*, *emptying Mines*, *drayning Fens* and *Marshes*." These activities would lead to a restoration of "the Empire of Man over Nature, and bring plentiful accession of Glory to [the] Nation, making Britain more justly famous then the once celebrated Greece; and London the wiser Athens."[85] John Ray made a similar claim, maintaining that the diligent naturalist, by making "large Additions to Natural History" could "benefit [the] Country by encrease of its Trade and Merchandise."[86] Yet these kinds of justification for natural history were less common than we might expect. This was partly because the pursuit of practical outcomes solely with a view to commercial gain was not thought to be consistent with gentlemanly studies. But, more prosaically, the economic benefits thus far produced by these researches had been meager, to say the least. Both the projectors of the mid-century and the Fellows of the Royal Society were particularly vulnerable to the claim that, in spite of their aspirations, they had actually accomplished very little in practical terms. "What useful thing have they done?" was a common, and not completely unjustified, question put to the promoters of the new sciences.[87] For the seventeenth century, the mooted economic rewards of scientific activity were at best promissory notes.

Natural history and the science of nature

The arguments advanced by Francis Bacon and defenders of the Royal Society pertained to both natural history and natural philosophy. To the extent that these arguments were successful, it can be said that the whole family of disciplines connected with the study of nature would enjoy an enhanced profile. However, as we have already noted, within that family of disciplines natural history had been regarded as inferior to natural philosophy because of its descriptive nature. Something of this subordinate relationship can still be seen in Jean le Rond d'Alembert's famous "Discours préliminaire" to the *Encyclopédie* (1751): "The philosopher gives the reasons for things, or at least seeks them. Those who stop at reporting what they see are mere historians."[88] In spite of this dismissive pronouncement, however, for some time the role of "mere historians" in the production of natural knowledge had been recognized as vital to the production of a sound philosophy of nature. Again, Francis Bacon was a key figure, for he had taken up the challenge of reforming natural history and redefining its relationship to natural philosophy.

Bacon made the point that the existing deficiencies of natural history were due not to its subject matter but to the incompetent manner in which it had been conducted in the past. There were flaws in the original deposit of knowledge accumulated by classical authors, but worse, their successors had rested content with the bookish enterprise of simply copying, translating, and occasionally augmenting the observations of the ancients. The fundamental principle of a reformed natural history was to "admit nothing but on the faith of eyes, or at least of careful and severe examination." Only through the strict adoption of this principle could the discipline be purged of "fables and superstitions and follies."[89] First-hand observation, along with a healthy skepticism about third-person reports and unsubstantiated written relations, thus became important filters for raw information about the natural world. Naturalists who followed this new critical approach generally specified that they had witnessed the relevant facts themselves, or that they were attested to by a sufficient number of reliable witnesses.[90] The founding of scientific societies furthered this trend by facilitating the communal witnessing of relevant facts.[91]

The traditional problem remained, however, of how to derive the general lesson from the particular fact. As we have already noted, this was a problem of the relation of history (which dealt with individual facts) and philosophy (which dealt with general explanatory principles). Bacon argued that the flaw of past natural philosophers had been a tendency to draw conclusions prematurely, on the basis of a few commonsense observations. This deficiency could be remedied if philosophers could be persuaded to postpone the formulation of general theories until a sufficiently large number of observations had been compiled. In Bacon's understanding of the relation of the disciplines, then, natural history was to provide the indispensable factual foundation for natural philosophy.[92]

Bacon's insight that causal speculations and general conclusions must be grounded in large collections of facts was now widely acknowledged.[93] If the natural sciences were to be grounded in systematic and objective observations of the world, natural history provided the first, and arguably most important, stage of the science of nature. Gradually this principle came to be enshrined in formal accounts of the relationships among the natural sciences. The third edition of the *Encyclopaedia Britannica* (1788–97) thus stated that "classification and arrangement is called NATURAL HISTORY, and must be considered as the only foundation of any extensive knowledge of nature."[94] The new status of natural history was also reflected in institutional settings. John Gascoigne has estimated that the eighteenth century witnessed a significant rise in the number of university professorial chairs that were dedicated to natural history, a trend that gradually redressed the grievances of Ray and others.[95] Natural history also enjoyed an increasing vogue amongst the educated public. Louis Daubenton, in response to the *Encyclopédie*, remarked that "the science of natural history is more cultivated than it ever has been . . . at present natural history occupies the public more than experimental physics or any other science."[96] Daubenton had written some 900 articles for the *Encyclopédie* on topics relating to natural history. The total number of entries in this field was 4,500, which provides a clear indication of the profile and popularity of the field.[97]

In all of this, it may be said that, in the eighteenth century, natural history came of age. Originally contentious claims about its moral and theological utility gained credence. Works of natural history attracted a growing and increasingly broad audience, and the idea that such books had a salutary moral and religious effect on their readers was widely accepted. Although at times natural history could degenerate into a mere "culture of collecting," its scientific status also became more secure. The Baconian justifications for natural history were bolstered by the development of powerful systems of classification, which demonstrated that its practitioners were capable of more than simply amassing facts to pass on to their philosophical superiors. The Linnaean system, in particular, held out the promise of a complete catalog of living things, divided into types on the basis of rational principles. The eighteenth century also saw the emergence of the recognizably modern disciplines of botany, zoology, and geology. Together, such developments conferred upon natural historians a cultural authority that encompassed the realms of science, morality, and theology.

CHAPTER TWO

Biology, atheism, and politics in eighteenth-century France
Shirley A. Roe

In eighteenth-century Europe, many intellectuals (often called *philosophes*) worried about the specter of atheism. Referring to the question raised by the skeptic Pierre Bayle, "Could a society of atheists be a moral one?," major figures such as François-Marie Arouet de Voltaire (1694–1778) refuted atheism at every turn. One arena for these debates was that of living organisms, where questions about generation (reproduction) occupied center stage. Just as in the nineteenth century, when the evolution of living things became an explosive issue, or the twentieth century, with its controversies over when individual human life begins, how the process of generation could be explained in the eighteenth century led not only to biological but also to religious and political questions. How the natural world operates and how we understand it have been repeatedly at the center of major controversies that have spilled over into the realms of religion and politics.

One answer to the question of generation that was popular in the eighteenth century was preformation, the belief that "germs," pre-organized matter from which all organisms would ever be developed, had existed since the Creation, when they were directly created by the hand of the Divinity. Encased within one another (the theory of *emboîtement*), all germs existed in the first member of each species. By involving God's creative power in every future instance of reproduction, preformation came to be seen by many as a bulwark against the immorality to which atheism would inevitably lead. By contrast, the theory of epigenesis, which explained each instance of reproduction as the formation of an organism out of unorganized matter, was thought by many to open the door to materialism and atheism. Major naturalists at mid-century, such as Charles Bonnet (1720–93) and Albrecht von Haller (1708–77), combined scientific evidence in favor of the preexistence of germs with arguments concerning the existence of God. Even one of the century's major opponents of preformation, John Turberville Needham (1713–81), explained to Bonnet late in his life that he had undertaken to produce a new theory of generation not based on preexistence precisely because that theory, which he thought would collapse under the weight of new evidence, had provided one of the best defenses against atheism. Needham wanted to forestall the triumph of materialism and atheism by providing a new nonpreformationist foundation for religion and morality.[1]

In the eighteenth century, moral behavior was generally believed to rest on a belief in God and in a future day of judgment. Revealed religion, that is religious belief based on God's having revealed himself to mankind through revelation and miracles, was still very widespread. Natural religion, belief in God based on understanding and admiring the created world, was certainly increasing as a complement to revealed religion, as was deism, which based religious belief solely on the order and harmony present in nature. Deism was much more threatening to some because of its proximity to atheism. Yet whether as deists or as believers in revealed religion, naturalists and natural philosophers viewed nature as evidencing God's design and as providing a foundation for religious belief. Were nature to be separated from God as creator and explained solely in terms of natural or material categories, this source of religious belief would be seriously threatened. And thus the foundations of morality and social order could be undermined.

Furthermore, a self-creative nature, one based on active matter, would exhibit no preordained order, only an order born out of material interactions. Human beings, as part of nature, would now have to be understood as part of this material, continuously active nature, rather than as part of a divinely ordered system. Thus a new naturalistic basis for moral behavior, one resting on human nature rather than on divine rules, would have to be developed to replace religion. Those who supported materialism, the self-creative activity of matter, such as Denis Diderot (1713–84) and Paul-Henri Thiry d'Holbach (1723–89), welcomed the new biological evidence and set out to develop a natural basis for morality and society.

But why in the mid-eighteenth century did generation become the central arena in which these contradictory views clashed? Why did naturalists and *philosophes* such as Voltaire point to preformation in particular as a counteracting force to atheism? The answer is twofold. First, the rise of belief in the preexistence of germs in the late seventeenth century was directly tied to concerns over the limitations of the new mechanical philosophy. Whether matter and motion could be sufficient to produce new organisms at reproduction was doubted by many (although not by René Descartes). The complexities and regularities of living organisms seemed to be beyond the capabilities of mechanical interactions. Furthermore, mechanism, especially in its Cartesian form, skirted close to the edges of atheism. If God could have created the universe by simply creating all of the matter and adding the initial motion to start things off, was there not a danger that such a philosophy could easily lead to a denial of the necessity for God at all? Preformation by preexistence provided a solution to both of these problems.

The second way in which generation theory brought the atheism question to the fore was that new evidence was discovered for active matter and a self-creative nature. Observations of very small and microscopic creatures by Abraham Trembley (1710–84), Georges-Louis Leclerc, Comte de Buffon (1707–88), and Needham indicated that preexistence might not be the norm in the microscopic world. This new evidence directly challenged the hold preformation had maintained over generation theory for more than a half-century. The result was a resurgence of preformationist research and thinking, especially in the work of Bonnet, Haller, and Lazzaro Spallanzani (1729–99), as well as the rise of a truly materialist view of living phenomena, in the hands of Diderot. This placed generation theory at the heart of the debates over biological materialism that dominated the 1750s, 1760s, and 1770s. The implications for atheism and consequently for morality of a self-creative nature without preexistent germs was obvious to both sides—worrisome to the preformationist, provocative to the materialists.

But it was not only naturalists and *philosophes* who worried about the implications of material activity and generation. These concerns entered the political sphere as well, especially in France in the 1750s and 1760s, when controversies between King Louis XV and the Paris Parlement over political power, an attempted assassination of Louis himself, and increased censorship of radical or atheistic publications brought the problem of the generation of living organisms into the realm of political controversy. Not only was active matter seen as providing a foundation for atheism and immorality but for challenges to the political hierarchy as well. If all living organisms (including human beings) had been created as part of an ordained political order, preformationism provided good evidence for maintaining the status quo of a stable society with an unchangeable religious and political hierarchy. But if material activity produced all that we know in nature, without divine guidance, then the traditional view of a monarchical society could be undermined. Concerns over the implications of these new views of nature entered the political realm of debate in the storm of protest over the *Encyclopédie* and other scandalous works that erupted in the late 1750s.

In this chapter I look first at the rise of preformationist thinking in the late seventeenth century. I then turn to the biological evidence that challenged preexistence in the mid-eighteenth century and to the reaction this provoked among the preformationists. I also focus on the work of Voltaire and Diderot to illustrate the connections between the generation debates and the materialism question more widely. Finally, I demonstrate how these controversies about nature entered the wider political debate in France.

Mechanism and the preexistence of germs

The theory of the preexistence of germs, or preformationism, arose in large part as a reaction to Cartesian mechanism. René Descartes (1596–1650), whose goal was to explain all natural phenomena on the basis of matter and motion alone, extended his work on physical nature to human physiology, writing his treatise *L'Homme* in the 1630s (Figure 2.1).[2] Yet the key to generation eluded him for another decade, and it was not until shortly before his death in 1650 that he finally worked out his explanation for the mechanical formation of animal embryos from the mixing of semen from both parents, through a fermentation of particles. His resulting treatise, *De la formation du foetus*, appeared posthumously with his earlier physiological work in 1664. (Descartes, a Catholic, had feared reprisals from the Catholic church after the condemnation of Galileo and did not publish these works himself.) Descartes' mechanistic explanation of generation based solely on matter in motion was for him the capstone of his mechanistic view of life. The organism that was to result from reproduction was, he believed, determined by the matter out of which the two semens were

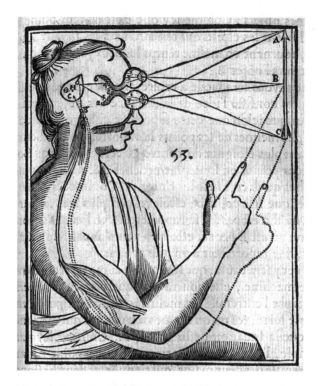

FIGURE 2.1 Mechanical coordination among eyes, brain, muscles, and nerves as the finger points to different spots on the arrow. (From René Descartes, *L'Homme*, 1664. Courtesy of Yale University, Harvey Cushing/John Hay Whitney Medical Library.)

made and the mechanical process that ensued upon their mixing. As he put it, "If one knew what all the parts of the semen of a certain species of animal were, for example, particularly of man, one could deduce from this alone, by reasons entirely mathematical and certain, the whole figure and conformation of each of its members."[3]

Descartes' theory, however, was seen by others as both insufficient and disturbing. Nicolas Malebranche (1638–1715), who first formulated a theory of the preexistence of germs in 1674, was responding directly to Descartes when he wrote, "The rough sketch given by this philosopher may help us to understand how the laws of motion are sufficient to bring about the gradual growth of the parts of an animal. But that these laws should form them and link them together is something no one will ever prove. Apparently M. Descartes recognized this himself, for he did not press his ingenious conjectures very far." As a Cartesian in physics, Malebranche believed that God's role in the physical universe was limited to his having imparted the initial motion that then would be communicated from body to body for as long as the world existed. The question regarding generation was: "Could this communication of motion from one particle of matter to another be sufficient to create a new organism?" Malebranche thought not and found evidence in the tulip bulb, in which he saw, using a magnifying glass, the folded up parts of the future tulip plant. Others believed they had seen the same thing in chicken and frog eggs. These observations made sense to Malebranche because "it is easy to see that the general laws of communication of motion are too simple for the construction of organized bodies." Rather, motion was the means by which God's initial creative act was to be carried out through all future time: "At the time of the Creation he constructed animals and plants for all future generations; he established the laws of motion that were necessary to make them grow. Now he rests, for he does nothing other than follow these laws."[4] As to the question of how all future organisms of each species could possibly be contained in their first representative created by God, Malebranche argued that because matter is infinitely divisible (another Cartesian position), this is at least conceivable. But what is not conceivable, he maintained, is that the laws of the communication of motion themselves could create new organisms at each generation. Thus preexistence combined mechanistic physics with the Cartesian view of God's initial involvement in our world. After creating the matter out of which it would form and imparting to it the initial necessary motion, God ceased to be directly involved in natural phenomena.

Malebranche was well aware of microscopic research being done by his contemporaries, and he referred to Marcello Malpighi's observations on chick eggs (Figure 2.2) and Jan Swammerdam's on frog eggs to support his ideas on preexistence. Swammerdam had proposed the preexistence of germs in a Dutch work of 1669, which became more widely available in Latin in 1672. Malpighi's observations on chick development, published in 1673, gave clear evidence that the rudiments of the embryo could be seen in a fertilized egg that had not yet been incubated (but not in an unfertilized egg). Even so, Malpighi's observations, along with Swammerdam's, were often cited as evidence for the preexistence of the tiny preformed organism in the chick egg or the frog egg, or in the plant seed.[5]

Yet the fact that these observations were immediately taken up by those making preformationist claims indicates that the concept of

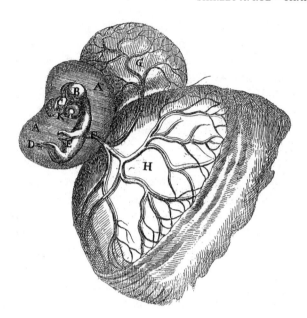

FIGURE 2.2 Chick embryo at six days. (From Marcello Malpighi, *Opera omnia*, 1686.)

preexistence was not one that grew out of observational evidence alone. It is clear, as several scholars have pointed out, that preformation through preexistence was a theory that responded more to philosophical than to observational needs. Jacques Roger was the first to explore the ties between preexistence of germs and mechanism, and to point out that, because the mechanistic view of the universe rested on passive matter and a non-interventionist God, preexistence made very good sense: "nature as a whole . . . , by becoming a complex of mechanisms, had lost all spontaneity and become pure passivity in the hands of God, the God who had created it and was now content with simply maintaining its existence and motion. In the last analysis, nothing appeared in nature that did not come from the original creation of all things."[6]

Furthermore, as in the case of Malebranche, one of the principal motivations behind preformationist theories was the need to combat the implications that trying to explain generation mechanistically without preexistence would entail. Although it was widely believed in the late seventeenth century that the universe must operate through mechanical laws, it was also felt that these laws were not sufficient to account for its origins and especially for the construction of living organisms. Claude

Perrault, another early proponent of development through preexistent germs, made explicit such a concern over the limits of mechanical laws. As he remarked in 1680, "I do not know if one can comprehend how a work of this quality would be the effect of the ordinary forces of nature . . . for I find finally that it is scarcely more inconceivable . . . that the world has been able to form itself from matter out of chaos, than an ant can form itself from the homogeneous substance of the semen from which it is believed to be engendered."[7]

The preformation–mechanism synthesis was very successful. Although there was some early opposition among a few mechanists, preexistence became the dominant theory by 1705 and remained so until mid-century. Among its supporters one can cite a number of figures who did not agree in many other aspects of their philosophies, including, for example, Hermann Boerhaave, Gottfried Wilhelm Leibniz, Bernard de Fontenelle, and René Antoine Ferchault de Réaumur. (Some believed in preexistence in the male spermatozoa, but most were ovists.) The notion that God must be fundamentally involved in each instance of reproduction, but only from his initial creative act and the mechanical laws that he had established, had widespread appeal.

As mentioned earlier, atheism was seen by many as implicit in the Cartesian universe. Descartes had shown how God could have created the world out of matter mechanistically had he so chosen. But if matter could form the universe by itself, who is to say that it had not? Likewise, if the material interactions resulting from the mixing of male and female semen could produce a living organism, how were any of us tied to God's creative act? The problem of origins, of the world and of living things was, for many, the weak point in Descartes' vision, and requiring simply that God give the universe the first motion was seen as insufficient by most to provide a bulwark against atheism. Preformation by preexistence of germs created a much more satisfying bulwark by ensuring that God was part of each and every generation of a living organism. Thus one had good mechanistic science without the specter of a godless universe.

Challenges to preexistence

The impact of Newtonian mechanism on generation theories was not felt until mid-century, when, at the hands of Newtonians such as Pierre-Louis Moreau de Maupertuis (1698–1759) and Buffon, new ideas on epigenetic development by attractive forces were proposed. The preformation–mechanism synthesis was thus challenged by the broadening of mechanism to include forces in addition to matter and motion. Might one be able to conceive of a theory of generation not based on preexistent germs but rather on matter and forces? This is what both Maupertuis and Buffon attempted to do; not wholly independently of one another, for the two apparently discussed issues relating to generation frequently in the mid-1740s. Maupertuis' anonymous work, *Vénus physique* (1745), presented an explanation of generation based on the notion that particles resembling all parts of the body are collected in the reproduction organs, there to form, via attractive forces, the new organism when seminal fluids from both parents are mixed together. Attractive forces had long been used in astronomy and chemistry: "Why," Maupertuis asked, "should not a cohesive force, if it exists in Nature, have a role in the formation of animal bodies?"[8] As one of the first proponents of Newtonianism in France, Maupertuis felt that attributing generation to attractive forces was far preferable to the absurdities of preexistence.

But it was not just Newtonian notions of attraction that led to these new challenges to preexistence theories. During the 1740s a series of biological discoveries seemed to indicate that matter is not totally passive in the generation process, at least among very small organisms. The first of these was Abraham Trembley's discovery in late 1740 that a little aquatic creature, the freshwater polyp, possessed remarkable regenerative capabilities. Unlike other creatures that could regenerate an injured limb, the polyp, when cut up into several pieces, regenerated into as many completely new organisms as there were cut-up pieces. Trembley's observations caused an immediate stir when they were announced to the Paris Academy of Sciences and to the Royal Society, for, as Aram Vartanian has remarked, "Trembley's contemporaries had the startling spectacle of Nature caught, as it were, *in flagrante* with the creation of life out of its own substance without prior design."[9]

Two other sets of observations from the 1740s were to prove equally controversial, those of Needham on revivification of eels in blighted wheat and those of Needham and Buffon on the generation of microorganisms in infusions. In the early 1740s Needham had observed that when he added water to whitish fibers found in grains of blighted wheat these fibers appeared to come to life as tiny worms, or "eels," that moved in a twisting motion for several hours. When he tried the experiment two years later on some of the grains he had saved from the original lot, the same phenomenon occurred. Similar, but somewhat more ambiguous observations were made by Needham and a colleague on "eels" found in mixtures of flour and water.[10] Although Needham made few speculations in these early years about the manner in which these organisms appeared to generate, his observations later became quite controversial in the hands of Voltaire (who ridiculed them) and d'Holbach (who used them as evidence for materialism).

The event that was to have the most impact was the publication in 1749 of the first volumes of Buffon's *Histoire naturelle*. The theory of generation and view of life that was presented in volume 2 was immediately controversial. Here Buffon argued that all animals and plants were composed of a type of matter that was both particulate and active. His *molécules organiques* [organic molecules] cycled through living systems in an unending process of generation, growth, and decay. Buffon countered preexistence of germs with the reorganization of organic particles into new creatures at each instance of reproduction. And he unequivocally promoted active matter as the basis of this process, proclaiming, "Living and animation, instead of being a metaphysical degree of being, is a physical property of matter."[11]

Buffon conceived of his new theory in the mid-1740s and began writing an account of it. Then in 1748 Buffon joined forces with Needham to perform a series of microscopic observations on seminal fluids and on infusions (Figure 2.3). These observations provided new evidence, both believed, for the operation of active matter at the microscopic level. Buffon concluded that freed organic particles could unite together spontaneously to produce microscopic organisms. Needham built his own theory of generation on the evidence these observations seemed to provide for the existence of a universal vegetative force.[12]

The publication of Buffon's and Needham's new views on generation, based on concepts of active matter and on evidence from the microscopic world, had a profound effect on biological thinking. The immediate effect was to change, or to begin to change, the theorizing of several figures who were soon to play key roles in the materialism debates. Diderot, for example, read the first volumes of the *Histoire naturelle* in 1749 while imprisoned at Vincennes for publishing the *Lettre sur les aveugles*.[13] The effect on him was to turn his attention toward biology as a fertile source of evidence for materialism. Haller, to take another example, reacted to the *Histoire naturelle* by beginning a shift in thinking that eventually brought him back from epigenesis to preformationism. Haller's encounter with Buffon's

FIGURE 2.3 Needham and Buffon (seated) performing observations with a microscope. (From Georges-Louis Leclerc, Comte de Buffon, *Histoire naturelle*, vol. 2, 1749.)

theory of generation brought him up against the materialist implications of epigenesis, only a few short years after he himself had adopted epigenesis largely in reaction to the discovery of the polyp's regenerative abilities.[14]

Bonnet, to cite a third example, was in the midst of composing his own treatise in support of preformation when he read the first volumes of the *Histoire naturelle*. The effect on him was to cause him to abandon his project altogether for over a decade. Only after Haller completed his own transition back to preformationism and published a series of important observations on chick development did Bonnet resuscitate his book, adding significant new material to it for its publication in 1762.[15] He later explained in his autobiography the reasons why he felt compelled to present his own preformationist theory: "to combat the different systems based on epigenesis, particularly those of MM. de Buffon and Needham; . . . [and] to oppose to these strange opinions a hypothesis that conforms more to the facts and to the principles of sane philosophy."[16]

Thus one of the effects of the new theories of generation proposed by Maupertuis, Buffon, and Needham was the production of the most clearly articulated theories of preformation in the work of Haller and Bonnet. Another effect was to promote materialism and to provide biological evidence for the existence of active and self-creative matter. This is most clearly seen in the works of Diderot and d'Holbach, and in the rising concern over materialism expressed by Voltaire. Both Haller and Bonnet found two principal areas of concern in the new biological theories. First, they did not see how forces could, on their own, be responsible for the generation of complex, seemingly designed, living organisms. Second, they were worried about the implications of active matter—not just in biology, where the old issue of spontaneous generation was given new life, but even more for religion, morality, and social stability.

With regard to the role forces were required to play in generating new organisms, the objections were quite similar among the preformationists. Réaumur was the first to make these worries clear, in his reaction to Maupertuis' *Vénus physique*:

> how will attractions be able to give to such and such a mass the form and structure of a heart, to another that of a stomach, to another that of an eye, and to another that of an ear? How will they fashion other masses into vessels, valves, etc. . . . . It is too evident that in order to succeed in forming such a complicated edifice, it is not enough to multiply and vary the laws of attraction at pleasure, and that it is necessary to give to this attraction the greatest amount of knowledge.[17]

Haller echoed these sentiments two years later when, in critiquing Buffon's new theory, he argued,

> I do not find in all of nature the force that would be sufficiently wise to join together the single parts of the millions and millions of vessels, nerves, fibers, and bones of a body according to an eternal plan. . . . M. Buffon has here the necessity of a force that seeks, that chooses, that has a purpose, that against all of the laws of blind combination always and infallibly casts the same throw.[18]

Buffon had argued that a "penetrating force" acted as part of the process whereby the *moule intérieur* [internal mold] organized the organic molecules in the mixed semen into the new offspring. But such a force, Haller believed, would violate the laws of Newtonian mechanism, where forces were considered to be invariable and automatic, and without any capacity to "choose" or to have a "purpose."

But it was not just the addition of forces to the generation process that caused problems for Haller, Bonnet, and other supporters of preformation. Buffon and Needham had based their theories on conceptions of matter that went beyond the traditional mechanist reliance on passive matter. The new generation theories—supported by the new biological evidence provided by the polyp and by microorganisms—proclaimed a nature in which matter was fundamentally active, not one that relied entirely on God-given automatic forces for any activity. And it was this new view of nature that was so provocative for individuals like Diderot and so disturbing to preformationists like Bonnet and Voltaire. Nature seen as active and self-creative directly challenged the conception of a divinely ordered, unchanging, and hierarchically arranged nature so fundamental to the preformationists. This clash of views of nature and the reactions to the possibility of defining a nature in which God was no longer necessary can best be seen by examining more closely the works of Voltaire and Diderot.

Voltaire and preformation

Having been born before the turn of the eighteenth century, Voltaire was one of the oldest of the *philosophes*; he was also the longest lived. One of the formative influences on his life was the years of exile he spent in England from 1726 to 1729; he returned to France a confirmed Newtonian. Along with Maupertuis, Voltaire was to greatly influence the introduction of Newton's ideas on the continent, in part through his *Éléments de la philosophie de Newton* (1738).[19] Voltaire adopted as well a deistic vision of nature, believing that the order and harmony of the universe testified to the existence of an intelligent cause, and he subscribed to the mechanist view of passive matter.

Although Voltaire indicated his belief in preexistence as early as 1740, in his *Métaphysique de Neuton* his preoccupation with biological topics, which is prominent in many of his later writings, did not begin until the 1760s.[20] He had learned of the new epigenetic ideas from reading Maupertuis' *Oeuvres* in 1752. He reacted quite critically to Maupertuis' theory, but it was what he learned about Needham's observations about finding live "eels" in formerly dried blighted wheat grains that piqued his satirical wit. In a series of pamphlets he wrote satirizing Maupertuis' work, Voltaire described a "séance mémorable," a parody of the Berlin Academy of Sciences, of which Maupertuis was president for ten years. Voltaire humorously portrayed the "galant" president serving the ladies present "a superb dish composed of pâté of eels all one within the other and born suddenly from a mixture of diluted flour" as well as "great platters of fish that were formed immediately from grains of germinated wheat."[21]

But it was not until the mid 1760s that Voltaire became alarmed at the implications of Needham's and Buffon's work for materialism. What seems to have prompted this concern was a visit Voltaire had from one of Diderot's friends, Étienne-Noël Damilaville (1723–68). Damilaville visited Voltaire in the summer of 1765, and he apparently presented pro-atheism and pro-materialism arguments to Voltaire, under the exhortation of Diderot (who was writing to him at Voltaire's house).[22] It was very likely from these conversations with Damilaville that Voltaire began to realize the connection between the new biological theories and materialism. By 1765 Diderot had completed his own transition from deism to atheism, and his growing interest in biological evidence for materialism had played a major role in this transition.

By coincidence, Voltaire had become embroiled, that same summer, with Needham in a pamphlet war over the subject of miracles. He did not discover Needham's authorship until August, and thereafter he included lampoons on Needham's biological work along with his arguments on miracles. "You have made a small reputation for yourself among atheists by having made eels from flour, and from that you have concluded that if flour produces eels, all animals, starting with man, could have been born in approximately the same manner." This was not in fact what Needham had been arguing, but Voltaire never bothered to learn much more about Needham's actual views. In a collection of his and Needham's pamphlets he published that same year, Voltaire added a preface to one of his pamphlets, which made his concern over the implications of Needham's work even more explicit. Falsely characterizing Needham as an Irish Jesuit disguised in secular clothing and roaming the countryside spreading papist dogma, Voltaire explained that he had also "meddled with experiments on insects" and believed he had seen the flour of blighted wheat change into small animals. Needham had been wrong, Voltaire claimed, as had recently been shown by experiments made by Spallanzani. Furthermore, Voltaire continued, Needham's claim was false for another even more superior reason, namely, that the fact is impossible.

> If animals are born without germs, there would no longer be a cause of generation: a man could be born from a lump of earth just as well as an eel from a piece of paste. This ridiculous system would moreover obviously lead to atheism. Indeed it happened that several philosophers, believing in the experiment of Needham without having seen it, claimed that matter could organize itself; and the microscope of Needham came to be seen as the laboratory of atheists.[23]

Among these duped philosophers Voltaire was to include Maupertuis, Buffon, whose *Histoire naturelle* Voltaire seems to have first read in 1767, and an anonymous translator of a new French edition of Lucretius' *On the Nature of Things*, which appeared in 1768 and which contained a positive report of Needham's work in a note. This translation had actually been produced by the tutor of d'Holbach's children (an individual named Lagrange) under the direction of Diderot and d'Holbach, and it is quite likely that the note came from Diderot. When d'Holbach's anonymous *Système de la nature* appeared in 1770 he too was added to the list. During 1767 and 1768 Voltaire unleashed a series of attacks on Needham and the philosophers misled by him. In *Des singularités de la nature* (1768), for example, he described Needham's experiments, then remarking, "Immediately several philosophers did their best to cry marvel, and to say: There is no germ; everything is formed, everything is regenerated by a living force of nature. It is attraction, said one [Maupertuis]; it is organized matter, said another [Buffon]; these are living organic molecules that have found their molds. Good physicists were deceived by a Jesuit."[24]

In 1770 Voltaire's fears about the foundation that Needham's work could provide for materialism were realized in d'Holbach's *Système de la nature*. Using Needham's observations, along with examples of chemical combustion, d'Holbach argued that matter is fundamentally active. Needham's work had shown "that inanimate matter can pass into life, which is itself only an assemblage of movements." Moreover, d'Holbach continued in a footnote, "would the production of a man independently from the ordinary means be more marvelous than that of an insect from flour and water?"[25] Voltaire was incensed by d'Holbach's book as well as its reliance on Needham's views. His reaction culminated in a pamphlet, *Dieu: Réponse de Mr. de Voltaire au Système de la nature* (1770), which formed parts of his entry "Dieu, Dieux" [God, gods] in his *Questions sur l'encyclopédie*.[26] Voltaire attacked Needham's work and d'Holbach's having been duped by it, and in response simply reaffirmed his belief in the incapacity of matter to produce life and intelligence. Thinking, sensing beings could only have been created by a power superior to mankind.

In many respects Voltaire was not the typical preformationist. He was not a naturalist himself, relying on what he had read or heard about the new biological discoveries. He was also atypical in being a deist, for Bonnet, Haller, and other preformationists were believers not only in the order and harmony of nature but also in revealed religion. (Both Bonnet and Haller, for example, wrote books on religious apologetics during the same years they were publishing on preformation and combating the work of Buffon and Needham.) Voltaire was always critical of organized religion and wrote much against the excesses of Christian fanaticism. Yet Voltaire's underlying concerns that led him to defend preformation were surprisingly similar to those of Bonnet and Haller, even though they never recognized him as an ally. For Voltaire, the rise of atheism would mean the demise of a moral society. As he explained in a letter written in 1768, atheists had never been able to answer the argument "that a watch proves a watchmaker." Mentioning to his correspondent the work of Needham, who had "recently furnished arms to atheistic philosophy in pretending that animals can form themselves all alone," and "another fool Maupertuis," Voltaire proclaimed, "May God preserve us from such atheists." Finally Voltaire concluded simply, "My dear Marquis, there is nothing good in atheism. This system is very bad in physics and in morals. An honest man can protest strongly against superstition and fanaticism; he can detest persecution, he renders a service to the human race if he spreads human principles of toleration; but what service can he render if he spreads atheism?"[27]

This concern was echoed two years later in another letter written in response to d'Holbach's *Système*. Pointing once again to Needham's "false experiment" as the foundation of d'Holbach's views, Voltaire continued, "Moreover, I think that it is always very good to support the doctrine of the existence of a rewarding and vengeful God; society needs this opinion. I do not know if you recognize this verse, 'If god did not exist, it would be necessary to invent him.'"[28] This verse was in fact Voltaire's own, written a year earlier in a poem intended as a reply to the infamous anonymous tract, *Traité des trois imposteurs* (1765).[29] Whatever Voltaire's own personal views were with regard to God, he felt that belief in God's existence was essential to preserving the social order, in addition to being demonstrated by the natural order. For Voltaire the proper society required a government based on an enlightened monarch and a moral code founded on justice, toleration, and a belief in God. Preformation, by involving God necessarily in the formation of each living organism and by laying out a lawful process whereby development could proceed, provided just the biological foundation Voltaire sought to ward off the onslaught of atheism and social disorder.

Diderot and materialism

Born nineteen years later than Voltaire, Diderot was one of the next generation of *philosophes* and was most active from the mid 1740s through the 1770s. His early work as a translator led to the *Encyclopédie* editorship, which began in 1747 and lasted until the final volumes of plates were published in 1772.[30] He was imprisoned for just over three months in 1749 for the publication of his *Lettre sur les aveugles* (1749); previously, his *Pensées philosophiques* (1746) had been condemned and burned by the Paris Parlement.[31] These experiences caused Diderot to leave many of his more radical writings unpublished during his lifetime.

Sometime between 1749 and 1751 Diderot made the acquaintance of the Baron d'Holbach; from then on he was a constant participant in the famous dinners at d'Holbach's house, and a friend and intellectual confidant of d'Holbach himself. D'Holbach contributed a number of articles to the *Encyclopédie* (many on mineralogy and geology, but also on more controversial topics), and Diderot very likely made at least an editorial contribution to d'Holbach's *Système de la nature*. He was actively involved in the series of anti-clerical works that issued from d'Holbach's pen in the 1760s, prompting Diderot to remark in a letter in 1768, "It is raining bombs in the house of the Lord."[32]

Diderot's introduction to the new biological theories occurred in 1749 with his reading of the first volumes of Buffon's *Histoire naturelle* while he was imprisoned. Previous to this he does not seem to have been particularly interested in biology, and in his *Pensées philosophique* (1746) he presented the classic deist argument for the existence of God: "It is not from the hand of the metaphysician that atheism has received its heaviest blows. The sublime meditations of Malebranche and of Descartes were of less use in shaking materialism than a single observation of Malpighi's." Diderot rejected spontaneous generation as well, claiming "The discovery of germs alone has destroyed one of the most powerful arguments for atheism."[33] Although a deist in the *Pensées*, by 1749 Diderot had moved much closer to atheism.

In 1751 Diderot published an article "Animal" in the first volume of the *Encyclopédie*, in which he reprinted sections from Buffon's *Histoire naturelle* with interspersed comments that indicate a willingness to go beyond Buffon's views (for example, by suggesting that animals possess the faculty of thinking). In his *Pensées sur l'interpretation de la nature* (1754),

Diderot began questioning Buffon's distinction between organic and inanimate matter, asking if there were really any differences between living and dead matter other than organization and self-movement.[34] In his letters in the late 1750s he continued to wonder how matter could pass from a dead state to a living one. In one of these he remarked that, by eating dead food, organisms grow, so that, as he put it "something dead put alongside something living began to live." Yet, he continued, "You might as well say that if I put a dead man in your arms he would come back to life."[35] By 1765 Diderot seems to have rejected Buffon's distinction and to have settled on the idea that there is no essential difference between dead and living matter; rather something he called *sensibilité* is a property of all matter—inert in inanimate matter but rendered active in living organisms and even producing thought in higher organisms. It was in this same year that, after hearing of Diderot's views from Damilaville, Voltaire became alarmed at the implications of active matter in biology and began his extended critique of Needham and Buffon.

The culmination of Diderot's materialist thinking in biology was his witty and provocative dialogue, written in the late summer of 1769, the *Rêve de d'Alembert*. While in the midst of its composition, he wrote to his friend Sophie Volland,

> It is the height of extravagance and at the same time the most profound philosophy. It is quite shrewd to have put my ideas in the mouth of a man who is dreaming. It is often necessary to give wisdom the appearance of folly to gain admittance for it. I like it better when it is said: But this isn't so mad as you might think, than to say: listen, here are some very wise things.[36]

In the second part of the *Rêve*, Diderot portrayed his fictional character d'Alembert in the midst of a dream. At one point d'Alembert dreamt he was looking through a microscope at an infusion. As one of the other characters described it, "The flask in which he perceived so many momentary generations he compared to the universe; he saw in a drop of water the history of the world. . . . He said 'In Needham's drop of water everything happens and passes away in the twinkling of an eye. In the world the same phenomenon takes a little longer; but what is our duration in comparison with the eternity of time?'" Depicting a universe of ceaseless activity and change, Diderot proclaimed, "You have two great phenomena, the passage from the state of inertia to the state of sensibility, and spontaneous generation; let them suffice for you: draw from them the correct conclusions."[37] For Diderot, Needham's microscopic observations provided the model for a world based on ceaseless activity and change, rather than preordained stability.

During the same years as he was developing his ideas on biological materialism, Diderot was actively involved in the anti-clerical and atheistic publication ventures of d'Holbach's coterie. Their program was not entirely negative. For one of their goals was to replace the traditional concept of morality based on future reward and punishment with a new naturalistic morality, one that would be founded on the laws of human nature and human society. Diderot developed a purely empirical moral theory, one that based moral ideas and the development of a moral sense on the individual's experiences of pleasure and pain. Yet Diderot did not make these views public; after the publication of the *Système de la nature*, it was d'Holbach who went on to publish a series of books in the 1770s expounding a materialist view of morality.[38]

The political stakes

A naturalistic morality not based on religious tenets also challenged the underpinnings of absolute monarchy. The threat that a ceaselessly active material world presented to a hierarchical, preordained political world was not lost on the French government. This is evident in the clash between the *philosophes* and the monarchy over the *Encyclopédie*, which began with the very first volume. Once again, we find Diderot at the center of the controversy.

Diderot's political views were most clearly presented in two of his *Encyclopédie* articles, "Autorité politique" [Political authority] and "Droit naturel" [Natural rights]. Diderot extolled the virtues of a monarchy limited by law, in which sovereignty would lie with the will of the people. A firm believer in natural rights, Diderot argued like others of his contemporaries that humans are by nature reasonable and sociable. He opposed both the divine sanction of monarchy and the concept of enlightened despotism. His views were thus by no means radical but they were clearly in opposition to the existing governing institutions in France.[39]

It was through his editorship of the *Encyclopédie* that Diderot found himself in direct conflict with the monarchy and the Paris Parlement. The first volume appeared in 1751, and by the time the second one appeared, Diderot and d'Alembert found themselves in trouble with the authorities. On 7 February 1752, the King's Council issued a decree suppressing the first two volumes of the *Encyclopédie*. In part, the decree read: "His Majesty has recognized that several maxims inserted in these two volumes aim to destroy royal authority, to establish a spirit of independence and revolt, and, under obscure and ambiguous terms, to promote the foundations of error, of moral corruption, of irreligion, and of unbelief."[40]

The article that provoked much of the controversy was Diderot's "Autorité politique." Coupled with growing tensions between the king and Parlement over who had the final authority to approve laws, Diderot's political views found ready critics. The crisis died down, however, and the *Encyclopédie* resumed publication. Yet the 1750s continued to be tumultuous years, with France entering the Seven Years' War (1756–63), the king exiling Parlement, and the shocking assassination attempt in 1757, in which Louis XV was fortunate to escape serious injury. Surveillance of authors and their publications by the police increased, and following the assassination attempt, a draconian law was passed about subversive literature. "Anyone who is convicted," the new law read, "of having composed and printed writings tending to attack religion, to stir up spirits, to endanger our [the king's] authority, and to disturb the order and tranquility of our state, will be punished with death."[41] Although this law was apparently never enforced, its passage indicates that the king's ministers had become extremely hostile to the *philosophes*.

After five more volumes had appeared, the *Encyclopédie* was completely shut down. The most controversial article appearing in the seventh volume was d'Alembert's "Genève" [Geneva] in which he praised the Genevan clergy for their supposed deism, which incensed not only the Calvinist pastors but the French authorities as well, for the not-so-subtle implied criticism of the French Catholic church by comparison could not be tolerated. Consequently in January 1759, the Attorney General Joly de Fleury rose before the Paris Parlement to condemn the *Encyclopédie*, Claude Adrien Helvétius' *De l'esprit* [Concerning the spirit] and six other books deemed to be too radical. Joly de Fleury's speech opened with the following dire pronouncement: "Society, the State, and Religion present themselves today at the tribunal of justice . . . their rights have been violated, their laws disregarded. Impiety walks with head held high. . . . Humanity shudders, the citizenry is alarmed." What was the source of this fear and disquiet? A conspiracy, claimed Joly de Fleury, one that existed for the purpose of destroying society: "there is a project conceived, a Society formed, to uphold materialism, to destroy Religion, to inspire a spirit of independence, and to nourish the corruption of morals." Remarking that it was sad to think what posterity would think of their century, Joly de Fleury claimed that it had fostered "a sect of so-called Philosophers who... imagined a project... to destroy the basic truths engraved in our hearts by the hand of the Creator, to abolish his worship and his ministers, and to establish instead Deism and Materialism."[42]

Joly de Fleury was particularly critical of the *Encyclopédie* and the way the editors used cross-referencing among articles to subtly bring out radical points hidden in articles seemingly on an innocuous topic. One of his prime examples pointed again to biology. In Diderot's article on "Éthiopiens" [Ethiopians], which does not actually say much about Ethiopians, we find an interesting passage about how animals develop from the earth through fermentation caused by the heat of the sun (which was quoted by Joly de Fleury). At one point, Diderot claimed,

> The Ethiopians take themselves to be more ancient than the Egyptians, because their country has been more strongly struck by the rays of the sun, which give life to all beings. Whence one sees that these people are not far from regarding animals as the development of earth put into fermentation by the heat of the sun, and to conjecture in consequence that species have undergone an infinity of diverse transformations, before becoming the form that we see them in.[43]

At the conclusion of the article, Diderot cross-referenced the article "Dieu" [God]. Although this article is not especially controversial, in one section, drawn from the work of Fontenelle, we find a discussion about whether the first animals of each species were formed by chance interactions of matter or by the will of God. Fontenelle concluded that it was the latter, but principally by arguing that if animals had been formed by chance, why was it not still happening? A paragraph claiming that generation from corruption, or spontaneous generation, had been shown to be false by modern experiments ended in a cross-reference to "Corruption." This leads us to another d'Alembert article where Buffon's theory of organic molecules was favorably discussed. D'Alembert highlighted not only Buffon's claim that microscopic organisms can be formed by the fortuitous combination of organic molecules but also that tiny "eels" (larvae) form in flour paste in the same manner. Although d'Alembert admitted that most cases that appear to be spontaneous generation are actually regular generation from eggs, he queried, "but is it demonstrated in all cases that corruption can never engender an animated body?"[44]

These articles in the middle volumes of the *Encyclopédie* thus presented a very strong argument for the material basis of life and called into question the need for God's involvement even in life's creation. That Joly de Fleury used them as his prime example of the dangers contained within the pages of the *Encyclopédie*, which were reinforced by its pernicious system of cross-referencing to even more radical arguments, is clear evidence that debates in the biological realm had become central to those in the political realm as well.

As a result of Joly de Fleury's denunciation, Parlement suspended publication of the *Encyclopédie* pending further examination. Then, in March 1759, a royal decree was issued that condemned the *Encyclopédie* and suppressed its further publication by revoking its royal "privilege." The decree declared: "The advantage to be derived from a work of this sort, in respect to progress in the arts and sciences, can never compensate for the irreparable damage that ensues from it for morality and religion."[45] Diderot feared arrest, but he vowed to continue the project, even hiding some of his manuscripts in the home of the monarchy's director of the book trade and censorship, Chrétien-Guillaume de Lamoignon de Malesherbes.[46]

A second example demonstrating the central role radical views about life and matter played in the political discourse of this period can be found in a satirical pamphlet on the *philosophes*, the *Nouveau mémoire pour server à l'histoire des Cacouacs*, written by Jacob Nicolas Moreau (1717–89). Moreau, who had been employed by the Ministry of Foreign Affairs to write pro-France propaganda during the Seven Years' War (1756–63), was regarded by the *philosophes* as being aligned with the court. The "Cacouacs," a fictional tribe of savages, were the creation of the abbé Odet Joseph de Vaux de Giry de Saint-Cyr (1699–1761), advisor to the Dauphin (the heir to the French throne), who noted the similarity of their name to the Greek word "cacos," meaning "malicious."[47] The Cacouac episode began just when the seventh volume of the *Encyclopédie* was published and was part of the anti-*philosophe* publications that followed.

In Moreau's *Nouveau mémoire* we again find the connection between materialist biology and the dangerous threat the *philosophes* represented to society. The memoir opened with the capture of the young hero of the piece, who was eventually able to tell his story after he escaped and returned to Paris. The Cacouacs, he reported, lived in tents to signify their freedom, had no government, regarded ethics as a matter of convention, and did not believe in the existence of God. Although I do not want to describe the young man's adventures in detail, the key episode for my purposes here was when he was interrogated by a group of Cacouacs in preparation for his induction into their society. His first question, from a venerable old man, was "If dead matter could combine with living matter? How does this combination come about? What is the result?" A woman continued, asking, "If molds are the principal forms? What is a mold? Is it a real and preexisting being, or is it only the intelligible limits of a living molecule united to dead or living matter . . . ?" Under the influence of incense, the young man began "to understand everything perfectly," and he was told he could now regard the Cacouacs as his brothers. Eventually our young hero was rescued; he returned home only to find "dangerous and ridiculous" Cacouacs there as well. "I found," he reported, "that they had been given the name *Philosophes*, and that their works were being printed."[48]

Giry de Saint-Cyr wrote a follow-up work, the *Catéchisme et decisions de cas de conscience, à l'usage des Cacouacs*. The catechism was supposedly to be used by Cacouacs for the proper inculcation of the young into their philosophy, which of course allowed Giry de Saint-Cyr to expose the most radical and dangerous doctrines of the *philosophes* through the ruse of instruction. The catechism is in the traditional question-response format, beginning with the question "What is God?" and proceeding through topics such as the creation of the world, the soul, the relationship between humans and animals, humans in their primitive state compared with humans in civil society, freedom of thought, the nature of morality, happiness, and free will. What is most revealing about the catechism is that the instructor's responses to the questions posed are compiled from direct quotations from works by the *philosophes*. One can imagine the abbé, with works by Diderot, Jean Jacques Rousseau, Voltaire, and Helvétius scattered about on his desk, earnestly searching for yet another scandalous quote to add to the catechism. The result is a somewhat disconnected pastiche, but effective nevertheless.

Giry de Saint-Cyr again used the question of living or dead matter as a key part of his critique. In the second section, on the creation of the world and the formation of beings, quotations from Diderot are used to claim that the universe had formed by the chance combination of atoms and that living organisms formed as well from particles that already possessed "desire, aversion, memory, and intelligence" (a reference to Maupertuis). In an even more revealing quotation, again from Diderot's *Pensées*, the Cacouac teacher explained, "the embryo, formed out of these elements, has passed through an infinity of organizations and developments; that it has had in succession movement, sensations, ideas, thought, reflection, conscience, feelings, passions, signs, gestures, sounds, articulate sounds, language, sciences and the arts."[49]

Although there are many more topics of ridicule in both Moreau's and Giry de Saint-Cyr's depictions of the Cacouacs, my purpose is to point out that here again we see biological ideas connecting life and matter, drawn from the work of Buffon, Maupertuis, and Diderot, being attributed as foundational to the *philosophes*' whole enterprise. It was only by understanding these issues that one could become a true Cacouac/*philosophe*, at least in the eyes of those who found them so dangerous. Of course, the use of these questions also heightened the satirical quality of the narrative, since they sound even more absurd out of context. Yet I do not believe that Moreau or Giry de Saint-Cyr used these simply because they would sound hilarious. Rather they were definitional and seen as providing a foundation for the *philosophes*' dangerous undermining of religion and society. Both authors were closely connected with the French court, and we can see, in this episode and in the attorney general's condemnation of the *Encyclopédie*, just how high up in the political realm worries over the implications of such radical ideas reached.

Although Diderot never wrote any political treatises, it is clear from his other writings that he favored a limited monarchy and feared the rise of a despotic one. His concern over the state of the monarchy is evident in his reaction to unfolding political events. As France's financial difficulties skyrocketed in the late 1760s, following the country's defeat in the Seven Years' War, the struggles between Louis XV and the Paris Parlement intensified. The Parlements were opposed to new taxes, and much of their bid for power coalesced around the taxation issue. When the new chief minister René Nicolas de Maupeou proposed in 1770 a new tax, the vingtième, the Paris Parlement refused to accept it; in a bold stroke Maupeou simply dissolved Parlement, exiling its members to the provinces. Diderot was shaken by this action. Although he was highly critical of the Parlement, it was really the only body in the French government that could serve as a limiting force against the monarch. Imperfect as it was, dissolving it was a step toward despotism. As he wrote in a letter: "This event has caused great emotion among all the orders of the State. . . . Heads are warming up and the heat is slowly spreading. The principles of liberty and independence, formerly hidden in the hearts of a few people who think, now take hold and are openly expressed." Diderot went on in this letter to claim what was perhaps the real basis for the fear of atheism expressed by Bonnet and Haller. As he explained, "Once men have dared in some way to attack the barrier of religion, the most formidable and most respected barrier that there is, it is impossible to stop. When they have cast a hostile glance over the majesty of heaven, they will not hesitate the next moment to cast one over earthly sovereignty."[50] In other words, anti-clericalism and the promotion of atheism went hand in hand, in Diderot's mind, with political reform.

Diderot's answer, then, to Bayle's question, "Could a society of atheists be moral?," was unhesitatingly "Yes." In fact he believed such a society could be far more moral insofar as it incorporated principles of toleration and justice, and was founded on a naturalistic concept of human behavior. What Diderot opposed was the rigidly hierarchical, unchangeable social order of French society, and, like others, he recognized the underlying ties between such a concept of society and a natural world built on pre-ordained order. Diderot's discovery of biology as a source of materialist arguments developed into an unbridled enthusiasm for any evidence that demonstrated material activity. Thus nature and society were both to be understood on the basis of natural laws.

Did Diderot, then, promote what Voltaire and the other preformationists feared? In one sense he did. Diderot's and d'Holbach's search for a definition of morality based on human nature grew out of a total rejection of religion as a foundation for moral behavior. Even though Voltaire was a deist he still claimed that society needed a notion of God's future judgment, at least for the masses. The other preformationists like Bonnet and Haller went even further in promoting organized religion based on God's revealed word. Yet in another sense Diderot was never as radical as atheists were feared to be. However, the potential for the breakdown of social order was there, according to the preformationists, in the loss of the natural order being necessarily tied to God. And this had to be prevented. As Needham wrote to Bonnet in 1768, "In spite of all that ... we have done to bring our unbelievers to reason, I have always hoped that somehow we will stop the contagion by warding it off from those it has not yet touched."[51] Needham was not a preformationist; but, as was mentioned earlier, he was just as worried about the impact on morality and society that the collapse of preformation would entail. It was the "contagion" of atheism that had to be stopped, because if the generation of living organisms could happen without God then so could the generation of human societies.

CHAPTER THREE

Eighteenth-century uses of vitalism in constructing the human sciences
Peter Hanns Reill

Disciplinary boundaries are never secure. In fact if they become secure, it signifies that the discipline or science involved no longer can be understood by the wider public or that it has lost its "relevance" for them. The more a science appeals to and energizes the people who come into contact with it, the more likely attempts will be made to translate the methods, assumptions, and conclusions of that science to disciplines removed from the original set of inquiries. Such sciences lend themselves to be used and also abused, especially if their relevance speaks to core beliefs concerning nature, humanity, politics, and morality, belief often comprised under the general heading of ideology. In the period of the high and late Enlightenment such an exchange took place, in which ideas, methods, and assumptions were drawn from the life sciences to reform the human sciences by sharpening their message and, by implication, making their ideological meaning more explicit. The goal was to improve the human sciences by naturalizing them, infusing them with the spirit that animated the search for the principles of life in what we today would call biology. This imperative to naturalize the human sciences was made clear by Johann Gottfried Herder (1744–1803), a major German thinker of the period, who during his productive life opened new ways to understand history, anthropology, and aesthetics, in the opening of his multivolume study, *Ideen zur Philosophie der Geschichte der Menschheit* (1784–91): "If everything in the world has its philosophy and science [*Wissenschaft*], why shouldn't that which lies closest to us, the history of humanity, have, on the whole, a philosophy and a science?"[1] The question served as both a challenge and a plea, a challenge to transform the history of humanity into a form of knowledge with systematic content and a plea to make the knowledge it generated comparable to natural science. Herder made this plea because he, like many of his contemporaries, believed that "the whole of human history is a pure natural history of powers and drives according to place and time."[2] For them the human and the natural sciences partook of the same intellectual operations and could produce similar results.

The general impulse of turning to natural science to guide the inquiry into the human sciences was not new. At least since the "Scientific Revolution" of the seventeenth century, generally characterized as the triumph of the mechanical philosophy of nature, thinkers who probed the humanities had turned to natural philosophy for their models of explanation and proof. At first, following the mechanists, they sought to transform contingent knowledge into certain truth, to reduce the manifold appearances of nature and humanity to simple principles. Their strategy was to incorporate the methods and assumptions of formal mathematical reasoning into their explanations for human phenomena, guided by the mechanists' definition of matter, their methodological and explanatory procedures, and their epistemology.

By the middle of the eighteenth century, however, this project had become contested, its implicit ideological assumptions open to debate. Many younger thinkers of the period no longer considered mechanism's language of nature satisfying or self-evident, especially those exploring the human sciences. Mechanism's very success had made it suspect. The brave new world of seventeenth-century mechanism, at first shockingly radical, easily lent itself in the early eighteenth century to supporting the status quo—to justifying political absolutism, religious orthodoxy, and established social hierarchies. The unease that accompanied such applications of mechanism to human affairs was marked by an increasing crisis of assent, expressed in a wave of mid-century skepticism directed against a one-sided reliance upon abstract reasoning. Many leading thinkers of the high and late Enlightenment deemed deductive philosophy incapable of accounting for nature's vast variety. A strong skeptical critique, formulated most forcibly by the Scottish philosopher David Hume (1711–76) but reflected to various degrees by many other thinkers, opened the way for a new approach to constructing nature.[3]

This skeptical critique, although attacking abstract reasoning, did not deny the need to turn to nature for a model upon which the human sciences could be constructed; it merely led to a shift in focus, replacing mechanical models with other alternatives. Responding to this shift, many mid- to late Enlightenment thinkers elevated the life sciences to the primary position in the hierarchy of the sciences. For in these newly emerging sciences they discovered principles that spoke directly to their needs and desires, in short to their ideological assumptions. Three of these new assumptions were based on a revised appreciation of what knowledge could or could not undertake and accomplish. These thinkers desired to elevate the contingent over the coherent. They assumed that human knowledge was extremely constricted, both because of its reliance upon sense impressions and because of its limited scope. If humans were endowed with reason, its ability to pierce the veil of the unknown was circumscribed by the weak tools at its disposal and modified by the passions, the habits, and the imagination, which animated all human endeavors. At the same time, these Enlightenment thinkers surrendered the idea that nature's operations could be comprehended under the rubric of a few simple, all-encompassing laws. They replaced uniformity and identity with ideas of variety and similarity. Nature not only was deemed complex, it also was considered to be a continuous movement in which old forms of existence are replaced by new ones. In short, nature had a history. This triple movement—the limiting of reason's competence, the expansion of nature's complexity, and the historicization of nature—set a new agenda for Enlightenment natural philosophers and those desirous of formulating the human sciences upon natural scientific models.

Recognizing the existence of this movement within the Enlightenment raises the controversial question of how to define the Enlightenment. Some scholars interpret the Enlightenment as a unitary project transcending all areas, though carried on by a few "great thinkers." Others prefer to localize it, usually in terms of "national context." In this chapter, I take a middle position. Though I accept the fact that there were many variants upon Enlightenment themes, there were also some more basic assumptions that united thinkers from different localized traditions. I argue that there were a limited number of Enlightenment discourses, often determined by the vision of nature they proclaimed, though these discourses had many dialects. This chapter deals with one of these larger discursive categories.

For many Enlightenment thinkers, the agenda they found most appealing had been set by a loose group of natural philosophers whom I call, for want of a better term, Enlightenment vitalists. Their interests usually centered on natural history, chemistry, the life sciences, and medicine, and on the interconnections between these fields, concerns that today would mostly fall under the general heading of the biological sciences. Eighteenth-century vitalists such as Georges-Louis Leclerc de Buffon, Pierre-Louis Moreau de Maupertuis, John Turberville Needham, Paul Barthez, Johann Friedrich Blumenbach, Charles Bonnet, Peter Camper, Alexander von Humboldt, William Cullen, John and William Hunter, Joseph Black, Horace-Bénédict de Saussure, and Toberg Olaf Bergman, to name but a few, sought to reformulate the concept of matter, along with those of force, power, and connection, in their construction of a natural philosophy that respected variety, dynamic change, and the epistemological consequences of skepticism. In some ways their theories resurrected the Renaissance and classical biological ideas rejected by the mechanists. But they placed them in a different context, designed to satisfy the truth demands of modern science and the ideological concerns of the modern age.

FIGURE 3.1 Adam Smith (1723–90), Scottish doctor, philosopher, and pioneering economist. He was a leading light in the Scottish Enlightenment. (The Granger Collection, New York.)

FIGURE 3.2 Johann Gottfried Herder (1744–1803), poet, literary critic, and notable German philosopher of the Enlightenment. He was a firm believer in founding the history of mankind in the natural sciences. (Portrait of Johann Gottfried Herder, painted by Anton Graff, 1785, Gleimhaus, Halberstadt.)

In this chapter, I focus on two examples of how biological vitalism enabled Enlightenment thinkers to formulate strikingly modern concepts of the human sciences, concentrating on the thoughts of Adam Smith (1723–90; Figure 3.1), the great Scottish thinker who pioneered the modern study of political economy, and on Johann Gottfried Herder (Figure 3.2). However, to do so requires me first to provide a sketch of what I consider to be the defining characteristics of Enlightenment vitalism.[4]

Enlightenment vitalism

Enlightenment vitalists sought to discover the secrets of living nature, defined as a dynamic entity in which active forces worked in harmony within a complex set of interrelations. They posited the existence of self-activating forces and "principles" in living matter, envisioning a living thing as a teeming interaction of drives, sympathies, and elective affinities, united in what they called "organized bodies." An organized body constituted a complex conjunction of energies and forces of varying intensities and functions that could not be reduced to a single dominating principle, yet in their interaction formed a unit with its own unique character, which, driven on by these forces, was forever subject to dynamic change. In other words, a "living body" was "self-organizing," guided by energies directing the body to reach its own perfection, or when faced with environmental challenges or by sickness to compensate as best as possible to keep the body alive and well.

The reintroduction into nature of active, goal-directed living forces led Enlightenment vitalists to reassess the basic methodological and analytical categories of scientific investigation and explanation. The new conception of matter dissolved the mechanists' strict distinction between matter and spirit and thus between observer and observed, since both were related within the larger conjunction of living matter. Relation, *rapport*, *Verwandschaft* [affinity], and interconnection replaced mechanical aggregation as one of the defining principles of matter. By emphasizing the centrality of interconnection, Enlightenment vitalists modified the concept of cause and effect. In the world of living nature, each part of an "organized body" was both cause and effect of the other parts. All forces were symbiotically linked, usually through the universal interactions of sympathy and synergy, forming a complex whole. Further, with the reintroduction of goal into living nature, Enlightenment vitalists made it the efficient cause of development. An explanation for something's existence took the form of a narrative modeled upon the concept of stage-like development or *epigenesis*, in which a body evolves through stages from a point of creation. Unique creation and true qualitative transformation, both positive and negative, were central to the vitalists' vision of living nature.

These shifts in natural philosophical assumptions challenged Enlightenment vitalists to construct a theory of knowledge capable of justifying and validating them. Vitalists agreed that active life forces could not be seen directly, nor could they be measured. At best they were announced by outward signs, whose meaning could be grasped only indirectly. This language of nature reintroduced the topos of locating essential reality as something lying hidden within a body. Understanding entailed a progressive descent into the depths of observed reality, using "signs" as the markers to chart the way. Thus Enlightenment vitalists reintroduced the idea of the "science of signs" or semiotics as one of the methods to decipher the secrets of nature.

The basic problem was to understand the meaning of these signs, to locate them within an interconnected "system," and to perceive the interaction of the individual yet linked active forces without collapsing one into the other. To resolve this problem Enlightenment vitalists called for a form of understanding that (a) combined the individualized elements of nature's variety into a harmonic conjunction recognizing both nature's unity and diversity and (b) required the use of imagination to apprehend this conjunction. The methods adopted to implement this program were analogical reasoning and comparative analysis.

Analogical reasoning became the functional replacement for mathematical analysis. With it, one could discover similar properties or tendencies between dissimilar things that approximated natural laws without dissolving the particular in the general. The fascination for analogies was strengthened by a general preference for functional analysis, in which outward form was subordinated to inner activity. Comparative analysis reinforced the concentration upon analogical reasoning. It allowed one to consider nature as composed of systems having their own character and dynamics, yet demonstrating similarities not revealed by the consideration of outward form.

In pursuing this program, a further problem arose. If nature was unity in diversity, on what grounds could one choose which element to emphasize? When should one concentrate upon the concrete singularity and when should one cultivate generalizing approaches? The proposed answer was to do both at once, thus allowing the interaction between them to produce a higher form of understanding than that provided by simple observation or discursive, formal logic. This new form of understanding—necessarily paradoxical in form—was referred to as divination, intuition, or *Anschauung* [intuition]. Its operation was based on the image of mediation, of continually moving back and forth from one level to the other, letting each nourish and modify the other, without attempting to dissolve one in the other. Buffon described this practice in the introduction to his *Histoire naturelle* (1749): "the love of the study of nature supposes two seemingly opposite qualities of the mind: the wide-ranging views of an ardent mind that embraces everything with one glance, and the detail-oriented laboring instinct that concentrates only on one element."[5] For him, as for vitalists inspired by him, the general and the particular were joined yet not identical. The understanding of the one was necessary for the understanding of the other, precisely because of the need to avoid homogenizing the two. Thus, it required the interpreter to keep the elements of these conjunctions intact, an approach that appeared to elevate paradox over identity.

In this seemingly paradoxical back-and-forth movement, many Enlightenment vitalists believed that understanding passed through a third, hidden and informing agent that constituted the ground upon which all reality rested. In eighteenth-century vitalist language, this hidden middle element was called variously the "internal mold" (Buffon), "prototype" (Jean Baptiste Robinet), "*Urtyp*" (Johann Wolfgang von Goethe), or "*Haupttypus*" (Herder). For others, that "middle" was defined as the inner core of reality, the essence of humanity—"the man within the breast" (Smith)[6]—to which subjective human understanding could be compared and authorized. For these vitalists the polarities that necessarily limited objects were most successfully resolved harmonically in the middle, the place where each extreme enhances and reinforces the other. Mediation and harmony constituted the core concepts and dominant metaphors guiding this language of nature.

This new vitalist language of nature combined in a unique way the concepts of connection, harmony, striving, sympathy, and mediation, justified by a theory of knowledge in which imagination supported and enhanced empirical research, expressed in the creative use of analogical reasoning. Those attracted to vitalism felt compelled to translate this language of nature into the humanities, for, as observed above, they believed a close analogy existed between nature and culture, that human history and natural history followed analogous courses. They considered it essential, as Wilhelm von Humboldt asserted, to "combine the study of physical nature with the moral in order to perceive the true harmony of the universe."[7] This was the "science" to which thinkers desirous of establishing the human sciences on a scientific basis, yet critical of mechanism, turned; they made this program even more appealing by embedding it as a model of explanation and proof within a commitment to freedom and humanity. Herder's *Ideen* and Smith's *Wealth of Nations* serve as striking examples of how vitalism enabled Enlightenment thinkers to construct a new vision of the human sciences. Both have become foundational texts in the human sciences.

The choice of these two works may appear arbitrary but I make it for a number of reasons. The most obvious is that Herder and Smith have traditionally been interpreted as being on opposite ends of the epistemological and ideological spectrum. Herder has been widely pictured as one of the most significant eighteenth-century opponents of the Enlightenment, a founder of anti-Enlightenment Romanticism and, along with the Neopolitan philosopher Giovanni Battista Vico, a major proponent of what Isaiah Berlin called the "Counter-Enlightenment."[8] In contrast, despite the important reevaluation now going on in Smith studies,[9] Smith's work and especially his *Wealth of Nations* is usually represented as one of the clearest expressions of mechanistic rationalism, a prime example of the Enlightenment's commitment to instrumental reasoning. In these evaluations, Smith emerges as the ultimate positivist, who sought to prescribe invariable economic laws governing all societies. Herder appears as the irrational dreamer who established the framework for the explication of nineteenth-century Romantic nationalism. Locating them both within the discourse of Enlightenment vitalism not only suggests new ways of conceptualizing the Enlightenment but demonstrates the extent and depth of vitalism's appeal.

This appeal is made even more evident because there was little or no influence of one writer on the other, though obviously they both turned to the same core of sources in pursuing their work. Herder, though very much aware of the major trends in Scottish and British thought, apparently had not read Smith, even though *The Theory of Moral Sentiments* and the *Wealth of Nations* were quickly translated into German. But Herder was acquainted with the British thinkers Smith most admired, including Hume, Hutcheson, and Bacon. In addition, Herder had a very thorough knowledge of Adam Ferguson's writings. Smith's knowledge of German sources was limited to works in natural philosophy (usually in Latin or in English or French translation), for he believed German writers too pedantic in most areas to offer anything new. Finally, both came from a peripheral region, Herder from the borderlands between northeastern Germany, Livonia, and Russia, Smith from Scotland, leading both to consider themselves outsiders to dominant culture. Herder's struggles with his outsider role are well known.[10] Smith's feelings of being an outsider wishing to enter the core were expressed in his letter to the *Edinburgh Review*, where he tried to associate himself with British thought. But he never fully integrated himself into English culture, as evidenced by his unhappy experiences in England and by his strained relations with those Scots such as Boswell who tried to fully conform to English patterns of thought and behavior. This perception probably made them extremely open to the new trends in thought being forged during the period and challenged them to employ these trends in a constructive and original manner. Thus, when one looks at Herder's *Ideen* and Smith's *Wealth of Nations* through the lens of Enlightenment vitalism, both Herder and Smith emerge in a different light, not only as translators of Enlightenment vitalism, in fact two of its most passionate advocates, but also as indicators of the appeal Enlightened vitalism held for intellectuals seeking to rethink how the human sciences should be conceived and how society should be established, a basic concern of ideology conceived in its broadest sense.

Establishing a vitalist human science

Both Herder and Smith had the knowledge and contacts to appreciate vitalism's message. Herder had immersed himself deeply in this language of nature. He had read widely in the most modern "scientific" literature, knew or corresponded with many of its leading figures and supplemented this knowledge by delving deeply into travel reports and anthropological treatises. Herder consciously incorporated this knowledge into the *Ideen*, directed by his assumption of the essential correspondence between the physical and the moral world. He argued that all human events be investigated by concentrating upon observable positivities and then generalizing from them, guided by ideas found in nature. When correctly done, the investigator will have "discovered a sound philosophy that is difficult to find elsewhere except in natural history and mathematics."[11] Clearly, Herder desired to establish a discursive realm that claimed equality with natural history and mathematics. In essence, he proclaimed, "spirit and morality are also physics,"[12] all ruled by the same laws. At the same time, Herder's explanatory model did not attempt to collapse the particular into the general, to adumbrate an essential identity between nature and morality. He sought to chart the interaction between the particular and the general that would lead to a deeper understanding of both the natural laws of history and history's unique and irreducible concrete manifestations; that is, to offer what he considered the ultimate goal of a causal explanation that answers the question "Why is a thing different than it could be?"[13]

Smith was equally at home with vitalistic thought. Glasgow and Edinburgh were two of the leading centers for the development of vitalism in the eighteenth century, as evidenced by the work of Robert Whytt, William Cullen, Joseph Black, John and William Hunter, and James Hutton. As a young man, Smith's "favorite pursuits while at the university were mathematics and natural philosophy."[14] He demonstrated these interests in his early writing: in his lectures, in his 1756 letter in the *Edinburgh Review*, and especially in his fascinating *History of Astronomy*,[15] composed some time around 1758. Smith had a strong knowledge of what was happening in the intellectual world of eighteenth-century Scotland, acquired through the many contacts he had in the leading clubs and societies of Edinburgh and Glasgow, which proliferated in the eighteenth century, the most important being the Edinburgh Philosophical Society, the Scottish Royal Society, the Select Society, and the Poker Club.[16] Later with Joseph Black and James Hutton, important figures in the explication of vitalism, Smith would play a major role in the newly formed and highly influential Oyster Club.[17] The holdings of Smith's library also confirm his interest in the emerging life sciences, though it is obvious that owning a book did not mean that he had read it.[18] Still, it appears that Smith's dissatisfaction with mechanistic assumptions was long-lived and entrenched.

Smith expressed this dissatisfaction in a 1775 university lecture. In it, he drew a distinction between mechanical and "natural" conceptions of humanity. "Man is generally considered by statesmen and projectors as the materials of a sort of political mechanics. Projectors disturb nature in the course of her operations in human affairs; and it requires no more than to let her alone, and give her fair play in the pursuit of her ends, that she may establish her own designs." He followed this indictment of political mechanics by remarking:

> Little else is requisite to carry the state to the highest degree of opulence from the lowest barbarism, but peace, easy taxes, and a tolerable administration of justice; all the rest being brought about by the natural course of things. All governments which thwart this natural course, which force things into another channel, or which endeavour to arrest the progress of society at a particular point, are unnatural, and to support themselves are obliged to be oppressive and tyrannical.[19]

Thus, twenty-one years before the publication of the *Wealth of Nations*, Smith had already formed the basic outlines of his approach, contrasting "nature" and her operations, characterized by nature's "fair play" implementing "her own designs," with the oppressive tyranny of political mechanics, which forced things into improper channels. As is clear from this thumbnail sketch, Smith's goal was to show how nature's goals were best to be achieved. To do so, however, required him to propose a new "system," in which these operations were made evident.

For Smith, the creation of a new system meant "combining together the powers of the most distant and dissimilar objects."[20] For him systems were human constructs designed to allay the emotions of "wonder" and "surprise," that is, feelings of unease and fear generated when one confronts things that appear at first solitary and incoherent, that disturb the "easy movement of the imagination."[21] Systems reintegrate these obscure, strange, and new objects into a pattern that allows the flow of the imagination to run undisturbed. New systems introduce "order into this chaos of jarring and discordant appearances." They "allay this tumult of the imagination and . . . restore it . . . to that tone of tranquility and composure, which is both most agreeable in itself, and the most suitable to its nature." As a result, they present humans with a new theater of nature, "a coherent, and . . . magnificent spectacle."[22] The task of any new system was to link as many disparate objects as possible together into a coherent whole in which their interrelationships were perceived and made evident. This often necessitated destroying older systems.[23] Smith's project in a larger sense was both to destroy and to create anew, a process he described in his *History of Astronomy*:

> For, though it is the end of Philosophy, to allay that wonder, which either the unusual or seemingly disjointed appearances of nature excite, yet she never triumphs so much as when, in order to connect together a few, in themselves, perhaps, inconsiderable objects, she has, if I may say so, created another constitution of things, more natural indeed, and such as the imagination can more easily attend to, but more new, more contrary to common opinion and expectation, than any of the appearances themselves.[24]

In essence, Smith's goals were similar, if not identical, to those of Herder. Both sought to create "a new constitution of things." To do so required them to propose a system that combined a larger number of seemingly unconnected events in a manner more convincing than had been done by political mechanics. In this sense, both the *Ideen* and the *Wealth of Nations* were designed to establish new fields of humanistic inquiry and explanation.

Nations as "organized bodies"

The *Ideen*'s title defined the field of inquiry as a systematic study of the philosophy and the history of humanity. The *Wealth of Nations*'s stated goal was to imagine a system that could explain how some societies acquired wealth and well-being while others did not. But what form did these new systems take? Herder's answer was clear. All of the units he investigated and explained were "organized"; that is, they were composed of a conjunction of living forces acting in space and time. Simply said, the system was analogous to a vitalistic one. The unit's character or constitution was formed by the interaction of all of its cultural activities at a specific time and place, each linked to the other, defined and limited by its specific cultural and physical boundaries. These relations took place in space, in a nation's *habitus* [mode of life]. The *habitus* shaped humans, gave them direction, brought out certain drives while minimizing others. For example, the Greek *habitus* made it possible for Greeks to develop their language—the "best formed in the world"—to evolve a culture in which song, dance and history were its primary expressive forms, to live in a world where ideas and goods circulated freely, and to generate their central passion, the love of freedom.[25] But nations or cultures were not self-contained entities, possessing an unchanging character or identity. Greek culture and its political forms did not remain the same; their character was modified with changing conditions, especially as the society's boundaries (in the largest sense of the word including its space and culture) were either expanded or constricted. Boundaries were flexible and subject to change, which implied that an organized body's character was malleable, differentiating it from what we today would call an organism. Thus Herder's new philosophy of humanity systematically strove to analyze the activities of organized bodies seen as a complex conjunction of culture, language, political organization, artistic forms, and economics all located in time and place. In so doing, Herder also sought to differentiate this philosophy of history from abstract metaphysics or purely conjectural history. "I believe . . . that . . . speculations, devoid of empirical observations and analogies drawn from nature, are airy ones that very seldom lead to the goal."[26] Every analysis had to begin and end with a deep commitment to empirical validation and explication.

Smith never gave his system a simple name, but it is clear, I believe, that it too was modeled upon the vitalist idea of an organized body. Indications of this are to be found everywhere in his writing, for Smith continuously resorted to comparing a nation to a living body, using images drawn from it to describe what happens in a "body politick." An example was his characterization of Great Britain in his discussion of her monopoly of the North American commercial trade and the threat posed by its possible loss:

> In her present condition, Great Britain resembles one of those unwholesome bodies in which some of the vital parts are overgrown, and which, upon that account, are liable to many dangerous disorders scarce incident to those in which all the parts are more properly proportioned. A small stop in that great blood-vessel, which has been artificially swelled beyond its natural dimensions, and through which an unnatural proportion of the industry and commerce of the country has been forced to circulate, is very likely to bring on the most dangerous disorders upon the whole body politick.[27]

In Smith's account, bodies politick have properties similar to those of living bodies; they are attacked by "disorders" (frequently occasioned by the "mercantile system") that are "often difficult to remedy."[28] Some of its "vital parts" may grow well out of proportion, thus endangering the body's health, which he defined as the just proportion of all its parts. Yet living bodies also have their own inborn abilities to regenerate themselves; they can compensate for mistakes and sometimes even thrive under very "unwholesome" conditions.

For these reasons, it was, Smith argued, folly to apply speculative systems to describe the body politick when discussing "political oeconomy." This was François Quesnay's error. Quesnay (1694–1774), a trained physician and a leader of the French physiocrats, whose writings are seen as precursors to Smith's economic theories, was, in Smith's view, a "speculative" physician who, though proposing a much better system than that of mercantilism, failed to understand correctly how bodies operate in the real world, the world of experience.

> Some speculative physicians seem to have imagined that the health of the human body could be preserved only by a certain precise regimen of diet and exercise, of which every, the smallest, violation necessarily occasioned some degree of disorder proportioned to the degree of violation. Experience, however, would seem to show that the human body frequently preserves, to all appearance at least, the most perfect state of health under a vast variety of different regimens; even under some which are generally believed to be very far from being perfectly wholesome. But the healthful state of the human body, it would seem, contains in itself some unknown principle of preservation, capable either of preventing or of correcting, in many respects, the bad effects even of a very faulty regimen.[29]

Clearly Smith believed that natural bodies and social ones were directed by similar hidden, unknown, principles, assumed by vitalists to be essential to a body's operations. And Smith, like Herder, argued that charting the effects of these principles could be achieved only through consulting experience and "real" history.

Smith's system was as wide ranging as Herder's, combining land, labor, exchange, politics, and culture into an interlocking whole, where activities in one sphere had important consequences in the others. The system, at least in its developed stage as "commercial society," was organized around the "three great orders" of land, labor, and capital. Their interactions were "strictly and inseparably connected with the general interest of society."[30] All three depended upon the primary force or principle that animated them, namely labor, which was employed in agriculture, manufacture, and trade. Labor, in Smith's vitalistic system, was "the only universal."[31] As an active principle, it imparted life to all of society's institutions. Each of the three orders and each form of labor's activities joined men together in such a manner that "man has almost constant occasion for the help of his brethren."[32] In civilized society, because of the "division of labour" necessary to such a system, "man" stands at all times in need of the cooperation and assistance of great multitudes."[33] Thus all groups in society were interdependent, and when left to follow their own passions and interests would "naturally . . . divide and distribute the stock of every society, among all the different employments carried on in it, as nearly as possible in the proportion which is most agreeable to the interests of the whole society."[34]

The metamorphoses of nations

Neither Herder nor Smith conceived of living bodies or bodies politick as being static. Rather, adopting the vitalistic model, they argued that living bodies were directed by processes that forged a natural pattern greater than anything humans could plan or foresee; vital bodies flourished and withered not as the direct result of human intention but due to larger impersonal forces. The history of civil society followed its own "fair play," which humans could at best chart and observe and in so doing apprehend the beauty of what Herder called "the great drama of active forces."[35] According to Herder, this drama followed eternal laws impressed upon primary matter at the beginning of time.[36] Herder interpreted creation in the same manner as Kant. The world was not created perfect, but rather was embued with specific principles that led matter to develop in the way it has and will continue to do so, all according to a plan. In one of his typical rhetorical flourishes, Herder asked: "Is not time ordered as space also is ordered? And are not both twins of a single fate?"[37] Time's order was not a property of physical nature alone; history too showed an order. "In physical nature," Herder continued, "we recognize laws, which we discover to be eternally efficacious, regular, and invariable; What?, And the realm of humanity with its powers, transformations and passions should be excluded from this chain of nature?"[38] For him, creation and dynamic progression or "progress" were guided by the eternal laws that bound everything in an analogous dynamic system.[39] Though Smith was not prone to such rhetorical exuberances, he would have agreed with Herder's major point; systems have a history, and it is necessary to show how they evolve according to their own internal principles or driving forces, acting on and from the environment in which they subsist.

Given these assumptions, it is clear that both Smith and Herder considered space, time, and conjunction as root organizing ideas, incorporating the vitalist attempt to merge structure and process without submerging one in the other. Both Herder and Smith strove to construct a dynamic system of events that proposed law-like generalizations yet emphasized unique individual content, due in part to the real effects of the environment in which process takes place. Here they shared the Enlightenment's fascination for the effects of the environment on social bodies, made popular by Charles de Secondat, baron de Montesquieu, and reinforced by the revival of Hippocratic thought in the last half of the century. Process, often referred to by Herder as metamorphoses, was, for both, an essential element of a living system.

Herder, in accounting for these metamorphoses, concentrated upon the interlocking elements of space, time, and conjunction. Space delineated the effect wrought upon humans by the "habitus" in which they existed; it shaped and limited process. These "physico-geographical" causes were modified and opposed by the action of "free organic powers," which Herder characterized using the metaphors of generation and activity. Generation was a spontaneous process in which a hidden formative drive [*Bildungstrieb*] is ignited that separates a specific quantity of matter from the mass and gives it a unique form.[40] Herder expanded this analogy to cover all creation, ranging from the single organism to the nation and including the generation of language and ideas. According to Herder, creation is both immediate and inexplicable; it is mysterious, the instantaneous result of inner active forces forming matter. Creation cannot be part of a Great Chain of Being, which proceeds continuously through minute shadowings, blurring all real distinctions. Rather, there are leaps in nature,[41] testified to by the appearance of new species, new languages, and new nations. Though Herder denied the existence of the Great Chain of Being, he did believe in what he called a ladder of organization—a regular pattern of development from the simple to the more complex, guaranteed by the constant action of internally active forces, following simple laws, on the given environment.

Smith's explanatory strategies were similar to Herder's, though in his discussions he refrained from resorting to the metaphor of generation. Smith too sought to establish the rules governing the way nations change, especially as these changes applied to the nations' wealth and well-being, though he, as did Herder, modified these laws by reference to the specific conditions in which the principles operated. That is, Smith saw a symbiosis between active principles and given conditions in which each modified and acted upon the other. When proposing a general statement concerning the dynamics of historical change, Smith always included the modifier that "in every society this rate varies according to their circumstances, according to the riches or poverty, their advancing, stationary, or declining condition."[42] He too envisioned change as occurring in stages, proposing "four distinct stages which mankind pass thro:—1st, the Age of Hunters; 2nd, the Age of Shepards; 3rd, the Age of Agriculture; and 4th, the Age of Commerce."[43] And he also sought to meld images of radical change with those of long-term development, in which certain principles act "silently and gradually" to change the condition of a nation or civilization.[44] In large-scale social changes, revolution, according to Smith, occurred as a result of the accumulation of unintended effects wrought by humans following their own private interests. Such was the case, Smith argued, with the undermining of the feudal system, accompanied by the rise of cities and commerce, an event unintentionally generated by the activities of greedy merchants and vain landholding aristocrats:

> A revolution of the greatest importance to the publick happiness, was in this manner brought about by two different orders of people, who had not the least intention to serve the publick. To gratify the most childish vanity was the sole motive of the proprietors. The merchants and artificers, much less ridiculous, acted merely from a view to their own interest, and in pursuit of their own pedlar principle of turning a penny wherever a penny was to be got. Neither of them had either knowledge or foresight of that great revolution which the folly of one, and the industry of the other, was gradually bringing about.

> It is thus that through the greater part of Europe the commerce and manufactures of cities, instead of being the effect, have been the cause and occasion of the improvement and cultivation of the country.[45]

For Smith, process constituted the central consideration of his analysis, leading him to emphasize not the end or goal of development but the very activity of change itself. In this sense, Smith's analysis of wealth and well-being, of happiness in its larger sense, focused upon where a nation and its people stood in the process of change, whether it be "progressive," "declining," or "stationary." Well-being and happiness were found in advancing societies, hardship in stationary ones, and misery in declining societies.

> It . . . is in the progressive state, where the society is advancing to the further acquisition, rather than when it has acquired its full complement of riches, that the condition of the labouring poor, of the great body of the people, seems to be the happiest and the most comfortable. It is hard in stationary, and miserable in the declining state. The progressive state is in reality the cheerful and the hearty state to all the different orders of society. The stationary is dull; the declining, melancholy.

Despite the general opinion that Smith cared little for the condition of the "labouring poor," he believed that "[n]o society can surely be flourishing and happy, of which the far greater part of the members are poor and miserable."[46] Only in a "flourishing" society would it be possible to satisfy the material and spiritual needs of all the people The satisfaction of these needs with its accompanying moral component was only a possibility, not the inevitable result of such an expansion. Smith was well aware of the dangers that the division of labor could entail, namely reducing humans to virtual automata, devoid of culture, thought, or reflection. It was incumbent upon such a society to ensure that all of its citizens received a liberal education, one that made them both laborers and citizens.

Smith's idea of process was complicated. Process, whether positive or negative, was complex and tortuous. Things could not change overnight, for all change, even if it followed its "natural" course, encountered great obstacles. Smith thus proposed what we may call two different types of analysis for change, both vitalist in form. One was an "ideal type" that sought to make clear the principles of the system, the other a "real type" that took account of historical realities, of how the system actually reacted to its environment. The first could never be totally realized, nor did Smith expect it to be. Rather, it was to serve as a guideline for action in the world, to help achieve as healthy a body politick as its social, cultural, and economic environment would allow. This ideal type posited the existence of "perfect liberty," which would allow nature to follow its course unimpeded, since the society would be in "its natural and free state." These ideal conditions would produce effects that could be anticipated. All that was needed was the understanding of how bodies politick flourished without any interference from outside; that is, interference from their given historical environment. Smith provided an example of this type of thinking in his discussion of the colony trade.

> The effect of the colony trade in its natural and free state, is to open a great, though distant market for such parts of the produce of British industry as may exceed the demand of the markets nearer home.... In its natural and free state, the colony trade ... encourages Great Britain to increase the surplus continually, by continually presenting new equivalents to be exchanged for it. In its natural and free state, the colony trade tends to increase the quantity of productive labour in Great Britain, but without altering in any respect the direction of that which had been employed there before. In the natural and free state of the colony trade, the competition of all other nations would hinder the rate of profit from rising above the common level either in the new market, or in the new employment. The new market, without drawing any thing from the old one, would create, if one may say so, a new produce for its own supply; and that new produce would constitute a new capital for carrying on the new employment, which in the same manner would draw nothing from the old.[47]

But it is imperative to emphasize that these dynamic "laws" were limited by the unlikely possibility that a nation could achieve "perfect liberty," or ever realize "a natural and free" state in history. These laws served the same function as the prescriptions of physicians who imagine how a body could best achieve health if that body functioned in an ideal state; that is, in an environment devoid of any dangers of disease and disorder.

But disease and disorders exist and cannot be ignored. For that reason, each society at its given moment of development has its own unique set of forces that direct action, making the achievement of perfect liberty impossible. In Great Britain, for example, Smith believed that "to expect ... that the freedom of trade should ever be entirely restored ... is as absurd as to expect that an Oceana or a Utopia should ever be established in it. Not only the prejudices of the publick, but what is more unconquerable, the private interests of many individuals, irresistibly oppose it."[48] Private interests and prejudices, passions found in all communities, are but two of the forces hindering change. Another, according to Smith, was the force of established institutions and conditions, which constituted a habitual system having its own "nature." Thus change can be retarded by "ignorance and attachment to old customs, but in most places to the unavoidable obstructions which the natural course of things opposes to the immediate or speedy establishment of a better system." Transforming an existing system is an arduous task requiring effort and time. "These natural obstructions to the establishment of a better system cannot be removed but by a long course of frugality and industry; and half a century or a century more, perhaps, must pass away before the old system, which is wearing out gradually, can be completely abolished through all the different parts of the country."[49] Human institutions, necessary for the existence of society, therefore make changes in society more difficult; for they "frequently continue in force long after the circumstances, which first gave occasion to them, and which could alone render them reasonable, are no more."[50] In serious cases, when very "dangerous disorders" have been introduced to the body politick, they are "often difficult to remedy, without occasioning, for a time at least, still greater disorders."[51] Progress then was not automatic, not the result of inevitable laws guided by an "invisible hand" to a fixed teleological end.[52] For Smith, there were always limits beyond which change could not reach, set by the *habitus*—to use Herder's term—in which action took place.

History, mediation, and harmony

Smith's moderate stance concerning the manner in which change occurred and the difficulties encountered when seeking progressive transformation illustrates an important feature of Enlightenment vitalist thought, namely its fixation with the concept of the harmonic mediation between competing extremes. Both Herder and Smith propose such a scheme. For Herder "progress" always took place between the extremes of maxima and minima, limiting poles between which healthy life is led. In the history of nations, those two extremes were represented by "patriotism" and *Aufklärung* [enlightenment], whose interaction could, under favorable conditions, lead to the harmonius blending of the competing elements.

Harmony was a marriage of extremes founded upon complicated interactions of living forces, a marriage in which each partner retained its own character within the joined unit, "twins of a single fate."[53] The greatest harmony entailed increasing the number of interacting forces and widening the distance between the extremes limiting the active forces, leading to an expanded middle realm in which life was experienced. Herder condemned one-sidedness, seeing it as a distortion of nature. He vested the term "center" or "middle" with great symbolic importance, signifying a location linking freedom with stability.

For Herder, the turn toward the middle was part of nature's law, the place where everything good and beautiful was produced, either in the geographical sense as the middle realm, or more importantly in the realm of freedom, in which freedom is attained only by the juxtaposition of opposites. But every attained level of freedom contained its own limitations, for, given the relativity of all cultures, nothing absolute could be achieved. Each age or nation had to work to realize its own form of freedom, which was always threatened with degeneration. Hence the progress of culture, which Herder affirmed, took place in fits and starts; the achievements of one culture were never totally transferred to others but were capable of serving as analogs to what later can be achieved. However difficult progress may be, Herder, in good Enlightenment fashion, assumed the goal of all development to be the expansion of freedom and humanity. Freedom meant individual self-determination and personal autonomy, which Herder defined as "reason and justice in all classes, in all human activities."[54] Herder believed this goal to be dictated by nature, for our "whole being was organized by nature to this end."[55] Thus freedom's expansion was possible because it constituted an implicit part of our natural condition, but it could be achieved only by action, by consciously pursuing the desire to expand reason and justice, according to one's time and place.

Smith's vision of harmony and mediation was equally powerful and also directed to enhancing human freedom in a complex world of competing interests. His idea of the manner in which harmonic mediation proceeds, creating "an order of things," was determined by his definition of human inclinations. "That order of things which necessity imposes in general, though not in every particular country, is, in every particular country, promoted by the natural inclinations of man."[56] Following Hume, Smith assumed that humans were motivated by the passions. These passions were primarily guided by self-interest, characterized by two opposing desires. The first was "the passion for immediate fulfillment; which, though sometimes violent and very difficult to be restrained, is in general only momentary and occasional."[57] In many societies, this passion is made manifest by what we would call conspicuous consumption, the desire to spend what one has for enjoyment and to demonstrate one's position and elevation to others. This passion often is institutionalized in orders such as the aristocracy, wealthy landowners, and successful merchants. It also supports the creation of art, luxuries, and the like. Hence its effects are many, though most do not lead to the creation of new wealth and therefore are not in the general interest of society.

The second passion "is the desire of bettering our condition, a desire which, though generally calm and dispassionate, comes with us from the womb, and never leaves us till we go to the grave."[58] This principle prompts humans to save money. As such, it is the engine driving the accumulation of wealth, which, when invested, converts "dead stock into active and productive stock,"[59] leading naturally to society's progression. Smith considered the first passion negatively; the second embodied the positive and productive side of human nature, which society should nourish and foster, since it alone was capable of accounting for human progress even in the face of extreme vicissitude. Using an analogy drawn from life, Smith described the effects of this passion and its ability to compensate for the negative effects of society and government.

> The uniform, constant, and uninterrupted effort of every man to better his condition, the principle from which publick and national, as well as private opulence is originally derived, is frequently powerful enough to maintain the natural progress of things toward improvement, in spite of both the extravagance of government, and of the greatest errors of administration. Like the unknown principle of animal life, it frequently restores health and vigour to the constitution, in spite, not only of the disease, but of the absurd prescriptions of the doctor.[60]

In Smith's view, the human passions had to be channeled; the violent and momentary passion for immediate enjoyment held in check in order to allow the desire for bettering our condition ample room to work and create.

But the tension between the passions can never be resolved, since it forms a basic part of human nature. The play between violent activity based on immediate impulse and calm and dispassionate commitment to long-term goals forms the nodal points around which the drama of human activity gravitates, a point Smith already made in his letter to the

Edinburgh Review, when he contrasted the views of the philosophers Bernard Mandeville and Jean Jacques Rousseau concerning "primitive man," where Mandeville emphasized the violent nature of savage society and Rousseau its "profound indolence."[61] This tension between the positive and negative poles of activity runs like a red thread through Smith's *Wealth of Nations*, where progressive life is represented as the harmonic mediation between both, as the condition when the vital parts of a body are "properly proportioned." Hence, we have Smith's continual invocation of the image of a natural balance between maxima and minima, which creates a center of "repose and continuance."[62] Though Smith employed the term "balance" to represent this movement between extremes, it was not meant in its purely mechanicist connotation as a simple equilibrium. The balance was not a static one. It varied with the progress or decline of society and hence represented either an expansion of liberty or, in the negative sense, its restriction. Thus the gravitation between extremes took place in a system of active matter analogous to that envisioned by Enlightenment vitalists. "Once started," Campbell and Skinner observed, "the process of capital accumulation may be seen as a self-generating process, indicating that Smith's model of the flow should be seen, not as a circle of given size, but rather as a spiral of constantly expanding dimensions."[63]

If such an expansion occurs, and Smith was aware of the contingent nature of that proposition, then society would be propelled toward greater freedom, toward a stage where humans could exercise their own autonomy. In this, as in many other things, Smith's hopes paralleled those of Herder. Both assumed that freedom could be expanded, not by any return to a mythical state of nature or by the implementation of some system of rational control, but by letting "things follow their natural course." If done, then a living system, approximating nature's, could be instituted "where every man was perfectly free to chuse what occupation he thought proper, and to change it as often as he thought proper."[64] This regime of liberty would help us to realize the human potentialities we possessed due to our natural organization.

Analogy, semiotics, and understanding

To end this chapter, I would like to briefly raise three additional points that illustrate the manner in which Enlightenment vitalism inspired Herder's and Smith's conception of a science of humanity, namely their use of analogy, of semiotics, and of the epistemology that supported it. All reveal the realignment of scientific thinking that vitalism encouraged. As mentioned, both Smith and Herder sought to apprehend what Herder called "the great drama of active forces."[65] But it was not an easy drama to observe, for the major players were hidden from view, performing on a stage located within the depths of living matter.[66] As vitalists, they were critical of mechanism's concentration upon the immediately apparent, which often appeared to them superficial. For them, reality's essential nature could not be apprehended directly, especially not by the tools of abstract reason. Therefore, some method had to be evolved to discover and prove the intimate connection of things not usually joined. The path they chose was the use of analogical comparisons.

According to Herder, analogies gave us the "key that allows us to pierce the core of things."[67] All great scientific thinkers, Herder argued, had relied upon them. Newton, Leibniz, and Buffon had arrived at their great and daring discoveries through "a new picture, an analogy, a striking similarity."[68] Smith was equally convinced of the importance of analogies; only through them were great advances in thought made, as the examples of Johannes Kepler and René Descartes in the history of astronomy proved. Analogies were crucial because each new system had to bridge the gap between "distant objects" that had "stopped and interrupted" the natural movement of fancy. Only imagination could accomplish the task of incorporating singular and strange elements, of smoothing "the passage from one object to the other," into a new chain that reintroduced repose and harmony.[69] This was possible only by imaginatively associating the unexplained elements with things already known in other spheres of life. Simply said, the imagination used analogies to domesticate the strange and singular occurrences that had generated wonder and surprise. Analogies "became the great hinge upon which everything turned."[70] For Herder and Smith analogical reasoning functioned as a crucial instrument of discovery in developing new systems.

But they went even further, for they also employed analogical reasoning as a tool of proof or explanation. Analogies served as semiotic figures carrying explanatory burdens of proof far surpassing that of simple metaphor. The most obvious and powerful explanatory analogy Herder and Smith deployed was that of the living body. As I have tried to demonstrate, it was central for their representation of a nation's operations and dynamics; Smith's extensive use of the idea of sympathy in his *Theory of Moral Sentiments* anticipated and underscored the centrality of the body analogy. Closely related to the analogy of the living body was the analogy of growth from infant to adult, which Herder and Smith employed for their representation of society's progression from the simple to the complex. Herder used it to explain the emergence of poetic genius in different cultures without proposing a classical definition of beauty, truth, or eternal norms. Early societies are analogous to youth. They are poetical by nature. As they grow older, they turn to prose.[71] Smith deployed the analogy of growth to describe what he called the progress of society from barbarism to civilization that proceeded through the four stages mentioned above.

Though both turned to analogies in their proofs, Herder was far more exuberant in the use of analogies than Smith. Herder roamed freely in the world of physical phenomena to explain specific moral, social, historical, and religious truths as well as to characterize universal occurrences. Smith was more concerned with finding signs for specific phenomena whose activity was hidden from view. His primary objective was to discover a way to represent value and wealth without resorting to the mechanistic idea of measuring value by the amount of gold and silver a nation possessed, indicators Smith considered "dead." "The gold and silver money which circulates in any country, and by means of which, the produce of its land and labour is annually circulated and distributed to the proper consumers, is, in the same manner as the ready money of the dealer, all dead stock."[72] Smith sought rather for an active agent that produced wealth and argued that the value of any commodity was, therefore, equal to the quantity of labor expended to acquire it.[73] But what signified this quantity of labor and how can we represent it? Certainly not by simply asking how much it costs, for prices are nominal; they vary. Smith therefore searched for one "sign" or symbol that expressed the money price of labor and concluded that the price of grain (corn) could represent it.[74] Using this new sign, Smith then could evolve an analysis that seemed to run counter to all of the precepts of established economic wisdom, creating "another constitution of things, more natural indeed . . . but more new, more contrary to common opinion and expectation, than any of the appearances themselves."[75]

According to Herder and Smith, their new systems, based upon images and models drawn from the analogy of living nature, were guided by the imagination and spoke to the ever-changing forms of imaginative apprehension. But if science's major task was to satisfy the imagination, what were its truth claims? What guaranteed their validity as being "scientific," as speaking to real relations? In their answers, both Herder and Smith illustrate one of the defining characteristics of Enlightenment vitalism, its modesty concerning the absolute veracity of its scientific claims. Neither Smith nor Herder claimed that their system was objective in the sense that the relations they uncovered expressed evident truths, transparently true and eternally valid. Rather, they related the truths' efficacy to their ability to account for a larger number of phenomena in a more convincing manner than had other systems. One reason for their success, they believed, was that their systems did not submerge the particular in the general; they retained the paradoxical relationship between individual occurrences and the course of nature intact. In short, they assumed that their systems spoke more directly to nature and to the human heart.

For Herder, the efficacy and truth value of his representations were guaranteed by two things. First, the universal circle of similarities and interconnections that linked life-forms validated our projection of self upon the other. The human was, Herder argued, a composite of all the types of living things that have existed; its history had recapitulated the history of organized life. The analogies he drew had, for him, the substratum of a universally lived experience. Second, this projection was true because it reflected the hidden core of living reality constituting our human nature, what he called the *Haupttypus*.[76] Smith's answer was more circumspect, though similar in structure. He too tied all knowledge to human sentiments, and specifically to the interplay between human beings linked together through universal sympathy, a feeling based upon the generation of "analogous emotions" that "springs up . . . in the breast of every attentive spectator."[77] And though spectators never experience exactly the feelings of the persons with whom they empathize, the sentiments linking them "have such a correspondence with one another, as is sufficient for the harmony of society."[78] Though these connections were in one sense subjective, based on the reciprocal interchange between feeling individuals, they were made valid by the fact that they also appealed, as Campbell and Skinner wrote, to "a still higher tribunal, namely that of 'the supposed impartial and well-informed spectator, the man within the breast, the great judge and arbiter' of conduct."[79] In many ways, Smith's idea of the "man within the breast" had a close affinity with Herder's image of the *Haupttypus*, both assuming a type of objective experiential *Urform* [archetype] underlying all subjective sentiments.

Neither Herder's *Ideen* nor Smith's *Wealth of Nations* was a paean to what is and has been, but rather to what could be, based upon the potentialities residing within humanity, suggested by the analogies of growth and development. These potentialities were, they believed, products of nature.

Humans were the paradoxical animal *par excellence*—composed of a fragile harmony between animal sensibility and spirit, forever unstable, yet capable of achieving an expansion of Enlightenment essential to humanity's organization. In the *Ideen* and the *Wealth of Nations*, Herder and Smith sought to chart the "progress" humanity had achieved; they provided, in their own way, a method of analysis and a system of proof that elevated their belief that the study of the science of humanity will encourage us to cultivate the "natural" powers that would make our future progress toward humanity, Enlightenment, and liberty less tortuous. And they did this by incorporating the methods and assumptions of Enlightenment vitalism into their research and writing.

CHAPTER FOUR

Biology in the service of natural theology: Paley, Darwin, and the *Bridgewater Treatises*

Jonathan R. Topham

In *The Blind Watchmaker*, Richard Dawkins quotes from the eighteenth-century Anglican theologian William Paley's *Natural Theology* (1802) the famous analogy between a watch and objects in nature. "[E]very indication of contrivance," wrote Paley, "every manifestation of design, which existed in the watch, exists in the works of nature; with the difference, on the side of nature, of being greater or more, and that in a degree which exceeds all computation." Since in the one case the manifestations of contrivance lead us to conclude that this is an artifact that must have had an artificer, so should they also in the other. This argument provides the foil for Dawkins's own. "Paley's argument is made with passionate sincerity," Dawkins asserts, "and is informed by the best biological scholarship of his day, but it is wrong, gloriously and utterly wrong." The analogy between watch and living organism is specious, he claims. In the wake of Darwin's *Origin of Species* (1859), we know that the appearance of design is the outcome of natural selection, which, "[i]f it can be said to play the role of the watchmaker in nature, ... is the *blind* watchmaker."[1]

Dawkins's argument here is not new. Indeed, the perception that Darwin turned Paley on his head is as old as Darwin's theory itself. Darwin read Paley's *Natural Theology* as a young man and was imbued with a sense of the perfect adaptation of structure to function in living organisms. In the 1830s, it became one of the objectives of his work on a theory of species' origins to account for such functional adaptation. Natural selection seemed to Darwin to provide an alternative, naturalistic explanation of Paley's *explanandum*: the appearance of design in nature. In his autobiography he famously wrote: "The old argument of design in nature, as given by Paley, which formerly seemed to me so conclusive, fails, now that the law of natural selection has been discovered." Yet this contrast between Paley's watchmaker and natural selection provides a very impoverished characterization of the role of biology in natural theology—and of natural theology in biology—in the intervening years.[2]

First published more than half a century before the *Origin of Species*, Paley's *Natural Theology* was hardly the height of contemporary theological or biological sophistication. Its presupposition of a created order that had come into being a matter of a few thousands of years earlier as a result of divine fiat was scientifically outmoded, as was its attempt to explain animal and plant morphology solely in terms of functional adaptation. In the interim, scientists and theologians had striven to develop natural theologies more in keeping with the progressive accounts of the history of creation emerging within the new science of geology and in astronomy, and with the new "philosophical" approach to natural history, which sought to identify the laws of life, rather than merely to describe and classify living beings. While Darwin used the phrase "Paley and Co." to refer to natural theologians, many of Paley's successors diverged from him in their approach to the living world.[3] It will be the object of this chapter to examine some of these alternative approaches and to explore the extent to which the literature of natural theology had trodden the path between Paley's watchmaker and Darwin's natural selection before the publication of the *Origin of Species*. In particular, I will focus on the *Bridgewater Treatises*, a series of eight extremely popular works published in the 1830s, which nicely illustrate the far-reaching development that took place in natural theology in early nineteenth-century Britain.

Introducing natural theology

Since the Paleyan stereotype is so prevalent, it will be helpful to begin by defining natural theology and briefly considering its place in British culture in the first half of the nineteenth century. In its strictest sense, the term describes the attempt to identify truths about God and human religious duties by the use of unaided (or "natural") reasoning—that is, without the aid of any supposed revelation of God by himself. This form of theological discourse has a long history and, by the Middle Ages, the arguments of natural theology had achieved considerable philosophical sophistication. The arguments took three main forms: the *ontological argument*, in which the existence and attributes of God were deduced from the very possibility of conceiving of such a perfect being; the *cosmological argument*, in which the existence and attributes of a divine first cause were inferred from the existence of the cosmos; and the *teleological argument*, in which the existence and attributes of a divine designer was inferred from the appearances of design in nature. Of course, we recognize Paley's argument as the last of these, but it is worth remembering that even in the nineteenth century natural theology extended beyond the argument from design.

It was in Britain—and later in the Anglophone world more generally—rather than on the continent of Europe, that natural theology became most strongly linked with the sciences from the late seventeenth century onward. Especially in the wake of the scientific achievements of Isaac Newton, numerous British works of natural theology were published that not only offered support to Christian doctrine but also thereby demonstrated to contemporaries the religious, and indeed the political, utility of the new mechanical philosophy. In the eighteenth century this developed into a resilient tradition, which historian John Brooke attributes particularly to the manner in which, in the distinctive sociopolitical conditions of eighteenth-century England, natural theology was valued as providing a means of mediating between individuals from different Christian traditions, and hence an integrative basis for both civil society and intellectual debate.[4]

For Brooke, the observation that natural theology served functions beyond its stated theological purpose accounts in part for its continued prevalence in nineteenth-century Britain, despite the devastating philosophical critiques published at the end of the eighteenth century by David Hume and Immanuel Kant. Natural theology continued to offer a means of mediating between scientific practitioners from different religious traditions, but it also provided an opportunity for those who were religiously committed to integrate their scientific researches with their individual religious commitments. As such, the project of natural theology undoubtedly shaped both the practice and the cognitive content of science. In addition, natural theology provided scientific practitioners with a means of manifesting the religious value and safety of their work at a time when fear of a revolution similar to that recently experienced in France was genuine, and concern was widespread that the sciences might provide support for revolutionary materialism. Indeed, in the hands of some, natural theology became a powerful means of enlisting science against political reform, since, it was argued, the existing state of affairs was natural, and thus designed by God. As we shall shortly see, however, natural theology was politically ambivalent.[5]

While natural theology certainly continued to fulfill important functions in early nineteenth-century Britain, it is important to observe that by no means all the references to design in nature that are to be found in nineteenth-century writings on the sciences were intended to be read as contributions to natural theology. The unwary might readily consider a reference to the creator having employed coral polyps to "construct and rear mighty fabrics in the bosom of the deep" to be a specimen of natural theology. Yet the author of such an account might be one of those for whom, while the study of nature must always be combined with reference to divine agency, knowledge of the creator must always come first from the Bible, rather than from any form of natural theology, strictly so called. Certainly, that was the position of William Kirby, the naturalist from whom the quotation is taken.[6] References to design in nature of this sort could function in many different ways. Often, indeed, their purpose was merely to link the sacred and the secular, so that those engaged in reading about the sciences would not find their minds taken away from the life of devotion to God. They might also, however, serve other purposes—for instance, clarifying Christian doctrine concerning God's role as creator—that were still far from being a natural theology as such.

In the light of this, it is worth reflecting that perhaps the backward projection of the Darwinian antithesis between natural selection and Paley's version of the design argument has had a tendency to incline historians to overemphasize the prevalence of natural theology in early nineteenth-century science. Certainly, early nineteenth-century Britain witnessed a significant retreat from natural theology among many Christians. The evangelical revival, for instance, contributed to such a reassessment. Associated first with John Wesley and the Methodists and subsequently with older dissenting groups and with Anglicans such as William Wilberforce, evangelicalism brought a new emphasis on the transforming power of the Gospel found only in the Bible. With their natural reason perverted by Adam's fall from grace, evangelicals claimed, humans were ill equipped to understand God by themselves, but the message of the Bible brought both understanding and salvation. Such a scheme seemed to leave little scope for a natural theology, independent of revelation, though some evangelicals still found a limited place for it. This did not, however, mean that evangelicals were reluctant to identify instances of design in nature. On the contrary, they particularly emphasized that a "Christian tone" needed to be infused into all secular learning, the sciences included, and references to design in nature were often part of this.[7]

There were other reasons, too, for Christians to be critical of natural theology. Particularly in the years following the French Revolution of 1789, various forms of religious radicalism became a matter of concern to Christians in Britain. Yet, while the radical political activist Thomas Paine (1737–1809) inveighed against Christianity and the absurdity of the Bible in his widely read *Age of Reason* (1794–95), he also used the arguments of natural theology to defend his deistic belief in a remote but real creator God. Many Christians, notably evangelicals and members of the Anglican High Church party, were not only quick to see that natural theology was no answer to prominent religious dissidents such as Unitarians or Paine's deists, but also that it could actually lead skeptics to rest contented with a religion far removed from the Bible or Church tradition. The truth of Christianity needed to be fought on these grounds, not on the grounds of natural theology.[8]

It was in this climate, however, that William Paley's *Natural Theology* was published to great acclaim. As we shall see, it was intended to be a natural theology in the strictest sense of the term, forming the logical foundation of Paley's whole theological system, prior to the introduction of the evidences of the Christian religion. Designed to complete his set of theological treatises for the use of Cambridge undergraduates, it was written accessibly and with great clarity. Partly in consequence, it reached a far wider audience, running to ten editions in the first four years. Moreover, it continued to sell well for more than half a century, with over 90,000 copies in print in Britain by the publication of *Origin of Species* in 1859. By dint of its continued production and use, it thus became something of a classic—read by succeeding generations in different ways, but widely familiar within British culture.[9]

In this chapter, however, I will place Paley's work in contrast with the much less well-known *Bridgewater Treatises* (1833–6). This series of eight works was published in fulfillment of a bequest entrusted to the president of the Royal Society by the eighth Earl of Bridgewater, Francis Henry Egerton (1756–1829), for the production of a work

On the Power, Wisdom, and Goodness of God, as manifested in the Creation; illustrating such work by all reasonable arguments, as for instance the variety and formation of God's creatures in the animal, vegetable, and mineral kingdoms; the effect of digestion, and thereby of conversion; the construction of the hand of man, and an infinite variety of other arguments; as also by discoveries ancient and modern, in arts, sciences, and the whole extent of literature.

The eight authors selected, with the assistance of the Archbishop of Canterbury and the Bishop of London, were leading men of science, who between them produced detailed accounts of the several sciences, considered in relation to the question in hand. Despite being expensively produced, the treatises sold well, with over 60,000 copies of the several titles in print by 1850. After Paley's work, they were thus the most widely circulated books on natural theology in the period prior to the publication of the *Origin of Species*, and they were widely read and discussed.[10]

The *Bridgewater Treatises* appeared to a generation of historians as mere adjuncts to Paley's work. This view received its canonical form in Robert M. Young's suggestion that there was, during the first six decades of the nineteenth century, "a relatively homogeneous and satisfactory natural theology, best reflected in William Paley's classic *Natural Theology* (1802)," which formed an important component of a "common context" of intellectual debate. On this view, the *Bridgewater Treatises* were merely "an attempt to codify this tradition in the light of detailed findings in the several sciences."[11] As we will see, however, the *Bridgewater Treatises* not only developed new perspectives on natural theology quite distinct from that of Paley, but their authors by no means shared a common approach. On the contrary, they developed a range of approaches, and my object here will be to outline some of their divergences from Paley and to indicate the growing biological sophistication of their natural theologies.

William Paley's *Natural Theology*

In the dedication to his *Natural Theology*, William Paley (1743–1805) announced that it was the completion of a comprehensive system of theology, but that his works had been written "in an order, the very reverse of that in which they ought to be read." The son of a Yorkshire headmaster, Paley had shone as a student at Cambridge, and for a decade he had been a well-regarded college fellow and lecturer. It was by drawing on his Cambridge lectures in ethics that, in 1785, having taken up a clerical career, he published his *Principles of Moral and Political Philosophy*. This became a set text in Cambridge almost straight away, and continued to be used in university examinations until 1857. Paley had also instructed his Cambridge undergraduates in the evidences of the Christian religion, and the historical credibility of the New Testament became the subject of his *Horæ Paulinæ* (1790) and *A View of the Evidences of Christianity* (1794). The latter was studied by Cambridge undergraduates, Charles Darwin included, for almost a century. When Paley completed his *Natural Theology* in 1802, he had thus completed a systematic introduction to Christian theology. The student might now encounter in sequence "the evidences of natural religion, the evidences of revealed religion, and an account of the duties that result from both."[12]

Paley's system was theologically orthodox, but both his theological principles and his friends were latitudinarian (that is, tolerating latitude in doctrine and placing confidence in reason over and above Church tradition or biblical authority), and his project to found a system of Christian theology on the philosophical arguments of natural theology was expressive of this. It was this theological rationalism that later led some High Church and evangelical critics to express serious reservations about the tendency of *Natural Theology*. Although the work was published in 1802, Paley's latitudinarian mind-set was somewhat at odds with the theological temper of the new century. While Darwin himself read *Natural Theology* as an undergraduate at Christ's College, it was very much less used in Cambridge education than Paley's other works. Nevertheless, as we have seen, the work was immensely successful. In part, this must be attributed to the easy and attractive style of Paley's prose. *Natural Theology* was based on sermons he had delivered over the preceding two decades,[13] and its accomplished use of rhetorical forms rendered it a compelling read. Moreover, while its account of the design argument merely restated a familiar analogy in a persuasive manner, the substance of the work illustrated the argument with lively examples drawn in many cases from the sciences.

Paley's statement of the argument from design as an analogical argument is sufficiently well known not to require much description. Starting with the *disanalogy* between a stone and a watch, Paley spent two chapters describing the inferences that any reasonable person would draw on encountering the latter while crossing a heath. The rational inference from the obvious contrivance in the watch was that it had been made by someone who had designed it for its use. This would be the case whether or not one had ever seen a watch being made, found that the watch did not work properly, failed to understand the use of some of the parts, or even found that the watch contained within it a mechanism to produce another watch just like itself. Only after considering these potential objections, did Paley introduce the analogy between the watch and the works of nature. Could anyone, he asked, reasonably doubt the inference from the design in the watch to the existence of a maker? Yet this was logically equivalent to the absurd position of the atheist, he concluded.[14] Natural objects are like human artifacts in exhibiting design, he claimed, but since human artifacts exhibit design because they are made by a designer, and since like effects have like causes, one can infer that natural objects too are likely to be the product of a designer.

It is important to note, at this point, that the inductive inference Paley urged his readers to make was necessarily only a probable one. Amassing additional evidence could contribute to the strength of the argument, but ultimately conviction was not merely a matter of reasoning. Recent commentators have consequently emphasized that natural theologians were adept in the use of rhetoric, which appealed to "the imagination and the emotions as well as to the faculty of reason."[15] As Matthew Eddy has shown, a key part of Paley's rhetorical approach was his presentation of "a long chain of related examples," which was intended to lead the reader to "a series of inductive assents." The argument was, Paley famously asserted, "cumulative."[16] Moreover, Paley selected his examples so as to maximize the probability of his argument. In particular, his preference was for examples drawn from the living world, both because here the analogy with human mechanism was strongest, and because the large number of individual parts on which the proper functioning of the whole depended increased the probability of their not having arisen by chance. At the start of his short chapter on astronomy Paley argued that, while the subject raised the mind that was convinced of God's existence to "sublimer views of the Deity" than any other subject, it was "not so well adapted, as some other subjects are, to the purpose of argument." The movements of the heavenly bodies were quite unlike human mechanism, and they seemed not to involve a mechanism involving carefully adjusted parts. Much more powerful in argument was an organic mechanism such as the eye, which so much resembled a human contrivance (such as a telescope) and involved a visible adjustment of parts to achieve a particular end.[17]

Natural Theology was thus dominated by examples drawn from the sciences of life. Paley introduced his analogical argument in six chapters, and a further five at the end were given over to considering the divine attributes and to concluding remarks. Of the remaining sixteen chapters, fourteen related to living organisms (almost exclusively animals), with only two devoted to the physical sciences. Paley's approach to living beings was to conceive of them in mechanistic terms, as functioning wholes, the various parts of which were mutually adapted. Usually implicit, but sometimes explicit, in his account were the mutual adaptations of whole organisms and their surroundings. Moreover, underpinning the entire analysis was a belief in the unchanging character of living forms. Current species had been created by God at the outset, some few thousand years previously, and had remained unchanged since. In this regard, of course, Paley was in good company. While he was clearly familiar with the evolutionary theorizing of Darwin's grandfather Erasmus (1731–1802) and made some reference to the impossibility of species transmutation in *Natural Theology*, Paley was hardly at odds with scientific orthodoxy in considering the order of living beings to be static. Moreover, while he did not draw heavily on the very latest scientific findings, his account relied on such familiar sources as James Keill's *Anatomy of the Humane Body Abridg'd* (1698) and Oliver Goldsmith's *An History of the Earth, and Animated Nature* (1774).[18] Even as Paley was writing, however, the scientific debate about the history of living beings was entering a period of rapid change. By the time the authors of the *Bridgewater Treatises* contributed to the literature of natural theology some thirty years later, they had to respond both to significant developments within the sciences and to a culture in which both the sciences and natural theology were viewed rather differently.

The *Bridgewater Treatises*

Given the classic status that Paley's *Natural Theology* had acquired by the 1830s, with around 40,000 copies in print, it is perhaps not surprising that historians have been inclined to see the *Bridgewater Treatises* as mere extensions and amplifications of Paley's *Natural Theology*. Moreover, there is some justification for this view. Several of the authors of these works expressly stated that they saw themselves working within a Paleyan tradition. Yet, of course, classic works have a tendency to be appropriated by those who come after, and an obeisance to the great master can hide a more revisionist intent. Certainly some of those who first read the *Bridgewater Treatises* saw in them important departures. One reader of the *Bridgewater Treatise* on astronomy and physics by Cambridge man of science William Whewell (1794–1866) considered it "surprising how much new light ha[d] been thrown" on the subject of natural theology since Paley's time. And, as we shall see, even the most Paleyite of the Bridgewater authors modified Paley's approach to take account of important developments in the sciences of life.[19]

The author apparently most in Paley's thrall was the distinguished Edinburgh-educated surgeon, Sir Charles Bell (1774–1842), who received a "handful of gold" for a work on Lord Bridgewater's favorite topic, "the human hand." In the early 1830s, Bell also produced with the Whig Lord Chancellor Henry Brougham an extensively annotated edition of Paley's *Natural Theology*. Here, the very first sentence, asserting the disanalogy of watch and stone, was annotated with a page-long footnote, beginning:

> The argument is here put very naturally. But a considerable change has taken place of late years in the knowledge attained even by common readers, and there are few who would be without reflection 'how the stone came to be there.' The changes which the earth's surface has undergone, and the preparation for its present condition, have become a subject of high interest; and there is hardly anyone who now would, for an instant, believe that the stone was formed where it lay.[20]

Not surprisingly, the author of the *Bridgewater Treatise* on "geology and mineralogy," the Oxford reader in these subjects William Buckland (1784–1856), was quick to make the same point. The inception of the new sciences of geology, and more especially paleontology and comparative anatomy, provided a new set of phenomena from which instances of design might be derived. Of course, these could be seen as mere extensions of Paley's approach. However, as we shall see, Bell and Buckland not only developed Paley's notion of functional adaptation in new ways, but they also introduced a temporal element strictly absent from the latter's *Natural Theology*.[21]

If these authors are to be found diverging from their acknowledged master, how much more so were those who were altogether more circumspect about Paley's approach? Bell and Buckland sought to historicize and to refine Paley's emphasis on adaptation to function as the primary determinant of animal and plant structures. By contrast, the author of the treatise on "animal and vegetable physiology"—London physician and secretary of the Royal Society Peter Mark Roget (1779–1869)—drew on the latest, controversial researches in comparative anatomy to move far beyond Paley's teleology. Roget suggested that there were higher morphological laws at play in nature. God had made species according to a "general plan of creation," and by attending to this, Roget meant to give the subject "that unity of design, and that scientific form, which are generally wanting in books professedly treating of Natural Theology, published prior to the present series; not excepting even the unrivalled and immortal work of Paley."[22]

Roget's morphological laws were not the only laws featuring prominently in the *Bridgewater Treatises*. In his treatise on physics, William Whewell observed that the "peculiar point of view" of modern science was "that nature, so far as it is an object of scientific research, is a collection of facts governed by laws." Whewell's aim in his *Bridgewater Treatise* was to "to show how this view of the universe falls in with our conception of the Divine Author." In so doing, he developed an approach that was by no means entirely original, but was certainly at odds with Paley's. Others of the authors, too, wrestled with the extension of scientific naturalism, seeking to establish in chemistry, geology, and physics that the attribution of phenomena to natural laws did not undermine the evidence of divine design. The mathematician and designer of calculating engines Charles Babbage (1791–1851), who wrote an unofficial ninth *Bridgewater Treatise*, took this approach furthest of all, extending the notion of divinely instituted natural law even to the origin of new species.[23]

As we shall see, these developments make the simple antithesis between Paley's watchmaker and Darwin's natural selection rather inadequate as a characterization of history. However, a further aspect of the *Bridgewater Treatises* is deserving of discussion at this point; for it is questionable to what extent these were meant by their authors to be works of natural theology at all. Certainly, the wording of Lord Bridgewater's bequest placed a relatively untaxing requirement on the authors: they were to write on the attributes of God as "manifested" in the creation. However, no prescription required that the authors attempt to infer the attributes, much less the existence of God, independently of the Christian revelation. Indeed, noting that the bequest referred to using arguments based on "the whole extent of literature," the septuagenarian parson naturalist William Kirby (1759–1850), who wrote the *Bridgewater Treatise* on the "history, habits and instincts of animals," considered that this gave him free reign to indulge his habit of uniting "the study of the *word* of God with that of his *works*."[24] When the Hebrew scriptures were deciphered according to an almost Kabbalistic system of interpretation, he argued, they revealed fundamental truths about nature. Kirby's treatise was admittedly unusual in this regard, drawing on the theological system of his fellow High Churchman John Hutchinson (1674–1737), which by this time was generally viewed as a historical curiosity. Yet, as we have already seen, there were many in the churches who were dubious about the epistemological validity of natural theology and about its apologetic usefulness in converting deists or radical Christians. These sentiments were more widespread among the Bridgewater authors than might at first appear.

The more theologically sophisticated among the authors were explicit about their concerns. Thomas Chalmers (1780–1847), a noted writer on political economy and a leading member of the Evangelical party in the Church of Scotland, had caused controversy in the 1810s by arguing that natural theology was at best useless, and at worst a distraction from the evidences of Christianity. By the time he wrote his *Bridgewater Treatise* on "the adaptation of external nature to the moral and intellectual constitution of man," he had become convinced that a limited natural theology was possible, but that it was only such as to lead people "to conceive, or to conjecture, or to know so much of God, that, if there be a profest message with the likely signatures upon it of having proceeded from Him—though not our duty all at once to surrender, it is at least our bounden duty to investigate." Whewell also showed signs of a serious discomfiture with a rationally constructed natural theology, which manifested themselves in his frequent recourse to rhetorical formulations when referring to indications of design. "Does not this imply clear purpose and profound skill?" he asked. "On any other supposition such a fact appears altogether incredible and inconceivable." Moreover, he viewed his obligations as an author in terms far removed from those motivating Paley's *Natural Theology*. "The subject proposed to me was limited," he announced: "my prescribed object is to lead the friends of religion to look with confidence and pleasure on the progress of the physical sciences, by showing how admirably every advance in our knowledge of the universe harmonizes with the belief of a most wise and good God."[25]

Even some of those authors who seem most Paleyite on first inspection appear to have fallen short of Paley's high aspirations. John Kidd (1775–1851), the Oxford Professor of Physic, had some flattering things to say about Paley's *Natural Theology* in his *Bridgewater Treatise* on "the adaptation of external nature to the physical condition of man." Yet he nonetheless considered that it was unnecessary to develop a formal natural theology, since unbelief was intellectually absurd and arose from immorality. His limp and unimpressive conclusion after 300 pages of examples of anthropocentric adaptations was that "whether from chance, (if any philosophical mind acknowledge the existence of such an agent as chance,) or from deliberate design—a mutual harmony does really exist between the corporeal powers and intellectual faculties of man, and the properties of the various forms of matter which surround him." Similarly, the arch-Paleyite Charles Bell confessed to his readers, somewhat shamefacedly, that "from at first maintaining that design and benevolence were every where visible in the natural world," circumstances had gradually drawn him "to support these opinions more ostentatiously and elaborately than was his original wish."[26]

None of the *Bridgewater Treatises*, with the possible exception of William Buckland's on geology, exhibited the confident Enlightenment rationalism of Paley's *Natural Theology*. Moreover, with the exception of Whewell and Chalmers, the Bridgewater authors were theologically ill equipped to provide a detailed or satisfactory alternative analysis. As several of the authors expressed it, they saw their brief as being to demonstrate the compatibility of modern science with orthodox Christianity at least as much as to use modern science to support a rationally constructed natural theology on which to base Christian belief. Similarly, few within the mainstream churches now felt quite so sure as Paley had been of the power of reason, unaided by revelation, to discover the existence and attributes of God. Rather, religious reviewers of the *Bridgewater Treatises* often valued them as providing a compendium of "safe" science, free from radical taint and combined with religious views.[27] Nevertheless, in treading their sometimes ambiguous paths between natural theology and a theology of nature, the Bridgewater authors did, as we have seen, make important departures from Paley's approach, and it is to these that we now turn.

Cuvierian teleology and progressive creation

As we have seen, Paley's version of the design argument gave fundamental importance to the functional adaptation of the parts of animal bodies both to each other and to external conditions. Yet in the thirty years between the publication of *Natural Theology* and that of the *Bridgewater Treatises*, this ultra-teleological approach to anatomy had been advanced far beyond Paley's account by the commanding researches of the French comparative anatomist Georges Cuvier (1769–1832) and his disciples. Paley was not a naturalist, but a theologian, and, although he included a chapter on comparative anatomy in his *Natural Theology*, the focus there had been chiefly on the more superficial resemblances of organisms. By contrast, Cuvier had transformed natural history by showing how the deep structure of animals revealed patterns of relationship between them, based on common solutions to the problem of functional adaptation. In so doing, he gave to teleological anatomy a vastly more developed scientific form and status. Cuvier began to codify his new approach around the time that *Natural Theology* was published, and by the 1810s it was being incorporated wholesale into British natural history. Thus, when such Bridgewater authors as Charles Bell and William Buckland emphasized functional adaptation in their treatises, it was Cuvier's work and principles, not Paley's, on which they drew. For this reason, it would be more accurate to describe teleologists of the type represented by Bell and Buckland as "Cuvierians" than as "Paleyites."

The application of the teleological principles of Cuvierian comparative anatomy to Paley's argument is well exemplified by Bell's *Bridgewater Treatise* on the hand. Bell explained that the human hand could not be studied in isolation, as a "superadded part," since "the whole frame must conform to the hand, and act with reference to it." This was Cuvier's principle of the *correlation of parts*, which maintained "the adaptation of any one part to all the other parts." Furthermore, the human frame itself ought properly to be studied in comparative perspective, in order to comprehend "a greater design." Cuvier had posited that only four fundamental animal body plans could conform to the *conditions of existence*, in terms both of their internal consistency and of their adaptation to the environment. His four "*embranchements*" or divisions of the animal kingdom represented the only viable solutions of a functional conundrum, and they had no larger significance than that. Yet in the work of Bell the existence of such types was itself seen as an indication of divine agency. Bell clearly considered that the existence of a "great plan of creation" provided evidence of design that was to some extent independent of functional adaptation. Indeed, even Paley, whose argument was so largely teleological, found in the common body plan of vertebrates an argument in favor of the unity of the deity. Nonetheless, for Bell, the real beauty of the vertebrate system—on which his account focused—was that "by slight changes and gradations hardly perceptible, the same bones are adjusted to every condition of animal existence."[28] In this emphasis on the perfect adaptation of organs to their function in the body, and of organisms to their function in the environment, Bell's *Bridgewater Treatise* presented contemporaries with one of the most accessible English accounts of Cuvierian comparative anatomy.

Despite the extent of their debt to Cuvier, Bell and Buckland can, in their continuing emphasis on functional adaptation, reasonably be characterized as using recent developments in natural history to extend Paley's basic perspective. In one key respect, however, they diverged very markedly from Paley. The natural history of Cuvier involved a comprehensive historicization of the living world. In his pioneering work on vertebrate paleontology and stratigraphy Cuvier had concluded that the physical history of the earth was punctuated by a series of catastrophic "revolutions," which resulted in the extinction of species now known only through fossil forms. In Britain, both Cuvier's account of geological revolutions and his use of characteristic fossils to identify strata had, from the early 1810s, served to underpin an extensive program of stratigraphic and paleontological research. Moreover, while some (including Buckland in the 1820s) were anxious to identify the most recent of Cuvier's catastrophes with the biblical deluge, the new geology clearly presupposed an earth and a living creation much older than that implied by conventional biblical chronology.

In one respect, the historicization of nature presented a new opportunity for natural theologians, and it was an opportunity that Bell and more especially Buckland readily seized. Cuvier's paleontology enshrined teleology as the primary principle of fossil reconstruction. Just as at present, he claimed, the extinct forms represented by fossils would always be found to be functionally adapted, both in terms of the correlation of parts within individual animals and in terms of their overall adaptation to their conditions of existence. To Buckland, who was dubbed the "English Cuvier" for his exceptional ability to conduct fossil reconstructions on teleological principles, this was a godsend. The new science of geology had faced significant opposition from various religious quarters, and Buckland took the opportunity of his *Bridgewater Treatise* to demonstrate that, while not to be found in Paley, the subject had an important and distinctive contribution to make to natural theology. In his preface he asserted that the subject extended "into the Organic Remains of a former World" the same approach that Paley had pursued in regard to living species.[29] Using Cuvier's meth-

FIGURE. 4.1 "Megatherium." (From William Buckland, *Geology and Mineralogy Considered with Reference to Natural Theology*, 2 vols. [London: William Pickering, 1836], vol. 2, pl. 5. Courtesy of Leeds University Library.)

odology, he reconstructed individual fossil specimens with the intention of displaying their functional adaptation. His *pièce de résistance* was his account of the fossil ground sloth *Megatherium*, which even Cuvier had considered to be ill adapted. Buckland's widely applauded reconstruction of the habitat and habits of this huge, ungainly creature demonstrated that the animal's structure was actually perfectly adapted to function (see Figure 4.1).

While it was very much in Buckland and Bell's interests that this development of a paleontological version of the functionalist argument should appear to be a mere extension of Paley's work, their substitution of a historicized account of creation for Paley's static one was a major departure. Buckland's announcement in his *Bridgewater Treatise* that "millions of millions of years" might have elapsed between the original creation and the first day in the Genesis narrative became, as one commentator put it, "quite as much a newspaper subject as would an horrid murder or a glorious victory."[30] Since Buckland's work provided the first and most prominent synthesis of a kind of progressivist geology that had been developed in England in the 1810s and 1820s, he prefaced it with a lengthy chapter in defense of the congruence of progressive creation with the Christian scriptures. The main body of Buckland's text began with the earth as a nebulous cloud of gas and dust, and moved through a series of geological epochs to the present era. Within each epoch, the characteristic fauna and flora were, Buckland asserted, perfectly adapted to their environment. Geological catastrophes brought each epoch to an end, wiping out many species, as evidenced by the abruptness of the mineralogical and paleontological changes between geological formations, but these had natural causes. It was only at the commencement of each new epoch, when the planet was repopulated with new, perfectly adapted forms, that the "direct agency of Creative Interference" came into play.[31] Thus, while Buckland still relied on miracles to account for the origin of new species, God's creative activity had been radically historicized and placed within a context in which much of the progressive history of the earth was to be understood as a consequence of natural laws—a theme to which we will return below.

Geoffroyan morphology and the idealist argument

Bell and Buckland's treatises clearly had much in common with Paley's earlier approach, notwithstanding their endorsement of a progressive account of creation. However, it is misleading to see this as the exclusive, or even the predominant, outlook of the *Bridgewater Treatises*. Some years ago historian Dov Ospovat identified what he called the "Bridgewater variety" of the concept of perfect adaptation, typified by Bell, which emphasized the strict adaptation of structure to function. Yet Ospovat's contrasting variety of the concept of perfect adaptation, the "doctrine of 'limited perfection,'" found one of its earliest exponents (as Ospovat himself recognized) in Roget's *Bridgewater Treatise*. This was the notion that, while organisms are perfect,

> they are created by laws, and they are only as perfect as is possible within the limits set by the necessity of conforming to these laws. Since the creator's laws are good and well-conceived, adaptation is the general rule. But an animal may possess an organ that serves no function, though it is useful to another animal of the same type.[32]

As this implies, Roget still made common cause with Paley in his *Bridgewater Treatise* in citing many instances of functional adaptation in the organic realm as evidence of design. In his opening chapter on "final causes" he argued that animal and vegetable morphology, physiology, and development could not be accounted for by material laws, but only in terms of means and ends. In his second chapter, however, he developed a morphological, idealist argument that was to run alongside the teleological, functionalist argument of the first. He was prepared, in explaining the variety of animal form, to go beyond the narrow confines of functional adaptation:

> Even when the purpose to be answered is identical, the means which are employed are infinitely diversified in different instances, as if a design had existed of displaying to the astonished eyes of mortals the unbounded resources of creative power. While the elements of structure are the same, there is presented to us in succession every possible combination of organs, as if it had been the object to exhaust all the admissible permutations in the order of their union.[33]

This "law of variety" was counterbalanced by another law, that of "conformity to a definite type." By this, Roget meant that "the formation of all the individual species comprehended in the same class, has been conducted in conformity with a certain ideal model, or *type*, as it is called." Roget strove to relate this to teleological concerns, arguing that "amidst endless diversity in the details of structures and of processes, the same general purpose is usually accomplished by similar organs and in similar modes." Specifically, he linked it to Cuvier's principle of the correlation of parts, which he called the "laws of the co-existence of organic forms." Yet in reality, Roget had gone far beyond Cuvier's strictly functionalist *embranchements* in positing the existence of an "ideal model, or *type*." The taxonomic classes represented "parts of one general plan" which "emanated from the same Creator." The morphological resemblances of organic forms thus provided an argument for the existence of God that was quite independent of their functional adaptation.[34]

In developing this new, idealist approach to natural theology, Roget was responding to contemporary developments in the sciences quite as much as Bell and Buckland. In the 1820s, the London medical schools were buzzing with radical new approaches to comparative anatomy that British students had encountered on visiting Paris after the end of the Napoleonic Wars. In particular, Cuvier's great Parisian rival, the deist Étienne Geoffroy Saint-Hilaire (1772–1844), had urged the need for a more philosophical approach to natural history. Animal form needed to be accounted for not in terms of final causes (what is it for?) but in terms of efficient causes (how was it made?). Geoffroy's view was that animal morphology was not determined by function, but obeyed certain fundamental laws of organization that were to be explained materialistically. In London's medical schools, the importation of such views had a political edge. Those engaged in supporting the "philosophical anatomy" of Geoffroy were often those marginalized by the Anglican-dominated medical establishment—the corrupt medical corporations and the Anglican universities of Oxford and Cambridge. The naturalistic morphology of Geoffroy Saint-Hilaire, like the transmutation theory of Jean-Baptiste Lamarck (1744–1829), provided a radical alternative to the top-down view of creation associated with Cuvier and his acolytes. God allowed lowly species and the socially excluded alike to rise naturally; in neither case had he fixed the scale by divine fiat.[35]

Roget's appropriation of Geoffroyan anatomy was extensive, but by wielding his idealist argument in place of Geoffroy's materialism he was able to disarm it of those radical connotations that led Bell and others of the Bridgewater authors to attack Geoffroy in intemperate terms. Furthermore, this radical departure in the literature of natural theology was to become more than a momentary one. In particular, as the London-based comparative anatomist Richard Owen (1804–92) rose to prominence in the 1840s, he gave greater sophistication and authority to the idealist version of natural theology. Owen was even prepared to consider that the historical development of life on earth, though expressive of underlying divine ideas, might take place according to natural laws. Moreover, while Roget was the only one of the Bridgewater authors to offer a developed idealist natural theology, it is notable that William Kirby joined him in invoking the "quinary" system of classification developed by the entomologist William Sharp MacLeay (1792–1865) as evidence of a divine plan. According to MacLeay, the plant and animal kingdoms were naturally divided into five basic divisions, each of which fell into five classes, and so on, in turn, through the subsidiary taxa. Each of MacLeay's assemblages of five groups was naturally arranged in a circular pattern, with those having the greatest morphological similarity being most closely connected in the taxonomic arrangement. Kirby had been one of the leading advocates of this quinary system in the 1820s, and in his *Bridgewater Treatise* he suggested that the divine plan of creation revealed by such morphological resemblances manifested design quite independently of teleological considerations.[36]

While the works of Bell and Buckland provided clear expositions of Cuvierian teleology, it would be more accurate to represent the *Bridgewater Treatises* as a point of departure between teleological and idealist perspectives than as the apotheosis of the former. That the series has generally been seen in teleological terms perhaps reflects the extent to which contemporary radical morphologists, seeking a clearly identifiable target, tarred all the volumes with the same brush. The Edinburgh surgeon and doyen of radical morphology, Robert Knox (1791–1862), coined the epithet "Bilgewater Treatises" in order to ridicule the "ultra-teleological school," but this memorable phrase has perhaps inclined historians to overemphasize the extent to which the series embodied the teleological approach, at the expense of the idealist innovations of Kirby and, more particularly, Roget.[37]

Natural law and the nomological argument

As we have seen in the preceding sections, several of the *Bridgewater Treatises* referred certain of the designed phenomena of nature to the action of natural laws, rather than to miraculous action on the part of the creator. Throughout the series, the expanding scope of naturalistic explanation since the time of Paley was clearly seen to be a potential source of public concern, and the authors responded in several ways. Thomas Chalmers, for instance, gave strictly mathematical form to Paley's observation that phenomena that could be explained by a few simple laws did not present such a strong argument for design as when a large number of independent phenomena were seen to have been separately adapted to achieve a given purpose. His answer, however, was to point out that, even when phenomena could be explained in terms of general laws, there were still the initial "collocations of matter" to be explained.[38] A similar distinction was made by Whewell in his *Bridgewater Treatise* on physics, but he went much further, seeking to make natural laws central to the project of natural theology.

As we have seen, Paley's God was a watchmaker—an artisan who could be known through the analogy between his creation and human artifacts. Consequently, Paley produced a catalog of God's mechanical handiwork, dominated by the contrivances in living organisms. Whewell's was a far less anthropomorphic argument. He insisted that in our conceptions of divine purpose and agency we must "go beyond the analogy of human contrivances." We are, he continued, "led to consider the Divine Being as the *author of the laws* of chemical, of physical, and of mechanical action, and of such other laws as make matter what it is;—and this is a view which no analogy of human interventions, no knowledge of human powers, at all assists us to embody or understand." Despite his protestations, however, Whewell did apply an anthropomorphic analogy to the deity, while cautioning that it should not be pushed too far. In place of Paley's artisan God he substituted a divine legislator. The world was governed by general laws, he explained, and just as a stranger in an unknown land might learn about the nature of its human government from the laws that were in force, so the cosmic stranger might learn about the nature of divine government by studying natural laws. This nomological conception of design had the advantage that it pushed Whewell's own subject matter into the limelight. Only in two departments of research—astronomy and meteorology—had science been able "to trace a multitude of known facts to causes which appear to be the ultimate material causes, or to discern the laws which seem to be the most general laws." These, then, were the parts of natural philosophy "in which we may hope to make out the adaptations and aims which exist in the laws of nature; and thus to obtain some light on the tendency of this part of the legislation of the universe, and on the character and disposition of the Legislator."[39]

Whewell thus seemed remarkably sanguine about the advance of naturalistic explanation. Most importantly, he applied his argument to the naturalistic account of the formation of the solar system developed by Pierre-Simon Laplace (1749–1827), which he christened the "nebular hypothesis," since it posited the condensation of nebulous matter under the influence of gravity. If the hypothesis were true, Chalmers had written, it would weaken "the argument for a designing cause in the formation of the planetarium." Whewell, by contrast, considered that "it only transfers our view of the skill exercised, and the means employed, to another part of the work." Both the laws, and the "primordial conditions" on which they operated, would equally evince design. Whewell's self-proclaimed purpose in his treatise was, after all, to show "that the notion of design and end is transferred by the researches of science, not from the domain of our knowledge to that of our ignorance, but merely from the region of facts to that of laws."[40] His influential treatment of the nebular hypothesis, while eschewing any consideration of the merits of the theory itself, still served to turn its atheist potential in favor of design, and thus to domesticate it.

A similar approach was also to be found in Buckland's geological treatise, which, as we have seen, provided a strikingly naturalistic account of the physical history of the earth. Following Whewell, Buckland considered that, if true, the nebular hypothesis would only "exalt our conviction of the prior existence of some presiding intelligence." Later he observed of the material components of the earth:

> If the properties imparted to these Elements at the moment of their Creation, adapted them beforehand to the infinity of complicated and useful purposes, which they have already answered, and may have further still to answer, under many successive Dispensations in the material World, such an aboriginal constitution so far from superseding an intelligent Agent, would only exalt our conceptions of the consummate skill and power, that could comprehend such an infinity of future uses under future systems, in the original groundwork of his Creation.[41]

Yet, as we have seen above, Buckland attributed the repopulation of the planet at the start of each epoch to "the direct agency of Creative Interference." Moreover, he considered one of the primary objectives of his *Bridgewater Treatise* to be the demonstration that geological evidence was utterly opposed to all existing theories of species transmutation. In this respect, too, Buckland found himself in agreement with Whewell. In an anonymous review of the first volume of Charles Lyell's *Principles of Geology* (1830), Whewell had written that by uncovering the "distinct manifestation of creative power transcending the operation of known laws of nature" that was to be found in the repopulation of the earth at the start of each epoch, geology had "lighted a new lamp along the path of Natural Theology." Whewell's position on the possibility of a law-like account of the origin of species is notoriously complex, but both he and Buckland were

FIGURE 4.2 "Argument in Favour of Design." (From Charles Babbage, *The Ninth Bridgewater Treatise: A Fragment* [London: John Murray, 1837], 36–7.)

reluctant to extend their nomological arguments in this direction, and, indeed, the other Bridgewater authors were also in practice implacably opposed to the transmutation of species (see Rupke, Chapter 6, this volume).[42]

The possibility was, however, confidently articulated as being consistent with divine creation in the unofficial *Ninth Bridgewater Treatise* (1837) of mathematician Charles Babbage. Piqued by comments in Whewell's treatise regarding the sceptical tendencies of "deductive reasoners," Babbage wrote a bizarrely fragmentary work to demonstrate that mathematicians could also contribute to natural theology. The first and most important of his arguments used his famous Calculating Engine to suggest that natural laws might be predetermined by the creator to change at specified intervals. Drawing on the enormous prestige of his Engine, Babbage described how he could set it to count up to one hundred million in intervals of one, before suddenly changing to a different number series involving different intervals (see Figure 4.2). He then applied the notion of changing laws from his Calculating Engine to the phenomena of nature, particularly the origin of new species, observing:

> To call into existence all the variety of vegetable forms, as they become fitted to exist, by the successive adaptations of their parent earth, is undoubtedly a high exertion of creative power ... But ... to have foreseen all these changes, and to have provided, by one comprehensive law, for all that should ever occur, either to the races themselves, to the individuals of which they are composed or to the globe which they inhabit, manifests a degree of power and of knowledge of a far higher order.[43]

This assertion was certainly radical, and Babbage's work received extensive criticism, but it was one that led to extensive discussion among men of science about the law-like origin of new species.

Putting Darwin back in the picture

Within seven short years, Babbage's invitation to offer a naturalistic theory of species change within a theistic framework had been answered by the anonymously published *Vestiges of the Natural History of Creation* (1844). Written by an outsider to specialist science—the Edinburgh publisher Robert Chambers—the work not only used Babbage's theological rationalization to justify a theory of species transmutation, but also excerpted to the same end the paragraph quoted from Buckland above. Nor was Chambers the only one to take up the challenge. Darwin's notebooks from the 1830s suggest that, as a still-devout theist, he too was convinced that "the Creator creates by laws" as he formulated his transmutation theory. Indeed, despite the subsequent erosion of his belief in God into a somewhat vacillating form of agnosticism, Darwin used an excerpt from a *Bridgewater Treatise* in a prominent position opposite the title page of the *Origin of Species* in support of this view. "But with regard to the material world," he quoted from Whewell, "we can at least go so far as this—we can perceive that events are brought about not by insulated interpositions of Divine power, exerted in each particular case, but by the establishment of general laws." Of course, the strategic nature of Darwin's quotation is plain to see. Yet the notion that natural theology, or at any rate a theology of nature, might in this way be found consistent with natural selection motivated a number of Christian evolutionists in the later nineteenth century.[44]

The binary opposition between Darwin's theory of evolution and Paley's static and miraculously created living world manifestly obscures the new forms of natural theology that were developed in the *Bridgewater Treatises* and other works to accommodate progressive creation and naturalistic explanation. In addition, it obscures the development of an idealist natural theology in Roget's *Bridgewater Treatise* and elsewhere that was intended to replace Paley's narrow emphasis on functional adaptation as the sole determinant of the anatomy and physiology of living beings. While some of the Bridgewater authors, such as Bell and Buckland, used Paley's strictly teleological notion of perfect adaptation to oppose naturalistic theories of species origins, Roget's move away from strict teleology to some extent opened the door to such theories. In the philosophical anatomy of Roget, Richard Owen, and others, the search was on for the general laws governing animal and plant morphology, including its functional adaptation. This was also Darwin's quest, as indicated by the heading "Zoonomia," or laws of life, in his first transmutation notebook. Moreover, while Darwin's approach was in many ways very different from that of the philosophical anatomists, it has been argued that he shared certain key ideas—such as the belief in the "limited perfection" of functional adaptation—with such naturalists.[45]

My object in this chapter has been to explore some of the divergences and developments within natural theology in early nineteenth-century Britain (particularly in relation to the sciences of life) that are obscured by the simplistic Paley–Darwin dichotomy. I have focused on the *Bridgewater Treatises* both because they were widely circulated books of great significance at the time and because their diversity serves to undermine facile generalizations about the literature of natural theology. What we have seen is that several of the works contributed in a limited but nonetheless significant degree to the development of a historical and causal science of life—what later came to be called "biology"—and to reassuring a large public that such developments were consistent with Christian belief. Historians have long known that Darwin in a curious sense owed a debt to Paley for clearly stating his *explanandum*. In another sense, however, Darwin also owed a debt to the *Bridgewater Treatises*. By adapting natural theology to a progressive and law-like view of the history of the creation, the series unwittingly helped to prepare an audience to receive Darwin's theory. Perhaps, after all, it is not so ironic that Darwin described his 1862 book on the adaptations of orchid flowers to ensure cross-pollination—a book greeted by the president of the Linnean Society as a remarkable exemplar for the new field of "biology"—as being "[l]ike a Bridgewater Treatise."[46]

Acknowledgments

I would like to acknowledge the helpful comments and suggestions of Denis Alexander, Geoffrey Cantor, Jon Hodge, Sandi Irvine, Ron Numbers, and the other contributors to this book.

CHAPTER FIVE

Race, empire, and biology before Darwinism[1]

Sujit Sivasundaram

On 26 November 1818, about a mile away from the temple that is said to house Buddha's tooth, in the highlands of present-day Sri Lanka, at a site that became known as "Hangman's Hill," two chiefs were executed by their British colonial overlords.[2] Before being executed, one of them, Kappitipola, washed his face and hands and begged to be left for a short time to perform religious observances. From within the folds of the cloth wrapped around his loins, he took out a small Buddhist prayer book. He was still reciting the Nine Attributes of the Buddha in Pali, the formal language of Buddhist scripture when his neck was struck with a sharp sword. "At that moment he breathed out the word Arahaan, one of the names of Boodhoo. A second stroke deprived him of life, and he fell to the ground a corpse. His head being separated from his body, it was, according to Kandyan custom, placed on his breast."[3] Henry Marshall (1775–1851), who was later hailed as the father of medical statistics, was at this time serving as a military surgeon in Ceylon. The two chiefs' skulls fell into Marshall's hands.

Eventually, Kappitipola's skull made the long journey to Scotland, and was presented by Marshall to the Phrenological Society of Edinburgh. Phrenology was by all accounts—even at the height of its amazing popularity—a controversial science. It asserted that the brain was the organ of the mind, and that the brain was composed of a set of organs which were spread across the head. This meant that temperament could be read off the skull.[4] Phrenology was founded out of the work of the Austrian anatomist Johann Franz Gall in 1795, and then popularized by others who made their name from phrenology, such as Johann Spurzheim and George Combe. Combe's text *The Constitution of Man* (1828) was one of the most widely read books of science in the early nineteenth century.

The September 1832 issue of *The Phrenological Journal and Monthly Miscellany* carried an article entitled "On the Character and Cerebral Development of the Inhabitants of Ceylon." By examining nineteen skulls from Ceylon, including that of Kappitipola, this article urged a typology of races, differentiating aborigines and naturalized foreigners on the island. The article asserted that the race of the "Cingalese" had generally "well filled up" heads in the region of Benevolence and Veneration, which was unsurprising given their devotion to Buddhism. The great size of the organ of Cautiousness was explained by the terror in which the Cingalese held demons and devils. The separation of Sinhalese and "Malabars," later termed Tamils, became entrenched in the racial classification of colonial Sri Lanka and continues to plague the country. Yet the fate of Kappitipola's skull shows that this racial divide did not emerge merely from colonial policies or indigenous ideologies; it was interlocked with practices of measurement and collection, which were scientific.

In the first half of the nineteenth century, when science became recognizably modern and new scientific disciplines such as phrenology were arising out of the former natural history, race and science were bedfellows. This chapter aims to unravel the complicated story of how this came about. The emergence of a science of race is sometimes mistakenly dated to the middle of the nineteenth century and aligned with the spread of Darwinism across the imperial realms. Because the sciences led to an identification of racial and national types they were an important part of the framework that upheld empire, both at home, amongst the populace in Europe, and abroad, amongst those who found themselves governing the colonies. Biology neutralized the question of whether empire was moral, by showing how races and peoples could be "improved," and so a justification was provided for rule by the supposedly superior colonizers. At the same time, empire provided the raw materials for science and helped to define the stereotypes. Kappitipola's skull would never have arrived in Edinburgh but for British colonization in Ceylon. These fundamental synergies between race and empire suggest that the study of the human body was vital to European expansion from the very start.

To make sense of the interlinking of science, race, and empire, it is important to begin not with the high point of modern empires, after the middle of the nineteenth century, but with the seventeenth and eighteenth centuries. The early modern period provides an important context for the science of race; there were continuities throughout this long period rather than intermittent revolutions in thought. This does not mean that the science of race was uncontested: for its hegemony should not be assumed in any historical period.

It is also crucial to keep view of developments in both Europe and the colonies. Race was made into a scientific concept out of new interconnections in thinking across a newly emerging world. The first part of this chapter concentrates on what might be called the Atlantic world of Europe, America, and to a lesser extent the Caribbean and West Africa; the second part considers the greater Indian Ocean, stretching from the South Pacific to South Asia. Today biology is a global discipline with practitioners everywhere. From one perspective, this chapter shows how and why biology was globalized.

The Atlantic triangle

The study of race in the early modern period was first and foremost a theological enterprise. Theologians interpreting the Bible considered scriptural geography, natural history, and geology, and sought to encompass the origins of newly discovered peoples within their explanations. Seventeenth- and eighteenth-century Protestant ethnologists hoped to interpret the differences between people in relation to such biblical events as the expulsion of Adam and Eve from the Garden of Eden, Noah's Flood, and the confusion of human language at the Tower of Babel. They believed that their study of people would lead them to an intricate set of family relationships, which could be traced back to the original man and woman in the Genesis story.[5] Central to the orthodox theological view was a commitment to what was termed monogenesis, the belief in a singular creation of human beings.

But from the intellectual margins there was some sparring: the Frenchman Isaac La Peyrère (1596–1676) created an uproar after publishing, in Latin, his *Prae-Adamitae* (1655), which was translated into English as *Men Before Adam* (1656). Peyrère held that there were two creations, first that of the Gentiles, and then that of Adam, the father of the Jews. The question of whether there was a singular creation or multiple creations was crucial to the character of the biology of race in the nineteenth century.

Though Peyrère's book was publicly burnt in Paris by the Parlement, its legacy remained. It was reprinted eight times before 1709.

Another ancient theory that came to have an important bearing on race biology in the nineteenth century was that of the Great Chain of Being. According to this scale, all of the created realms could be organized, ranging from rocks, to trees, to animals, humans, angels, and then the deity. In addition to gradation, the Great Chain of Being highlighted the plenitude of nature, and its inherent unity, for higher organisms were shown to be related to those below.[6] It was with this view of interrelation that the Swedish natural historian Carl Linnaeus (1707–78), who established a system of botanical nomenclature, wrote: "I cannot discover the difference between man and the orangoutang, although all my attention was brought to bear on this point."[7] Linneaus combined the concept of the Great Chain of Being with the idea of a divine economy of nature, where nature awaited improvement by human intervention, for instance by way of acclimatization. Georges-Louis Leclerc, Comte de Buffon (1707–88), the author of the voluminous *Histoire naturelle* (1749–89), which eventually ran into more than forty volumes, also placed humans and animals in the same system of natural history (see Roe, Chapter 2, this volume). The Great Chain of Being was certainly a pliable idea, and its tidiness lent support to Christian views of origin. Yet in the eighteenth century, it was perceived to be a threat to monogenist understandings because of the way it placed humans right next to beasts.

One of the detractors from this scale, the German doctor and physiologist Johann Friedrich Blumenbach (1752–1840) insisted on a separate taxonomic category for humans in his book *On the Natural Variety of Man*, first published in 1776.[8] Blumenbach held that human variety was explainable by climate, modes of life, and upbringing. But Blumenbach's scheme still took on the nature of a hierarchy and so harped back to the Great Chain of Being. Indeed, it was he who invented the term "Caucasian" because he thought that the people living near the Caucasus Mountains, where Noah's Ark may have landed, possessed "really the most beautiful form of the skull." He also held that white was the "primitive" or original color of humanity.[9]

The Dutch anatomist Pieter Camper (1722–89) sought to rebut the Great Chain of Being but in the end inadvertently lent support to its claims. Camper was renowned for his charts of skulls, and was the holder of a professorial chair at Groningen by 1763. His charts placed humanity on a scale ranging from the animals to the gods, on the basis of what became the oft-cited facial angle, the angle between two lines, one drawn from the forehead to the projection of the upper teeth, and the other the line running through the ear and nose.[10] The classical Greek face has a perfect angle of 100° at the top end and the tailed ape ends up at the bottom of the scale with a facial angle of 42°. Curiously, Camper was a strident monogenist. He insisted, for instance, that regardless of whether Adam was created "black, brown, tanned or white" his descendants changed color as they spread across the world.[11] In these debates about human origin, those who took on board and those who opposed the Great Chain of Being were equally drawn to concepts of hierarchy, and all had in common the project of classifying humanity, in keeping with the organization of the rest of nature.

The attention to climate and topography evident in this work arose directly from a growing awareness of the world that came with new explorations. Buffon was a particular advocate of the impact of climate on human variation and physiognomy. He wrote stridently that "there was originally only one species, which having multiplied and spread itself over the entire surface of the earth, underwent different changes, through the influence of climate, differences in food, diversity in way of life, epidemic illness, and also the infinitely varied mix of more or less similar individuals."[12]

Yet Buffon was also conditioned by the same prejudices as Blumenbach. He used the word "race" imprecisely to signify people in the same climactic belt; the most beautiful race was cast in the temperate zones, between 20° and 30° to 35° latitude north, from "Mongol to Barbarie," and from the "Ganges to Morroco."[13] As a scientific term "race" here denoted a lineage or stock of peoples, affected by where they lived. Out of the bedrock of monogenesis came the increasing utilization of this new word to describe human diversity. This environmentalist understanding of race was deeply concerned with gender: for women were seen to contribute, together with other influences such as climate, to the making of their children's noses, lips, and other physiognomic features. Buffon assigned the flat "Negro" nose to the fact that African women carried their children on their backs, while Blumenbach held that Belgians had oblong heads because Belgian mothers wrapped their infants in swaddling clothes.[14]

Yet there were those who critiqued climatic theories, and perhaps one of the most notable was the Scottish philosopher David Hume (1711–76), in his essay *Of National Character* (1748). For Hume, human differences were not rooted in physical reasons, but rather in moral and social factors. The character of a nation was said to depend on "signs of a sympathy or contagion of manners, none of the influence of air or climate."[15] Hume's scepticism about climatic theory might be fitted alongside the racialist footnote in *Of National Character*: "I am apt to suspect the negroes and in general all the other species of men (for there are four or five different kinds) to be naturally inferior to the whites."[16] A later edition of 1753 dropped the argument about distinct "species" and the implication that non-Europeans were not human.

Anticlimatism's alliance with polygenism, the belief in multiple creations, is evident in the work of Henry Home, Lord Kames (1696–1782), one of the leading lights of the Scottish Enlightenment, who was a philosopher and judge. Kames asked how Amerindians and Europeans could be so different if they shared the same climatic belt.[17] Yet well into the next century the importance of climate in understanding variability was seemingly unassailable, and this meant that a science of race that stressed innate difference rather than variation induced by the environment was kept at bay. The orthodox position continued to be that which held to the unity of mankind.

However, by the end of the eighteenth century, the old view of monogenism as a synthesis of lineage and Scripture had been slowly and subtly reinvented. For within the later Enlightenment discussion of race, we see not a simple acceptance of Scripture, but rather a desire to observe, collect, and make sense of the new information that was arriving from colonized lands. Monogenism had become more a matter of measurement and verification than of scrutinizing the pages of the Bible. This swing to empiricism coincided with a new wave of the British Empire, with the loss of the American colonies and expansion to the east, and the traffic of "curiosities" that came in its wake.

The shows of people in London allowed interested gentlemen to stare and even interact with the early modern period's imagined savages.[18] This genre of exhibition carried right through the nineteenth century, through the Great Exhibition of 1851 and beyond. By the end of the century, whole villages of colonized peoples were on display in France.[19] It is important not to cast this mode of engagement as merely popular, and therefore divorced from the scientific theorizing of race. Perhaps the most telling piece of evidence to support this comes from the story of the display of Sara Baartman, the so-called Hottentot Venus, who arrived in London in 1810, from South Africa, four years before the annexation of the Cape Colony by the British. Baartman was made to pose in London, and members of the public were asked to pay two shillings to see her.[20] Her story, and its attendant fascination with her exaggerated figure, provides another example of the entanglement of definitions and ideas of race and gender in this period. In the spring of 1815 in Paris, Baartman's body was observed over three days by professors at the Muséum d'Histoire Naturelle.[21] When she died later in the year Georges Cuvier (1769–1832), the French comparative anatomist and paleontologist, dissected her body.

Whilst the real experience of the tropics, and increasing acquaintance with newly colonized peoples changed older ideas of human variability, so too did the vociferous debate surrounding slavery, which concerned a group of people already familiar in Europe and America, the slaves exported from West Africa. The popularity of the abolitionist movement in Britain, and its emphasis on the "one brotherhood," meant that the onset of a science of race was delayed there in comparison with what was happening in France.[22] Abolitionism stressed the role of European civilization and freedom in improving the backward peoples of Africa. William Wilberforce, abolitionist and politician, poured scorn on *The History of Jamaica* (1774), the work of Jamaican plantation owner Edward Long. For Long, blacks were a species intermediate between orangutans and Europeans; Long wrote of their "bestial and fetid smell," and noted that their children matured more rapidly, like those of animals.[23] Indeed, it was in honor of the British abolitionist movement's victory of 1833, the date that marked the abolition of slavery, that the Hiedelberg physiologist Friedrich Tiedemann (1781–1861) wrote a bold article, "On the Brain of the Negro, compared with that of the European and the Orang-Outang." This was published in the *Philosophical Transactions*, the journal of the Royal Society, in 1836. Tiedemann threw out the premises of race science and concluded: "there are no well-marked and essential differences between the brain of the Negro and the European."[24]

The abolitionist context in Britain was also vital to the work of James Cowles Prichard (1786–1848), who emphasized the capacity for improvement inherent in all human beings, and who was even sceptical of climatic theories.[25] His *Researches into the Physical History of Man* first came off the press in 1813. According to Prichard, certain qualities, which he termed "connate," could be passed on to offspring. Yet at the same time all human beings were said to have common ancestors; indeed, Prichard believed that the "primitive stock of men were Negroes," and those human beings who became white had over time taken on civilization and a better mode of life.[26] Prichard was an ardent Christian, and his theories saw fulsome disdain of polygenism. He became a central figure in the Aborigines Protection Society, which was both a scientific society interested in ethnology and a forum of protest against the plight of indigenous peoples across the British Empire. The society was founded in 1837 by the Quaker doctor Thomas Hodgkin, who was Prichard's close friend. So Prichard's science was firmly tied to early nineteenth-century British humanitarianism, and it took on board the legacy of the early modern enterprise of aligning Scripture with the study of diversity.

Yet it is important not to paint the Anglophone world in general in heroic terms as antiracialist, for abolitionism itself operated on the basis of a hierarchy of civilizations, with the virtues of British Christian identity at the top. At the same time across the Atlantic, in another context, slavery carried on amongst those who debated the virtues of new theories of human variation in the American South. Here, given the livelihoods of their communities, intellectuals had to hold to both slavery and monogenism as scriptural, in contrast to what was happening in Britain. The "Curse of Ham" provided the best route for bringing these strands together: Noah was discovered drunk and naked by his son Ham, who in turn told his brethren about it, and Noah cursed his son, telling him that he would be the servant of servants. Slavers held that slaves were descended from Ham.[27] Yet at the same time, Southern intellectuals, in a wish to be part of the international intellectual sphere, preferred at times to ignore the problem of slavery.[28]

In 1843, Josiah Nott (1804–73), a respected physician from Mobile, Alabama, published an article on the offspring of black and white unions, in the *American Journal of Medical Sciences*, that argued that "mulattos" lived shorter lives and tended toward infertility, which was basically an attempt to describe blacks and whites as separate species. This was just the start. Soon Nott was giving lectures on the subject, and with George Gliddon, who had resided in Egypt for several years, he published *The Types of Mankind* (1854), which became a best seller. It had gone through ten editions by 1871.[29] Nott, who was driven by an anti-clerical agenda, threw out the Bible along the way, urging that modern science had updated it. In Charleston, the center of natural historical research in the South, Nott's work was unpopular. The Lutheran minister John Bachman (1790–1874), who went on to become a leading name in Southern science, was quick to criticize Nott. Bachman, who wrote a four-part review in the *Charleston Medical Review,* defended the unity of mankind from Scripture, but knew that he had to watch the implications of his position. So he made clear his support for slavery: "The Negro is a striking and now permanent variety, like the numerous varieties in domesticated animals."[30]

The absence of slavery as a factor in the debate over Nott's work is striking. This is further borne out by the work of Louis Agassiz (1807–73), who had contributed an essay to *Types of Mankind*, and later came to preside over American science from his professorship at Harvard. In responding to Nott, from the abolitionist North the polygenist Agassiz stressed the spiritual unity of mankind, even as he urged that humans were distinct from each other in zoological terms. He held that slavery was unrelated to the question of the origin of races. Writing in 1850 in the *Christian Examiner*, he set aside "any connection of these inquiries with the moral principles to be derived from the holy scriptures, or with the political condition of the negroes."[31]

The question of slavery therefore had a complicated impact on the relationship between race and science. In Britain, where abolitionism became a platform for defining Britishness, it determined, for a short period at any rate, scientific accounts of race, so that the unity of humankind was tied to anti-slavery. Meanwhile, in the USA the ubiquity of slavery in the South meant that it was impossible for those on either side of the debate about human origins to take different views on the question of slavery: slavery could be linked with scriptural monogenism or it could be dispensed as a topic of importance to the debate by proponents of polygenism. Nevertheless, in all of these Atlantic contexts, experience borne of colonialism and slavery was the precondition that made it possible to theorize about the non-European human body in the early nineteenth century.

The widening horizons of scientific awareness in turn prompted the birth of a series of new disciplines and institutions in the early nineteenth century: out of natural history were born phrenology, comparative anatomy, geology, and anthropology, amongst others. Phrenology, which we encountered in the account of Kappitipola's skull, stood against the traditional emphasis on climate as contributing to human variability. Instead it was committed to heredity and innateness, and from here it was a short step to the belief that different human groups are differently endowed and so fit for different roles in human society.[32] In turn these types could be used to support particular agendas. According to the phrenologists, African slaves were well suited for freedom. Phrenologists recommended their release from bondage, after a period of years of education that developed their abilities to the fullest limits evident in their skulls.

Phrenology can be viewed as a secularizing science, even though there were Christians amongst its practitioners. Evangelicals in Britain deemed it to be too naturalistic, and George Combe himself wrote in 1847: "Science has banished from [people's] minds belief in the exercise by the Deity, in our day, of special acts of supernatural power as a means of influencing human affairs, and it has presented a systematic order of nature, which man may study, comprehend, and obey, as a guide to his practical conduct."[33] However, phrenology's liberalism, evident in its commitment to the freeing of African slaves, descended from the humanitarian context of abolitionism in Britain. In this sense it must be linked back to earlier religious notions of difference. Phrenology also played an important part in firming up earlier entanglements of race and gender: men of lower races and women of higher races had their skulls compared. This interest in comparison lived on in the work of the German Carl Vogt (1817–95), amongst others. Vogt wrote of how the adult man in the lower races resembled the Caucasian woman who had given birth to children, in that they both possessed an extended belly.[34]

In France, Georges Cuvier's career saw the institutionalization of another discipline, namely comparative anatomy, at the Muséum d'Histoire Naturelle, which emerged out of the Jardin du Roi, the royal botanic garden, in keeping with the antimonarchical ideals of the French Revolution.[35] Cuvier is often thought to bear "a heavy responsibility for the nineteenth-century confusion about the meaning of the word race," because he blurred the distinction between the older idea of race as lineage with the notion of permanent type or variety that had emerged in the sciences.[36] On this count, Cuvier marks the growing onset of innatism and fatalism, namely that bodily difference should be understood from its inner workings and that it was determined by nature. Cuvier was a monogenist, but only just, and his theories are often cast as polygenist, because even though he believed in a singular creation, he urged that his three major races, Caucasian, Mongolian, and Ethiopian or Negro, had diverged from each other after a major natural catastrophe, the biblical Flood, and had then developed in isolation.[37] He advocated the importance of anatomical study to understand racial difference; he believed that physical differences had produced cultural and social changes rather than the other way around; this was in keeping with his insistence on the dominance of function over form.

Having been disillusioned by the Revolution and the subsequent Reign of Terror, Cuvier sought comfort in the empirical facts of measurement and steered clear of the uneasy generalizations that he witnessed in the work of his colleague, Étienne Geoffroy Saint-Hilaire (1772–1844), especially after Geoffroy's return from a Napoleonic expedition to Egypt. His constant insistence on facts meant that he waged a battle against theories that he saw as too metaphysical, and phrenology ranked amongst these.

His most famous conflict with Geoffroy pivoted in simple terms on a clash between his own empiricist functionalism and Geoffroy's philosophical morphology.[38] It was in this context of factual observation and anatomical measurement that his racial ideas emerged; his collection of skulls was vital to his classificatory scheme as was his interest in dissection, which we have already encountered in his engagement with Sara Baartman. He sought for the hidden functions of the organism and their interrelation.

Despite Cuvier and Geoffroy's differences, both of them hoped to reshape the sciences of life by utilizing the rival methods of empiricism and natural philosophical theory, both of which in the end became central to biology. Meanwhile, in Britain, as humanitarianism ebbed away by the 1840s, Prichard's older framework of Christian ethnology was under attack. The Aborigines Protection Society had to tend to its reputation so as not be seen to be critical of the British Empire.[39] The twinning of science and humanitarianism proved difficult and eventually led to the birth of a new society from within its membership devoted to the scientific aspects alone. The Ethnological Society of London was founded in 1843. By the end of the 1850s it had expanded beyond its initially marginal status with the entry of a series of new members who were of a distinctly different kind, as they came with an interest more in physical anthropology and archaeology: Joseph Davis, Joseph Thurman, John Beddoe, Robert Knox, and—the most energetic of them all—James Hunt. In 1862, some "differences of opinion" over the publication of engravings of the inhabitants of Sierra Leone in the society's journal led to the resignation of Hunt and the institution of a rival society, the Anthropological Society of London. The role of Sierra Leone in this split is revelatory: for it was a colony of freed slaves, and therefore occupied a cherished place in abolitionist consciousness. The engravings depicted romanticized Africans, and so did not capture the imagination of the newly emergent scientific theorizers, who held a different view of the non-European bodily frame.

The new Anthropological Society was an immediate success: within two years there were 500 members. Paradoxically, even as the Anthropological flowered, the Ethnological was taken over by Darwinians. For members of the Anthropological Society, Darwinism seemed like a new religious orthodoxy. Hunt noted that there was little difference between "a disciple of Darwin and a disciple of Moses—one calls in natural selection with unlimited power, and the other calls in a Deity provided in the same manner."[40] When there was an uproar over a massacre of Jamaican black farmers by the island's Governor Eyre in 1866, the Darwinians, led by Thomas Henry (T. H.) Huxley, famously known as "Darwin's bulldog" for his enthusiasm for Darwin's theory, were extremely critical, whilst the Anthropologicals could air the view that the killing of "alien races" was "a philanthropic principle."[41] The changing patterns of the institutionalization of the science of human beings in France and Britain are complex and reveal repeated attempts to redefine what we call the life sciences in relation to religion, empiricism, and natural philosophy. The Darwinian takeover of the Ethnological Society, the organization that enshrined the legacy of monogenism and humanitarianism, means that it is important not to present a teleological or linear story about the emergence of scientific racism. Early proponents of Darwinism, in Britain at least, often stood against the oft-cited scientific racism of the Anthropological Society. In this way they returned to earlier ideas.

The extent of popular interest in the meetings and debates of the Anthropological Society also show how discussions of race and science had become matters of public interest by the early 1860s, not just in this scientific society, but in others such as the well-attended Royal Geographical Society.[42] This interest in the science of man had been evident some decades earlier, in the tumultuous reception of the anonymous *Vestiges of the Natural History of Creation* (1844), which went to fourteen editions in Britain. *Vestiges* served up evolutionary theory for the people. Its readers were confronted by a developmental hierarchy, which was what its author—identified as Robert Chambers—termed "the universal gestation of nature," a grand theory that encompassed the whole of nature. By this he meant that in the sky there was development from a swirling fire-mist to nebulae, solar systems, and planets; in the ground there was evidence of development as invertebrates, fish, reptiles, mammals, and man follow in order; and in society there was development in civilization as Negro, Malay, American, Indian, Mongolian, and Caucasian give way in that order, one to the other.[43] The scheme proposed by the author of *Vestiges* was absorbed into the imperial scale of civilization, even as it was open to multiple readings in different localities.

The meanings that were attached to biology were not only defined by the nature of its reception by the public. Within the circles of élite men as well there were practical constraints on who could write biology; these were linked also to race and gender. The "gorilla debates" of the 1860s provide striking evidence of this.[44] These also centered on a best-selling work, *Explorations and Adventures in Equatorial Africa* (1861) by Paul Belloni Du Chaillu (1835–1903). The image facing the frontispiece of that

FIGURE 5.1 "My first gorilla." The frontispiece showed a foldout of a large gorilla. (From Paul Du Chaillu, *Explorations and Adventures in Equatorial Africa* [London: J. Murray, 1861], facing 71.)

book, a foldout showing a gorilla, indicates what caused the greatest stir. Readers were gripped by the lurid descriptions of the gorilla hunt, which in turn fed a competition amongst men of science to get their hands on gorilla skeletons (Figure 5.1). In Britain there was also a heated exchange between Richard Owen (1804–92), the comparative anatomist, who in some ways took on board and modified Cuvier's legacy, and Thomas Henry Huxley, about the distinction between humans and the gorilla. Owen claimed that human brains had a lobe, the hippocampus minor, that was unseen in gorillas, but Huxley disagreed and this in turn depended on Du Chaillu's skulls. The eventual controversy hinged on the veracity of Du Chaillu's descriptions in *Explorations*.

A close reading of Du Chaillu's critics shows that they felt that he had overstepped his mark. Du Chaillu was seen as a fieldworker who had a defined place in the new sciences, below the rank of the producers of scientific knowledge. Indeed, explorers like himself, who spent too long in Africa, were thought to lose their ability to be rational men of science. The attempt to expel Du Chaillu from the intellectual community was summed up in October 1861 by George Ord (1781–1866), a member of the Academy of Natural Sciences of Philadelphia: "If it be a fact that he is a mongrel, or a *mustee*, as the mixed race are termed in the West Indies, then we may account for his wondrous narratives; for I have observed that it is a characteristic of a negro race, and their admixtures, to be affected to habits of romance."[45] It is important to note, however, that for others this sort of criticism was ill-founded.

In the context of the enormous popularity of books such as *Vestiges* and *Explorations*, Darwin's *Origin of Species* loses some of its significance. Indeed, the Darwinian idea of race in the later 1860s and 1870s had an earlier parentage; for instance, there was a coming to terms with evolutionary ideas around the debate between Cuvierian teleology and Lamarckian transmutation in the 1830s and 1840s (see Rupke, Chapter 6, this volume). At the same time, the mid-nineteenth century should not be characterized by simple scientific racism in Britain. Here, in the third quarter of the century, national consciousness, and racial ideas of superiority were seen as Continental excesses.[46] Britain had risen above such ideologies and had advanced up the ladder of civilization; it was her duty to promote civilization across the Empire. Racism became a dangerously anti-imperial sentiment because the British Kingdom and Empire were multinational. This is not to deny the twinning of race and science: it is to make the point that they did not come together for the first time at the middle of the nineteenth century. We should not assume either the relative absence or the dominant presence of a science of race in the period before Darwin.

Before we leave the Atlantic contexts of Britain, France, America, and the fleeting glimpses of the Caribbean and West Africa, it is important to note the divergent trajectories of the science of race in each of these localities, conditioned by religion, politics, and culture. Yet at the same time there were interconnections and reverberations. The debate over slavery and the rise and fall abolitionism, questions about the character of colonialism, and scientific discussions about teleology and empiricism crossed the seas of this Atlantic zone and took root elsewhere just as they had come to a resolution at their place of origin. There was a circulation of meanings and theories throughout this period, rather than a steady advance toward scientific racism. Scriptural traditions, the Great Chain of Being and the hierarchy of civilizations, and the role of the environment and culture in explaining difference, all returned to the intellectual scene just when historians might have expected their defeat. This persistence of older ideas does not preclude the slow move toward a more determinist view of race, or the greater public interest in questions of race. Yet it is clear that the public absorption of the science of race was not straightforward. The entanglement of race and science did not have a simple teleology; instead it encompassed a story of recapitulations.

Oceans to the east

It was not only in the triangle which linked the slaving states in Africa, Europe, and America that there were discussions about how to theorize race. Another related network operated on the shores of the greater Indian Ocean, where Europeans were forging new empires founded in equal measure on free trade and the imagination.

The popular voyages to the South Pacific in the late eighteenth century allowed travelers to come to terms with the non-European frame in radically new ways. These oceans served as "a laboratory" for the human sciences, even as their inhabitants were made into sexual icons in the European imagination.[47] On Captain Cook's first voyage he had on board the young and wealthy naturalist Joseph Banks, who on his return quickly emerged as a leading light of London science. The aims of this voyage included the search for the missing southern continent and the observation of the transit of the planet Venus across the sun, and Cook's ships were equipped with the best scientific instruments. Cook was not exceptional in combining science with empire; his continental rivals were also not averse to the potential of this set of interests. Paul De Brosses's *Histoire des navigations aux terres australes* [A history of the voyages to the southern lands] (1756) set the framework for these kinds of expeditions by stressing the importance of taking both natural historians and illustrators of natural history to the South Seas.[48] These Pacific explorations were carried by the winds of the optimistic Enlightenment hope of mapping all of the created realms, alongside other mundane interests in commerce and navigation. The naturalist Johann Reinhold Forster (1729–98), who, together with his sixteen-year-old son Georg, accompanied Cook on his second voyage, wrote: "My object was nature in its greatest extent; the Earth, the Sea, the Air, the Organic and Animated Creation, and more particularly that class of Being to which we ourselves belong."[49]

An early scientific debate about race in Europe drew on the information gleaned from travelers from the southern hemisphere. The naturalist Philibert Commerçon (1727–73), who was on board the voyage captained by Frenchman Louis-Antoine de Bougainville over the three years 1766–

69, wrote for instance of Tahiti as a new Utopia, and supported his view with a distinctly polygenist line: "I therefore do not see why the good Tahitians should not be the sons of their own soil, that is, descended from ancestors who were always Tahitians, stretching back as far as the people proudest of its ancestry."[50] But curiously, his scientific polygenism did not seek to denigrate the Tahitians; it hoped to show that the Tahitians were a superior race. Johann Reinhold Forster, in his 400-page "Remarks upon the Human Species in the South-Sea Isles," set in motion an important tradition of distinguishing two categories of Pacific peoples. He wrote: "We chiefly observed two great varieties of people in the South Seas; the one more fair, well-limbed, athletic, of a fine size, and a kind benevolent temper; the other blacker, the hair just beginning to become crisp, the body more slender and low, and their temper, if possible more brisk, though somewhat mistrustful."[51]

This sensibility to physiognomy lived on in Georg Forster (1754–94), who recast his initially sympathetic view of Pacific Islanders on returning to Germany and entered into a debate with the philosopher Immanuel Kant (1724–1804) about human origins, illustrating how the two oceanic regions of this chapter were intertwined.[52] Yet European explorers' attachment to Tahiti as a Utopia and their insistent physical representation of Pacific Islanders came together most forcefully in the work of another French explorer, Jules Dumont d'Urville (1790–1842), who traveled through the Pacific, first from 1826 to 1829 and then from 1837 to 1840.[53] It was d'Urville who presented a cartographic representation of the races of the South Pacific in an article in the *Bulletin de la Socété de Géographie* in 1832, distinguishing between Polynesia, Melanesia, and Micronesia. D'Urville wrote of the blackness of the Melanesians and their "hideous" women, as compared with the European-looking Polynesians.

These Pacific explorers were succeeded by evangelical missionaries who, far from being antiscientific, saw themselves as practitioners of science who contributed to the program of ethnology outlined by Prichard. Thomas Hodgkin, the founder of the Aborigines Protection Society, praised the ethnological work of South Pacific missionaries.[54] Two of the most important works that arose from these missionaries in the first half of the nineteenth century were William Ellis' *Polynesian Researches: During a Residence of Nearly Eight Years in the Society and Sandwich Islands* (1826) and John Williams' best-selling *A Narrative of Missionary Enterprises in the South Sea Islands* (1837). These books were concerned to fit the origin of Pacific Islanders into the biblical account of the dispersal of peoples. Without using the term "Melanesia," early missionaries made sense of the darker-skinned islanders by utilizing the "Curse of Ham." One of them wrote: "We see this curse lying in all its crushing weight. The Papuans, the poor descendants of Ham, are lying in the lowest state of degradation."[55]

In missionary ethnology, this exploration of racial differences sat together with an emphasis on the "one blood" of all humanity. By converting islanders to Christianity, missionaries believed, following Prichard, that their charges would move up the scale of civilization. In curious classical language, Makea, the converted chief of Rarotonga in the Cook Islands, was described thus by one missionary: "the idea of a Homeric hero."[56] Missionary accounts of the Pacific therefore did not serve as a precursor to scientific ideas, or indeed disrupt older views, which saw Pacific Islanders as noble and thus worthy instructors of Europe. They were attempts to contribute to the science of ethnology, and they were contradictory in their emphasis on difference and unity. Missionaries also forged their views in relation to those of the islanders themselves, and in turn tutored the islanders on human nature and science.

The need to bring in indigenous notions of race and nature is even clearer as we move across the Indian Ocean to another context of European theorizing, namely India. Central to late eighteenth-century and early nineteenth-century understandings of India was the paradigm of an oriental scholarship that was fascinated with ancient Sanskrit texts.[57] The East India Company's rule of the subcontinent depended in practical terms on the acquisition of Indian languages. The necessity of linguistic expertise arose in part from the nature of the company's "indirect rule" via indigenous rulers and princes.

With this agenda in mind, the company relied practically on the assistance of indigenous *pandits* or scholars, who provided the necessary knowledge in language, philosophy, and indeed science, so that the company could govern as a régime interested in progress. Orientalism had perhaps its best-known elaboration in the famous "anniversary discourses" presented by William Jones (1746–94) to the Royal Asiatic Society of Bengal, which was founded in 1784. Jones, who arrived in Calcutta in 1783 from Britain as a judge of the Supreme Court, developed a deep admiration for the grammatical structure of Sanskrit, which arose through concentrated empirical study of its forms, and he urged a close affinity between it and Greek and Latin. This account of the grand history of Sanskrit ran over into an exposition of human origins. Jones was a strident monogenist and scripturalist; he took the view that after the Flood three distinct races emerged: the sons of Ham were the "Persians and Indians" (including the Greeks, Romans, Goths, Egyptians, and the Chinese and Japanese), the sons of Shem were the "Jews and Arabs," and the sons of Japhet were the inferior "Tartars."[58]

Jones' work was strengthened in India by others such as Thomas Henry Colebrooke (1765–1837), who also became a judge in Calcutta, and to whom Jones' legacy fell. Colebrooke accentuated Jones' commitment to Sanskrit texts as a repository of Indian civilization. The rapturous reception of this orientalism in the German Romantic movement might be credited, rather intriguingly, to a figure whom we already know, Georg Forster, who in turn convinced the philosopher Johann Gottfried Herder (see Reill, Chapter 3, this volume) of how the origins of humanity lay in India. The Germanic reception of Indian orientalism lived on in the work of the philologist Max Müller, who, after his arrival in Britain from Germany, took up the newly created chair in comparative philology at Oxford in 1868.[59]

According to this story, in the first half of the nineteenth century, intellectual thought generated from the subcontinent held to a hierarchy of civilizations, where India's past rather than its present was glorified, and progress was tied not to the body or physiognomy but to language. In this view, it is difficult to write of the rise of biologized racism until much later. Indeed, one leading historian of India has this to say about the view of culture prevalent in this period: "It was fundamentally different in character from that 'scientific racism' which sought to measure anatomical features such as the size of the brain and the shape of the head."[60]

According to this narrative, the anti-British Indian Mutiny of 1857 served as a turning point for Aryanism, the ideology that allowed the claim that Europeans and Indians were of one stock; it made British eyes take note of the true savagery of Indians. Together with the British–Maori wars of 1845–72 in New Zealand and the Eyre massacre in Jamaica, the Indian Mutiny is cast as a point of pressure that mutates ideologies of difference, allowing the birth of biologized racism across the colonial realms. It is in this context that George Campbell (1824–92) of the East India Company's Medical Service wrote *Ethnology of India* (1866), which emerged as a handbook to guide the cataloguing of all the races and classes of India, in a new régime of surveillance. In Campbell's instructions, language is still given a place as a criterion of classification but it is subordinate to physical appearance. At the same time, a grand program of photographing India

FIGURE 5.2 "Korewah Group. Aboriginal. Chota Nagpoor." The photographs were accompanied by descriptions; this one noted that the "Korewah" were "a wild tribe occupying a portion of the water-shed of Central India." (From *The People of India* [London: W. H. Allen & Co., 1868], vol. 1, Plate 21.)

began in the 1850s and yielded an eight-volume work of 468 photographs, *The People of India* (1868) (Figure 5.2). These photographs created types of race, caste, and tribe, and connected these types not to the "peculiar characteristics of costume" but to "the exact tint of their complexion and eyes." One Indian, Sayyid Ahmad Khan, protested that his countrymen were portrayed as the "equal of animals."[61]

The story of how language gave way to fixed biological types as a benchmark for race in India by mid-century has a wealth of evidence to support it. Yet the early century was not a period without race science. The role of phrenology in India is especially revealing.[62] A significant line of earlier racial thinking allowed Indians to be cast as black, and also saw an engagement with the Indian head and physiognomy; it placed Indians' physique in relation to the climate, most especially through the practice of phrenology in the subcontinent. The Calcutta Phrenological Society was founded in 1825 by George Murray Patterson, a Christian surgeon attached to the Bengal Medical Service, who by the time of his death had studied over 3,000 "Hindoo skulls" and had also written *The Phrenology of Hindoostan*. Skulls passed between Patterson in India and George Combe in Edinburgh.

What is striking about this society is how it drafted a whole new class of contributors into the intellectual discourse of race and the body. For the society was run not only by Britons, but it was also managed by mixed-race men; its lectures were well attended, probably by a clientele who were not exclusively European, and phrenology in India was as controversial as it was in Europe. By 1845, a second Calcutta Phrenological Society was launched by the intellectual Kali Kumar Das, and this society's membership was composed of Bengalis, who belonged to the Westernized, missionary-educated world of Hindu reformism. For Das, the physical study of the head provided a route for the regeneration of Indian culture; it allowed a bridge between rational empiricism and philosophical religion in India. But at the same time, in the print media of the years following 1850, phrenology was absorbed into the Indian traditions of *samudrika* or *samudrikvidya*, a set of practices for predicting destiny by reading the body and the head, akin to astrology and palmistry.

In addition to phrenology, there were other topics that allowed race and science to meet; one such topic was the study of climate, which as we have seen played a crucial role in theories of human origin at the end of the eighteenth and early nineteenth centuries.[63] Europeans wondered whether it was possible for them to adapt to the tropics, and also what modes of life and diet they should adopt to remain healthy in that environment. A belief in the importance of "seasoning," the acclimatization of the body, was born to avoid the diseases to which newcomers from Europe to India were said to be particularly susceptible. Buffon's theory of climatic change gave rise to the fear that white colonists would become black over the course of time. Knowledge that was borne out of experience was central to James Johnson's *Influence of Tropical Climates on European Constitutions*, first published in 1813, which urged that there was little prospect of Europeans becoming "acclimatized" in India.

In the second quarter of the nineteenth century, in medical engagements with India, there is not an exclusion of climate, but rather an inscription of climate within a determinist model that characterizes both the subcontinental body and the environment as related to, and different from, those of Europe. These early ideas of tropicality carried through to the middle of the nineteenth century and coalesced around the idea of "tropical races." Climate then did not die off as a mode of explaining racial difference, in India, giving way to other criteria such as language and then physical measurements by the middle of the century; environmental conditions became tied to ideas of innate difference and heredity. Nature and culture were not that far apart in understandings of diversity.

Some other lines of inquiry on characterizations of difference in India also show that race science had a long history in the subcontinent. Orientalists were also very interested in botany, topography, and the land. In 1802, the Scottish physician and naturalist Francis Buchanan (1842–94) was commissioned by Lord Wellesley, then governor-general, to classify and illustrate all the animals and birds of South Asia. Though this project did not get off the ground, partly because of a falling-out between Wellesley and the company, by 1807 Buchanan had been promoted to the title of surgeon of the company and was dispatched to undertake a topographical survey of Bengal. This was supposed to range over all of the land and its people, from the castes and histories of local families to the archaeological ruins of the land. He had undertaken something like it previously in Mysore.[64] In Buchanan's surveys, as in the surveys of his Scottish contemporary, the military surveyor Colin Mackenzie, we see both ethnography and natural history, and in many ways these are forerunners of the grand photographic archives discussed above. The typology of race science at the middle of the nineteenth century descended from these sorts of enterprises and also from the cartographic regimen more generally.

In addition to surveys, the study of natural history in India also related closely to engagement with animals. In the absence of gorillas, the Europeans' place in nature was assessed in relation to other large mammals, such as the elephant and the tiger.[65] Elephants in India were given human character and were even said by British observers to have their castes; one young male Indian elephant, Chuny, who was shot in spectacular fashion in London in front of crowds in 1826, was taken in one engraving to represent "India." Chuny had been anthropomorphized in a myriad ways before his shooting. His skull was later phrenologized by Spurzheim to determine whether the animal was "noble" or "savage." The elephant's remains were also dissected by pupils of anatomy, after its flesh was tasted by surgeons.

This became one of the sensational tales of this period of London life, and shows how an Indian animal could sustain discussions of the nature of human character, and the relation of the living things of India to the rest of the world. Here we find Cuvierian anatomy and Spurzheimian phrenology combined. Putting the elephant in its place became a metaphorical route to understanding the human body and racial difference.

Even weirder for modern sensibilities is the science of mesmerism, which arrived in India more than a decade before the mutiny and contributed to a ruthless engagement with the Indian body as a dispensable vehicle of experimentation and for the trial of mesmeric anesthesia.[66] Its main practitioner was the Scottish surgeon James Esdaile (1808–59), who instituted the Calcutta Mesmeric Hospital in 1848 and returned home by 1851, leaving it to an unexpected demise. Mesmerism does not fit into the neat teleology of the rise of the science of race, but it shows again that there were earlier moments when the human body, science, and notions of difference came together before Darwinism and before more determinist anthropology.

This brings us finally to the man who undoubtedly made a very significant contribution to later nineteenth-century ideas of human origin—Charles Darwin. Yet Darwin's intellectual formation was profoundly affected by the Pacific Ocean too, during his voyages on the *Beagle*. It is rather surprising to find that the first publication that arose from Darwin's pen was co-authored by the captain of the *Beagle,* Robert Fitzroy, and sought to provide a defense of the missionaries of the South Pacific.[67] While traveling through the islands, Darwin had benefited greatly from the hospitality of the missionaries, who arranged guides for him and provided him with company.[68] Darwin's views of human origins and race were also deeply affected by his experience of the three inhabitants of Tierra del Fuego whom Fitzroy had taken captive on his first voyage. These Fuegians, and one other who died from smallpox, were "civilized" in Britain and became like Europeans in the space of two years.[69] Fitzroy had given them an education whilst they lodged with him and now hoped to return them to the islands together with a missionary, thus setting the context for the Christianization of Tierra del Fuego.

Darwin was taken aback by the contrast between these three Fuegians and the other inhabitants of Tierra del Fuego: "I could not have believed how wide was the difference between savage and civilized man. It is greater than between a wild and domesticated animal, in as much as in man there is a greater power of improvement"[70] (Figure 5.3). Darwin's

FIGURE 5.3 "Fuegians." This group of portraits was contrasted with another set, later in Robert Fitzroy's *Narrative*, showing the Fuegians who were being returned after their sojourn in London; the latter would have seemed well dressed and civilized to the readers. (From Robert Fitzroy, *Narrative of the Surveying Voyages of His Majesty's Ships Adventure and Beagle* [London: H. Colburn, 1839], vol. 2, facing 141. Whipple Library, University of Cambridge.)

encounter with these Fuegians and other Pacific Islanders are said to have been crucial to the emergence of an evolutionary sense of man, where there is no essential difference between man and beast. Yet matters are not that simple, for in writing in defense of missions and in documenting the role of civilization in improving Fuegians, Darwin was nailing his colors firmly to another mast. He was taking on board Fitzroy's Christian humanitarianism, which saw all mankind as needing civilization and conversion. If conditions changed, as suggested by George Prichard, then humans could be improved.

So Darwin himself was shaped by the South Pacific missionaries' project of ethnology and conversion. This point might be widened by taking account of a twinning of civilizational or cultural ideas of progress with biological notions of evolution in Darwin's own thinking (see Ruse, Chapter 10, this volume); the application of Darwinism to understanding cultural and social differences was not a later mutation of his thought.[71] Darwin and Fitzroy wrote their paper in the Cape Colony, in Southern Africa, where they were aggrieved at the conditions of missionaries. So here we see that this idea of how humans might be improved emerged from a process of comparisons of missionary work in the South Pacific and at the Cape Colony, across the greater Indian Ocean.

As in the Atlantic circuit of writing about race, in the greater Indian Ocean we need to be circumspect about the rise of a distinctly new kind of biologized racism at the midpoint of the century. This is not to deny that there was a slow move toward fixity or indeed that new technologies such as photography and new kinds of colonialism altered older notions of difference. Yet in this connected realm, religion, art, and science blended and reinforced each other. At the same time, the study of language, culture, and dress sat alongside more physical forms of measuring the types of people; the importance of climate as a means of coming to terms with the tropics lived on and became tied to more deterministic ideas about the body. The story from this sphere also shows how widespread the discussion of race had become, and there is yet more to be discovered about how indigenous ideas and Indian and Pacific peoples became contributors to thinking on racial and biological differences. There are of course differences between the South Pacific Islands and India, tied to the character of colonialism and the nature of local political and social traditions. Nevertheless, the movement of peoples and ideas across this vast oceanic realm makes it appropriate to bring these terrains of intellectual production into the same remit.

Journey's end

As a way of bringing our travels to a close, let me summarize what has been argued in the course of this chapter. Race was and continues to be a potent and disturbing ideology. Its resilience lies in the way it attaches itself to seemingly discordant realms of intellectual thought: science, religion, philology, anthropology, and art, to name but a few. Yet in order to understand it, we must not restrict our gaze to a chosen set of figures in the past, to a particular line of theorizing, or to a particular period of time. It is difficult to disentangle the biologization of race from the production of cultural ideas of difference; this makes the point that, during the period covered in this chapter, race could seek security in rival contexts, so that those who took on board one racial way of seeing the world could use the intellectual credibility of writers elsewhere to forge their theories. No one took exclusive ownership of race or indeed of race science, and yet it can be discerned across the oceanic realms discussed here. Racial theories were undoubtedly forged by the politics of colonization, slavery, and labor; but here again what emerges is how flexible race was. It adapted, for instance, to the different contexts of the American South and India. In arguing that a biological view of race emerged earlier and that a scale of civilization lasted longer, it is not my intention to excuse the mid-nineteenth century's scientists. Rather, it has been my aim to show how complicated the cultural heritage and legacy of an idea such as race can be.

This has not been a truly global history of race, for whole continents have barely been mentioned. Yet it has been an attempt to show how biological ideas were created even whilst empires expanded and Europeans had to come to terms with peoples and social conditions across the world. The participants in discussions of race were both heterodox and unexpectedly connected, in their locality and their views of religion, politics, and culture. This means that it is incorrect to speak of the diffusion of a science of race from Europe to the rest of the world. The resilience of the idea of race lay in its ability to travel over distance and in multiple directions; discussing diversity became a global phenomenon in the period under review, and it drafted in other cultural traditions of knowledge. Biologists based in Europe did develop racist ideas, and yet to leave it at that claim misses quite a lot of the power, contestation, and mutability of the idea of race.

CHAPTER SIX

Darwin's choice
Nicolaas Rupke

To many of us, "ideology," especially in the combination "science and ideology," has a negative connotation and is used in a pejorative sense. With respect to "biology and ideology" in particular, we immediately think of historical examples when totalitarian governments coerced biologists to adopt theories that served party political purposes. Notorious among these is the Stalinist imposition of Lysenkoism, the theory of Trofim D. Lysenko, who dictated to Soviet agriculture his socialist/Marxist belief in the heritability of acquired characteristics, in opposition to the theory of Mendelian genetics.

Yet the interrelations between biology and ideology are more pervasive and subtle than is indicated by such showpieces of ideological infamy as Lysenkoism. Over the past few decades, historians of science have begun arguing that all biology—the coerced and the free, the bad and good, the false and the true, the current and the historical—is and was embedded in sociopolitical ideology. Biology, at some level, is ideology, in the sense that its key concepts have grown out of and have been applied to the system of ideas that structures civil society. This particularly holds true for Darwinism with its notion of natural selection—it has been maintained—and publications have come on the market with titles such as *Darwinismus und/als Ideologie* (2001).[1]

Ever since Robert M. Young's teachings and writings at Cambridge during the late 1960s and early 1970s, we have come to think of Darwinism as shaped by a Malthusian common context of biological and social theory. Darwin's evolutionism, Young explained, was irreducibly anthropomorphic, social, political, and ideological.[2] Others, not least Adrian Desmond, James Moore, and Janet Browne have fleshed out Young's work and turned his common context, which was defined primarily in terms of ideas, into a more embodied sociopolitical one, extensively taking account of place, time, institutions, career, family, vested interests, rivalries, strategies, and more.[3] An inventory of much of this contextualist work, for the most part dealing with Britain, was presented in the collected volume *Victorian Science in Context* (1997).[4] Further studies, by David N. Livingstone, Ronald L. Numbers, and others, expanded the geographical scope of this work to include other parts of the English-speaking world, delineating spaces of science, religion, and politics in which Darwinism was reconstituted to suit the local or regional needs of different constituencies.[5]

As Moore observes, biography has proved a powerful contextualizing tool. In several new, substantive biographical studies of both Darwin and his followers, evolutionary biology has been dissected down to the bare bones of its varied ideologies, showing how basic notions in biological science were molded by the cultural, ideological, and socioeconomic forces that shape the rest of the world in which the biologists lived.[6] This contextualist work represented a fundamental revision of the mostly essentialist Darwin hagiography, spawned by the 1959 centenary celebrations, and has had a deconstructionist or, in a later fashion to enter the history of science, a constructivist, effect on our perception of Darwinism.[7] In other words, we see Darwin's theory as constitutively affected by social conditions and, what's more, we understand the literature on Darwin, too, as reflecting to a significant extent the worlds of its authors, each with a distinct set of concerns and interests.

Yet to many historians of the generation that founded the field of the history of science after World War II, as well as to many practicing biologists and traditional philosophers, this "biology as an ideological construct" approach is little more than an arrogant solipsism that misses a crucial point. As one of the stellar members of that generation, Charles C. Gillispie, comments in a recent retrospective on the history of his subject, it is a truism that science is a cultural and social product—that it is created. No one ever doubted this. Yet not all products of human creativity are the same. Scientific ones differ from those in the creative arts—painting, writing, and music—such as Hamlet, the Mona Lisa, and the Mass in B Minor, in that these would not exist if Shakespeare, Leonardo da Vinci, and Johann Sebastian Bach had never lived. Such contingency does not apply to science. Newtonian physics would have come into existence also without Newton, and Darwinian evolution without Darwin, Gillispie maintains.

> If Newton had . . . died in infancy, the planets would still move subject to the inverse square law of gravity. Although no one else would have written the *Principia*, it could be argued convincingly that others would then or soon have written down everything in it that really mattered to later physics. Much the same is true of nearly all the great contributions to modern science. For scientists find their problems but also problems find their scientists, and usually not only one of them. Evidence is the frequent occurrence of simultaneous so-called discovery of many an important phenomenon, often attended by disputes over who had it first. It is all very well to say that convergent social forces are the explanation of simultaneous discovery, but *pace* social constructionists, they do have to converge on something out there in nature.[8]

Thus if Darwin had missed his boat, had not sailed around the world on board *HMS Beagle* (Figure 6.1), and consequently the *Origin of Species* had never been written, this would not have made a fundamental difference to the progress of the life sciences. A theory of evolution similar to Darwin's inevitably had emerged—or so Gillispie's line of argument leads us to believe. After all, is there not the famous instance of Alfred Russel Wallace (1823–1913), who simultaneously came up with the theory of evolution by natural selection? For all the contextual factors that go into the making of science, the facts of nature to which it pertains remain unaffected and, in the end, scientific theories converge on those facts. Whatever ideological influences may swirl around scientific theories or temporarily disfigure them, sooner or later, truth will be out.

By and large, practicing biologists agree. None other than "the Darwin of the twentieth century," Ernst Mayr (1904–2005), has insisted that Darwinism was not an ideology; on the contrary, it represented an "anti-ideology" that for the first time in history attributed the origin of species to natural causes: "The adoption of evolution by natural selection necessitated a complete ideological upheaval. The 'hand of God' was replaced by the working of natural processes."[9] This insistence on the objective, realist nature of Darwinian evolution—in its original formulation and yet more in its twentieth-century reformulated shape of the evolutionary synthesis—gained

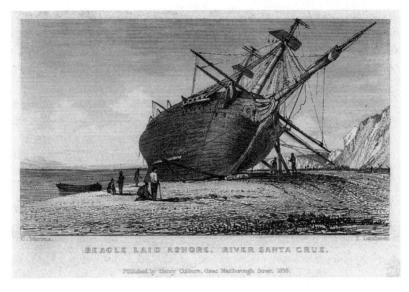

FIGURE 6.1 "Beagle laid ashore." If Darwin had been shipwrecked and the *Origin of Species* never been written, would a theory of evolution similar to Darwin's still have emerged? (From Philip Parker King and Robert Fitzroy, eds., *Narrative of the Surveying Voyages of His Majesty's Ships Adventure and Beagle* [London: Henry Colburn, 1839], vol. 2, opposite p. 336.)

additional importance in the context of legal action in several states of the USA to obtain equal time for the teaching in public schools of creation to that for evolution. Following a 1961 Supreme Court ruling that public schools are not the place for religious doctrine, the fundamentalists have emphasized that creationism is a science. Moreover, they have argued that evolution is a religion. As a result, the question about the ideological status of Darwinism has taken on more than academic significance. Michael Ruse, in an attempt at even-handedness that has brought his *Mystery of Mysteries: Is Evolution a Social Construction?* (1999) many accolades, allows that the theory of evolution has a dimension of religious ideology; yet he simultaneously stresses that Darwinian theory has a content of scientific validity that has increased over time (see Ruse, Chapter 10, this volume).[10]

A possible objection to Ruse's argument is that creationist biology, too, has produced valid science that during "the reign of creationism," which in Germany lasted until the end of the eighteenth century and in the English-speaking world until the middle of the nineteenth, showed a pattern of increase. Bearing witness to this are the successive contributions to botany, zoology, and paleontology by such creationist giants as Carl Linnaeus,

Georges Cuvier, and Louis Agassiz. The Darwinian way of coping with this problem has been to anoint a variety of creationist biologists as "forerunners" of evolution; they had tried but not fully succeeded in throwing off the shackles of religious dogma. A good example was Linnaeus, who in his younger years expressed the belief that species are created. However, having wrestled with the restrictive impositions of Genesis 1, toward the end of his life he conjectured that God's work of creation had gone no further than the ancestral form of each genus or even just of each order, and that natural diversification within the limits of these original types had produced today's many species. Draw a straight line from early to late Linnaeus, extrapolate this line and you arrive at Darwin—at least according to the conventional Darwin literature.[11]

Crucial to this discussion is the point made by Gillispie and others that the great contributions to modern science, including Darwinism, converged (and continue to converge) on "something out there in nature." That factual something is not contingent. The path Darwin walked was not a matter of choice but followed a course that was predetermined by the objective truth about the origin of species. An alternative road, trending in a different direction, was non-existent—even an impossibility. Right from the start, in 1859, Darwin himself insisted that the only available choice was between his theory and the non-scientific, religious doctrine of special creation. A scientific alternative to Darwinian evolution did not exist—or so many people continue to believe.

The purpose of this chapter is to argue that this view is a myth—one that goes back to Darwin himself and to the *Origin of Species*. I argue that: (1) long before Darwin, a naturalistic theory of the origin of species existed; (2) this theory represented a scientific alternative to Darwinism; (3) Darwin refused to engage with it for reasons of strategy; (4) his own theory produced a fork in the road of the development of the biology of origins; and (5) the Darwinian choice, in addition to the good it did to the life sciences, also caused damage as a result of scientific shortcomings that were integral to Darwin's positions. In other words, if Darwin had not boarded *HMS Beagle* and the *Origin of Species* had never been written, today's evolutionary biology could well have looked different, possibly fundamentally so. This does not imply that Darwinism is wrong, but does indicate that it is partial in the double sense of being incomplete and biased, being related to the author as much as to the subject matter. Putting it differently, I support the contention that Darwinian biology bears the hallmark of an ideology.

Here I restrict myself to documenting that a scientific alternative did exist. The question of what might have happened if the *Origin of Species* had never been published, I leave, apart from a single suggestion, to another publication.[12]

A contemporary alternative to Darwinism

Traditionally, in the classical literature from Aristotle to Lucretius, there had been the following four theories of the origin of species (plus subsidiary combinations of these): species are eternal, species are created, species originate spontaneously, and species have evolved. By the time the *Origin of Species* appeared, and within mainstream scientific thought, only the first of these four was no longer seriously considered. Species eternalism had all but vanished; one of its last few representatives was the military surgeon and philosopher Heinrich Czolbe, who in the mid-1850s was uncompromisingly taken to task for his views by, among others, Berlin's founder of medical pathology, Rudolf Virchow.[13] The second view—creationism—by contrast, continued to flourish, especially among Anglo-American biologists and paleontologists, famous among whom were Oxford's William Buckland, Cambridge's Adam Sedgwick, and Harvard's above-mentioned Agassiz. To be sure, they departed from the literalist interpretation of creation and Flood, known today as young-earth creationism, and instead envisaged repeated special creations in the course of a long geological history (see Topham, Chapter 4, this volume). The fourth view of the origin of species—evolution or rather the transmutation of species—found few adherents and remained marginal until Darwin was forced out of his evolutionary closet and published his magnum opus.

Wiped from the slate of our collective memory since shortly after 1859– the year the *Origin of Species* was published—is the fact that during the immediately preceding seven or so decades the majority of the great minds of what we now call the earth, life, and biomedical sciences, especially in the German-speaking world, believed neither in creation nor evolution but in autogenesis—the spontaneous origin of species, or *Urzeugung*. Following non-creationist speculations about species from the middle of the eighteenth century, especially by Georges-Louis Leclerc, Comte de Buffon, this theory was modernized during the period 1790–1860 and refitted to become the cutting edge of scientific thought about the origin of life and species during the first half of the nineteenth century.

The theory of autogenesis stated that species have originated, not by miraculous, special creation but by the spontaneous aggregation of their specific "germs" (see Roe, Chapter 2, this volume). The reader might be excused for objecting that the theory of spontaneous generation has not been quite so forgotten as here maintained. Do we not have—and one may well ask—a substantial body of secondary literature on the seventeenth- to nineteenth-century controversy over the spontaneous origin of "entozoa" (worms such as flukes living parasitically inside other animals) and "infusoria" (microscopic aquatic creatures), which controversy was settled to the satisfaction of most biologists by Louis Pasteur's classic experiments that appeared to disprove the phenomenon?[14] Is this story not well and widely known? However, two theories of spontaneous generation existed that, although related and in some publications interwoven, were distinct. One of these was the familiar theory about the contested observation that today, under our very eyes, primitive forms of life spontaneously originate from lifeless matter.

Yet spontaneous generation also had a second meaning. It could refer to the past origin of species, when the very first specimens of any fixed form of life, low or high, including humans, must have come into existence—it was postulated—by non-parental generation. Plants, animals, and humans had come into existence not by miraculous, special creation but by the natural generation of one, two, or many of the very first individual representatives. Various explanations were put forward as to why in taxonomically higher organisms the original spontaneous generation of its members had not continued but been suppressed to be superseded by a process of sexual reproduction in perpetuating the species.

Obviously, the first of these two theories could have a bearing on the second, but it was perfectly possible for biologists not to believe in the spontaneous generation of primitive life at present and still fully endorse the theory of the *generatio spontanea* of species in the past, and quite a few of them did.[15] If simple organisms originate spontaneously today, this enhances the likelihood that more advanced species came into being long ago in a similar way; but it is also imaginable that the origin of each permanent form of life had been unique to certain moments in geological history, because only at those moments, temporarily, the right conditions for the original aggregation of their germs or seeds had occurred.

About a dozen different terms were used to describe spontaneous generation (for example, *generatio spontanea, primigenia, cosmica, primitiva, originaria, automatica, aequivoca, heterogenea, Urzeugung, Urerzeugung, autogene Zeugung,* autogenesis). These terms were not always used synonymously and could have somewhat different meanings but, apart from *generatio heterogenea*, they all indicated a form of abiogenesis; in order to avoid confusing the two theories of spontaneous generation, I shall refer to the second one by using the synonym "autogenesis" or "autogeny" (and the related words "autogenist" and "autogenous"). It is this theory that has not been entered into the annals of the history of biology. Yet autogenesis gained widespread popularity through the period that ranged from the late eighteenth century to not long after the appearance of the *Origin of Species*, becoming an integral and organizing concept in several scientific fields: (1) the physiology of generation; (2) the paleontology of extinction and origins/renewal; (3) the geography of plant and animal distribution; and (4) anthropogeography. To a lesser extent, autogenesis was also discussed in the context of other subjects, for example the new discipline of physiological chemistry.

Let us briefly look at the place of autogenesis in each of these four fields of scientific theory and practice.

Autogenesis in biomedical physiology

First and foremost, the notion of autogenesis was part of what we would call the biomedical physiology, especially the study of generation, regeneration, and nutrition. More specifically, it was intimately connected with the theory of epigenesis (the notion that organisms develop from undifferentiated germ material) as opposed to the "nested boxes" view of "preformation" (the notion that organisms exist preformed in the germ material). After all, if the germ from which an individual member of a species grows is not a babushka doll containing in microscopical form all future individuals, but more simply is composed of a set of elemental substances from which the embryo develops afresh, it becomes possible to conceive of the very first such development—the incipient stage of non-parental origin of the first individual(s) of a species—as a process of aggregation of just the right elemental mixture under just the right physical conditions.

A central figure in the development of these speculations was the Göttingen professor of medicine Johann Friedrich Blumenbach (1752–1840), who as a leader of epigenetic thinking was also one of the early naturalists cautiously to express the likelihood of the autogenous generation of species. This he did as early as 1791, in a footnote to the fourth edition of his *Handbuch der Naturgeschichte* (1791); also later, in his *Beyträge zur Naturgeschichte* (1806).[16] One of Blumenbach's many students, Gottfried Reinhold Treviranus (1776–1837), the first person to introduce the term and the subject of biology as a separate discipline in Germany, in his main work *Biologie, oder Philosophie der lebenden Natur* (1802–22), substantially expanded upon autogenesis, combining a defense of the spontaneous origin of primitive organisms today with that of higher forms of life in the past and adding to that, as did Blumenbach, a belief in the variability of species as a further, subsidiary mechanism of the origin of new forms. Given the "analogy" of the spontaneous origin of simple life forms, we may assume "that also the original forms of mammals and birds once were generated in the same way that today for the most part only zoophytes [invertebrates such as sponges] are still formed."[17] This assumption is justified, Treviranus added, even if today spontaneous generation no longer takes place; but it does. "We can therefore no longer doubt that the very principle that is active in producing a microscopic world of plants and animals in infusions of putrified substances should also be capable of generating larger and more complex organisms."[18]

Others in biomedical physiology followed suit and treated the theory of the autogenous generation of species as integral to the physiology of generation and related phenomena. *Urzeugung* or *Urerzeugung* became a major topic, ranking substantial entries in the new encyclopedias and encyclopedic series of the period.[19] No less a figure than the Königsberg physiologist Karl Friedrich Burdach (1776–1847) advocated autogenesis, and he counted some of the greatest names of biomedical physiology and anthropology among his pupils, such as Karl Ernst von Baer and Martin Heinrich Rathke. In his physiological and anthropological works Burdach explicitly combined, as did Blumenbach and Treviranus, the autogenous generation of species with a certain modification through external conditions working on innate variability; yet he objected to the transformation of major forms into one another.[20]

Other leading lights such as Berlin physiologist Johannes Müller, in his *Handbuch der Physiologie des Menschen* (1844), and Virchow, in a major discussion of "Alter und neuer Vitalismus," used concepts and language of autogenesis in discussing the origin of species. Müller, however, indicated that the origin of species lay outside human experience and would better be dealt with by philosophy than physiology.[21]

Autogenesis in Cuvierian paleontology

Autogenesis was readily integrated into the new geology of the time, which recognized the extinction of many species in the course of earth history as well as the repeated origin de novo of new organic "worlds." Also in this field Blumenbach was leading with early speculations about repeated extinctions of old "worlds" and the appearance of new ones. Continuing to use the language of a "creator," he changed the meaning of "creation" from a miraculous event to a natural process. Following a geological revolution, nature had created new species that differed from the ones that had been wiped out because altered physico-chemical conditions made the *Bildungstrieb*—the morphogenetic force—deviate from its previous direction. Blumenbach stated

> that the morphogenetic force after the changes that such a total revolution is likely to have caused in the material substratum had to take a direction in generating new species that more or less deviated from the previous.... As a result, with these transformations creative nature no doubt also reproduced in part creatures of a similar type to those of the previous world, but in the overwhelming majority of cases was forced to substitute forms that were adapted to the new order of things.[22]

At the occasion of the fiftieth anniversary of the start of Blumenbach's legendary career as a Göttingen University teacher, his pupil K. E. A. von Hoff, by then himself a famous geologist in Gotha, saluted his mentor for having solved the problem of the origin of species.[23] Autogenous generation was readily brought to bear on a geological past marked by repeated "revolutions," accompanied by extinctions and new beginnings. In addition to a cautious Blumenbach, there was among the early advocates of autogenesis the outspoken Jean-Claude de Lamétherie (1743–1814). Taking his cue from the new epigenesis, he regarded generation as a process of true crystallization [*une véritable cristallisation*].[24] Following the general crystallization of rocks from the primeval waters, the generative germs of sea creatures had "crystallized" from dead matter and, after the gradual subsidence of the ocean waters and emergence of mountains peaks, so had the "germs" of land creatures.[25]

Given that the origin of species was a process of crystallization under specific physico-chemical conditions, or at least an analogous process, the same species could have originated multiply, in different parts of the globe. Yet new species had originated also as a result of organic variability; for example, as a result of hybridization. "The number of original species has therefore been truly far fewer than commonly believed."[26]

Georges Cuvier (1769–1832), a central figure in the (re)construction of a lengthy and repeatedly interrupted history of the earth and of life, seems to have stuck to a tradition of creationism, although he substituted repeated creations throughout geological time for the literalist, one-off event of young-earth creationism. Yet many Cuvierians and a majority of those who further developed the picture of "punctuated" earth history adhered to the theory of autogenesis. They assumed that, following a Cuvierian catastrophe, special environmental conditions may have existed that favored the spontaneous aggregation of germs and the autogenous origin of life and species in repopulating the earth. Among the influential names to work with this notion were Germany's leading geologist Leopold von Buch (1774–1853) and leading paleontologist Heinrich Georg Bronn (1800–62), as well as the prominent popularizer Hermann Burmeister (1807–92) or the notorious materialist Carl Vogt (1817–95)—the last two engaged at the time in left-wing, revolutionary causes. As Burmeister wrote:

> If we don't want to resort to miracles and incomprehensibilities, we have to attribute the origin of the first organic creatures on earth to the free, creative power of matter itself and deduce the reasons why this generative capacity today no longer continues in higher organisms, from general laws of nature that ordain only the necessary, not the superfluous.... If nature could create at some time a human couple, then she could generate also several of them, she even had to, if she wanted to know the existence of her creation for ever secured.[27]

In the course of the period 1790–1860 the notion of autogenesis was increasingly complemented by that of heterogenesis, a notion that included the belief that major life forms, especially taxonomically higher ones, do not immediately originate from inorganic matter but via intermediate organic stages. This process of descent of a higher from a lower form of life was not visualized, however, as a gradual transformation/transmutation of one species into another, but as a spontaneously generated leap taking place in the germ material of a particular individual. To help visualize this process of saltatory "evolution," the analogy was used of *Generationswechsel* [alternation of generations] or, what the British comparative anatomist and paleontologist Richard Owen (1804–92) called "metagenesis." A typical instance of metagenetic change is found in fluke worms. Their *Generationswechsel*

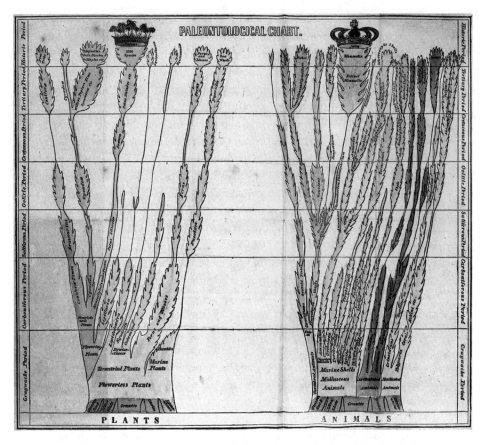

FIGURE 6.2 "Paleontological chart." Edward Hitchcock's early diagram of the distribution of life through geological time gave visual expression to the third theory of the origin of species. Characteristic are the following features: life originated from inorganic matter (the "trees of life" have roots of rock), plants and animals commenced separately (two "trees of life," not just one), all major plant and animal taxa began nearly simultaneously (even flowering plants and vertebrates go back to the earliest period), the major forms of life advanced as much convergently as divergently (depicted as shrubs of life, not a single, branching tree of life), and the development of life and species is teleological, culminating for the animal world in "Man" and for the plant world in "Palms." Yet "Man" is the highest form of life, wearing a kingly crown, whereas palms display a diadem of lower, ducal rank. Hitchcock believed that an origin of species by natural law need not be contrary to religious belief if such law be seen as divinely sustained. In places, he also used creationist language. (From Edward Hitchcock, *Elementary Geology*, 8th ed. [New York: Mark H. Newman, 1847].)

consists of a succession of three distinct forms—infusorial, worm-like, and tadpole-like—connecting the egg and the mature fluke-worm. One might imagine that under particular circumstances the cycle could be broken, and the separate stages go on reproducing. In this way wholly new genera or even orders might originate.[28] Taken together, the following pattern of life through time was envisaged (Figure 6.2): life from lifeless matter (the roots of life are rocks), the independent origin of plants and animals (each kingdom forms a separate "tree of life"), the multiple independent origins of major groups of plant and animal life ("shrubs" rather than "trees"), a certain consecutive and parallel development of life forms through time (shrub branches), and also occasional reemergences of life forms after extinctions (at the lower boundaries of geological periods/systems).

Autogenesis in Humboldtian biogeography

Like geology, Humboldtian physical geography was one of the cutting-edge fields of early nineteenth-century science. Probably more than any other field, biogeography adopted autogenous generation as a central organizing concept. Species were not just Linnean taxonomic entities but Humboldtian members of geographical communities. Plants in particular came to be thought of as species with a special area of distribution, vertically from low to high altitude in mountainous regions, and horizontally/latitudinally from equator to pole across the globe. Such provinces of distribution were interpreted as "centres of creation," meaning that the species that characterized a particular province had come into existence locally as a result of autogenous generation. In biogeography, autogenous generation equaled autochthonous (or aboriginal) generation.

In this context, distribution statistics and maps showing provinces of distribution took on special significance, and many were published during the first half of the nineteenth century. A particularly exciting discovery was that identical plant species occur in widely separated provinces of distribution, for example the Maldives and Finnmark. Such instances, in which migration/transport from one area to the other appears all but impossible, indicated that the same species can originate spontaneously in more than one location. That being so, the origin of a species in a single area may have happened by the autogenous generation of many individuals at the same time, increasing the chances of survival—a possibility referred to in the above Burmeister quotation.

Thus, within the context of biogeography far-reaching implications of the theory of autogenesis were discussed. Can one species have more than a single origin? And, related to this, does a species originate as a single individual/as a single pair, or as many? Looking still further: if there have been Cuvierian catastrophes, must the repopulating of the world be seen as a rather sudden spontaneous generation of whole ecological communities? Or should one follow Charles Lyell, abandon the catastrophist view, and consider extinction as well as autogenous generation as a gradual process of the coming and going of single species? And should one follow Edward Forbes, adding a geological dimension of gradual change over time to provinces of distribution?

An entire generation of famous Humboldtian biogeographers subscribed to the theory of autogenesis, among them Franz Meyen (1804–40), Alphonse de Candolle (1806–93), Ludwig Schmarda (1819–1908), and, until well after Darwin's *Origin of Species*, August Grisebach (1814–79). Like many subscribers to the theory of autogenesis, Candolle contended that the origin of species lay beyond our range of observation; but biogeography showed that most plant species had originated in the region where they occur, even though opinions diverged over whether or not new forms had come into existence as a single individual/pair or as many individuals. Species had not migrated across the globe. The facts taught us

> to regard each endemic species as *aboriginal* to the country where today it exists, and the more dispersed species (*sporadic*) either as transported by accident from one country to another during the time of their existence or as original to several countries at once.
>
> On the latter point, authors are divided. Some presume that each plant species comes from a single individual (or a single couple when it concerns dimorphic plants); others admit that from the beginning species must have had a considerable number of individuals, closely together or widely separated from each other across the surface of the earth.[29]

Autogenesis in anthropogeography

Most dramatically, Romantic anthropology adopted autogenous generation as a central concept. Nearly to a man, its star representatives in the German-speaking world believed in the spontaneous origin of *Homo sapiens*. Among them were Burdach and the polymath and Dresden professor of obstetrics Carl Gustav Carus (1789–1869). Also, the great Berlin anatomist and physiologist Karl Asmund Rudolphi thought of humans as aboriginals/autochthons. As in biogeography generally, in anthropogeography the phenomenon of provinces of distribution of human varieties was attributed to autochthonous origins and autogenous generation. Human races were autochthons, and polygeny (the notion that human races do not have a common origin—monogeny—but separate ones) was allied to autogenesis, well before it became part of the theory of evolution. Admittedly, more than in the case of plant and animal species, "monogeny" versus "polygeny" was controversial when applied to humans. Blumenbach famously argued for a unitary origin of mankind and J. C. Prichard, in his *Researches into the Physical History of Man*, followed suit by identifying a single "centre of creation" for humans. Burdach sided with Blumenbach in attributing to humans a single origin, putting the issue as follows:

> Because every region of our planet has generated those plants and animals that reflect each region's characteristics, and because the different human races, as far as our knowledge reaches back into antiquity, always have had the same characteristics that still today sets them apart from each other, it has been maintained that they were different from the start and came out of the hands of creating nature in different regions. . . . Human beings are not bound to their native soil like animals and plants, and the various human races can hold their own everywhere on earth far from their homeland, and are therefore not the products of distinct climates.[30]

Equally great names stood on the other side of the issue. Rudolphi argued that human races are true "aborigines" whose historical distribution is not a result of migrations but due to separate origins in situ.[31] A similar view was championed by Carus, who pleaded ignorance about the origin of species and of the human species in particular—suggesting that primordial origins lie outside the range of our observational possibilities—but who then specified nevertheless how humans must have come into existence, summing up the various ways in which this could *not* have happened. Humans had neither originated as told in the Bible nor had they evolved from apes. Specifically, *Homo sapiens* had not made its first appearance in adult form on land, as a single individual or a single pair, like Adam and Eve in the Garden of Eden; nor had it originated as some evolutionists speculated by the transformation of monkeys into men. The human species—he asserted—had its origins in primordial vesicles that had developed in enormous numbers in water under mild and stable climatic conditions:

> We may not think of the origin of humans as occasioned by a sudden appearance of one or more finished organisms; . . . we may think of a condition of the earth when, at a time of quite enormous generative activity, higher, even the highest, epitelluric organisms also arose from . . . primordial vesicles. In this case we should not think of their development in a dry but rather in a wet environment; and once one has clearly grasped this, then, . . . it should not be thought that this development took place only once or twice, but in enormously large numbers; . . . it may not be thought that this development was reached under the influence of anything other than . . . a mild climate. . . . Finally, however, and primarily, it should not be thought that human beings originated when an animal (an ape, for example) was elevated in his development to become a human.[32]

Thus human autogenesis could and would have taken place in different regions of the globe, and Carus was an advocate of polygeny, combining it with the explicit contention that the different races are unequal with respect to their intellectual abilities. In a famous *Denkschrift* [commemorative publication] of 1849 on the occasion of the centenary of Goethe's birth, Carus produced a map of the world on which he plotted the climatic zone that had produced the highest forms of life—unsurprisingly, this included his own part of the world—and zones that had given rise to lower levels of humanity.[33]

By no means all autogenists were species fixists. Several added the creative potential of organic variability to the autogenous origin of life and of a number of original types. Blumenbach did so in a modest way, for instance to explain the origin of human varieties. Others went much further. Jean-Baptiste Lamarck, Étienne Geoffroy Saint-Hilaire, and Robert Chambers, anonymous author of *Vestiges of the Natural History of Creation,* can be seen as autogenists who combined the spontaneous origin of basic organic forms with extreme species variability and change. In other words, we can think of Lamarck, Geoffroy, and others as advocates of autogenesis who went out on a variability limb, rather than as proto-Darwinists. Others, too, postulated the autogenous origin of life and of a variety of original forms, but added species variability and descent via major changes due to inner tendencies. The German philosopher Arthur Schopenhauer (1788–1860) advocated this *generatio in utero heterogeneo* [generation in the uterus of a different kind], whereby the egg of a new species originates in the uterus of a related but different species, not under the open sky on the bank of a river, in a pond, etc. Thus *Homo sapiens* could have spontaneously originated in the womb of an ape and yet not have evolved from apes. Owen came to defend this position, accepting that humans are descended from animals; this would explain why a major gap exists between apes and humans.[34]

Autogenists in context

The theory of autogenesis enjoyed its largest following in Germany; but in France, also, it had representatives.[35] In the German literature, the British authors Lyell and Forbes were cited in the context of autogenesis, but although they are likely to have sympathized with the theory, they stopped short of explicitly expressing support.[36] Bronn and Vogt cited Lyell.[37] In the second volume of his *Principles of Geology* (1832), Lyell had famously criticized Lamarck's theory of transmutation, upheld the constancy of species, and speculated that, on the basis of a gradual, uniform rate of extinction matched by an equally gradual origin of new species, in a region the size of Europe, only once every 8,000 years or more would a mammal disappear and emerge, making it a difficult process to authenticate.[38] Lyell did not specifically state that he believed in a natural origin of species but later admitted that he left this to be inferred. Bronn, for one, did infer just that and attributed to Lyell the view that species past, present, and future had been/were being/would be originated by primordial generation (*Urerzeugung*), in a slow and imperceptible process of change of the world's fauna. Moreover, autogenesis theory was familiar to British scientists either directly through Continental literature or via Ray Society translations into English, such as Meyen's outspokenly autogenist and autochthonist *Outlines of the Geography of Plants* (1846) and the more bizarrely holistic-vitalistic *Elements of Physiophilosophy* (1847) by Lorenz Oken.

Both Joseph Dalton Hooker and Thomas Henry Huxley, before they joined the Darwinian cause in the wake of the appearance of the *Origin of Species*, expressed doubts about species variability. Huxley, in his notoriously scathing review of *Vestiges*, strenuously opposed "transmutation" as well as divine guidance and intervention. Hooker, in a review of Candolle's *Géographie botanique raisonnée*, expressed agreement with Huxley but at the same time prevaricated on whether multiple creations or creation by transmutation had taken place.[39] It is likely that in the cases of Hooker and Huxley, as in the German instances of pre-*Origin* scientists, the combination of anti–miraculous creation and antitransmutation went hand in hand with an open-mindedness regarding the possibility of autochthonous generation.[40]

Yet it cannot be overlooked that autogenesis was first and foremost a German theory of the origin of species, just as natural selection was British in origin. Some decades ago now, John C. Greene observed:

> It is a curious fact that all, or nearly, all of the men who propounded some idea of natural selection in the first half of the nineteenth century were British. Given the international character of science, it seems strange that nature should divulge one of her profoundest secrets only to inhabitants of Great Britain. Yet she did. The fact seems explicable only by assuming that British political economy, based on the idea of the survival of the fittest in the marketplace, and the British competitive ethos generally predisposed Britons to think in terms of competitive struggle in theorizing about plants and animals as well as man.[41]

Greene's line of thought, linking evolution by natural selection to British national context, could, mutatis mutandis, equally be taken in the case of autogeny. Why should this theory have been primarily German? What in German culture predisposed biologists to develop the theory of a spontaneous origin of species and think in terms of autochthony? The spread of German naturalistic thought to Britain and France may well have been inhibited by the prevalence there of creationist belief, yet the question remains: why did autogenesis grow from German roots? Not just Darwinism, also Blumenbachian spontaneous generation, calls for the sort of analysis that Greene, Young, and a younger generation of Darwin scholars have carried out for evolution by natural selection.[42] Also, the theory of autogenesis will have been intrinsically ideological, its German ascendancy being most likely attributable to, among other things, a pervasive and deeply seated Germanic belief in environmental determinism.

Writing autogenesis out of the historical record

To sum up, a major tradition of non-creationist, naturalistic opinion on the origin of species existed by the time the *Origin of Species* was published, belying the essentialist contention, referred to above, that Darwinism was the first and only scientific theory that attributed the origin of species to non-miraculous, natural causes. Today, the pre-Darwinian paradigm of autogenesis has been forgotten. How can such forgetting have come about? The answer is that the theory was written out of the historical record by a memory culture that developed as part of the self-constituting of Darwinism as a cause. To many, the memory culture of Marxism may serve as "type specimen," but a Darwinian variety also came into existence that equally constructed a self-reconfirming perspective on the past. A collective memory formed, the content of which was determined by a process of sorting what seemed of interest from what seemed not, the former being highlighted and the latter, in some instances, being deleted and forgotten.

"Forgetting" is a particularly relevant aspect in the context of the present chapter, as this is what Darwin initiated with respect to the theory of autogenesis. His *Origin of Species* completely ignored the theory, even to the extent that Darwin belittled the problem of the origin of life to which he devoted no more than a few lines (see next section). Darwin's ridicule of the belief in an instantaneous origin of species was directed against the creationists, not the autochthonists, even though his words could, with all due reservations, just as well have applied to them:

> These authors seem no more startled at a miraculous act of creation than at an ordinary birth. But do they really believe that at innumerable periods in the earth's history certain elemental atoms have been commanded suddenly to flash into living tissues? Do they believe that at each supposed act of creation one individual or many were produced? Were all the infinitely numerous kinds of animals and plants created as eggs or seed, or as full grown? and in the case of mammals, were they created bearing the false marks of nourishment from their mother's womb?[43]

Following the early success of his book, Darwin added to it "An Historical Sketch" (prefixed to the third edition of the *Origin of Species* and, in an expanded form, to the fourth and later editions). Supposedly recording "the progress of opinion on the origin of species," the sketch omitted mention of the autogenist theory. Darwin failed to act like a "pure" scientist weighing the factual evidence and discussing the available theoretical options, but engaged in special pleading, selectively anointing forerunners and opponents. Thus he created a self-affirming historical framework for telling his story and arguing his case, the main line of persuasion being that there existed no scientific alternative to his theory because the choice was one of natural origins by means of evolution versus miraculous origins by creation. The latter was not science; it represented a religious doctrine.

> Until recently [that is, until the *Origin of Species*] the great majority of naturalists believed that species were immutable productions, and had been separately created. This view has been ably maintained by many authors. Some few naturalists, on the other hand, have believed that species undergo modification, and that the existing forms of life are the descendants by true generation of pre-existing forms.[44]

Darwin let those "few naturalists" have a share in his success by naming them forerunners of evolution; moreover, he appropriated several great names from the recent past of biology and geology by representing them as his precursors—even though these people had been autogenists of one kind or another. With this historical sketch Darwin retouched the map of "the progress of opinion on the origin of species," blotting out what was at the time the most commonly represented naturalistic theory. His omission was critically commented upon by, among others, the Heidelberg paleontologist Bronn, who translated the *Origin of Species* into German.[45] Another former autogenist who converted to Darwinism, the Utrecht zoologist Pieter Harting, in the introduction to his 1862 zoology textbook, like Bronn drew attention to autogenesis, discussing at length the pros and cons of this option before continuing with a plea on behalf of Darwinian evolution.[46] Darwin himself, however, stuck to his strategy of avoidance and did not discuss the theory of autogenesis, defending as late as the sixth edition (1876) of the *Origin of Species* his oversimplified scheme of creation versus evolution.

> As a record of a former state of things, I have retained in the foregoing paragraphs, and elsewhere, several sentences which imply that naturalists believe in the separate creation of each species; and I have been much censured for having thus expressed myself. But undoubtedly this was the general belief when the first edition of the present work appeared.[47]

This view, we have seen, was a distortion; yet it proved effective. By ignoring "the origin of life" Darwin circumnavigated a dispute with formidable, highly regarded representatives of the earth and life sciences (see next section).

The efficacy of the historical sketch as a template for later histories is apparent from the fact that the theory of autogenesis to this day has remained a largely forgotten chapter in the development of evolutionary biology. Scholars have contracted a blind spot that makes those who read the original sources fail to see the theory of autogenesis. Time and again they have reclassified autogenists as "forerunners of Darwin"[48] or simply "forgotten" them. Darwin's self-alignment with selected predecessors was approvingly repeated 100 years later on the occasion of the centenary of the *Origin of Species*. The great biologists and their scientific contributions from the eighteenth and nineteenth centuries were interpreted as foreshadowing the coming of the Messiah of organic evolution—an approach that received its almost hymnic rendition in Bentley Glass's preface to the classic *Forerunners of Darwin: 1745–1859*:

> By the middle of the eighteenth century, theories of evolution were being heatedly debated. As the evidence in favor of organic evolution grew, the struggle between two seeming alternatives, the one, Divine Plan and Providence in an Original Creation and the other, a godless mechanism of chance and of blind cause and effect, became not so much a debate that divided scientists and philosophers into two camps as a cleavage in the heart of each individual man. Thus, as the individual often resolves an implacable conflict by repressing it into the subconscious, so human thought in the first half of the nineteenth century stubbornly and blindly repressed the implications of the growing evidence in regard to the origin of species, including his own. Darwin was the outburst of those repressed conclusions, the victory of that submerged scientific conviction. His was the magnificent synthesis of evidence, all known before, and of theory, adumbrated in every postulate by his forerunners—a synthesis so compelling in honesty and comprehensiveness that it forced such men as Thomas Henry Huxley to say: How stupid not to have realized that before![49]

A further advantage of "forgetting" was that this made it easier for autogenists and especially the outspoken critics of species transformation among them to convert to Darwinism. It provided shelter for those who changed their minds and decided to join Darwin. They did not object when their earlier antitransformist views were not held up against them, when their public pronouncements in favor of the spontaneous origin of species were quietly passed over, and when they were praised for having helped along the cause of Darwinism before the appearance of the *Origin of Species*. Among the beneficiaries were Hooker and Huxley. A remarkable case was Vogt, an instant convert to Darwinism who previously had passionately argued for autogenesis and against both creation and evolution, but who later claimed not to have known what to think about the origin of species before Darwin's book.[50]

It is well known that Darwin admired Lyell and was deeply influenced by his *Principles of Geology*. A parallel can be drawn between Darwin's historical introduction and the historical chapters with which Lyell opened his *Principles of Geology*. On the occasion of the Lyell Centenary Celebrations of 1975, Roy Porter discussed Lyell's introduction to the *Principles*, arguing that it functioned less to provide a historical background than to formulate a polemic in support of his philosophy of geology. The introduction made uniformitarianism seem the inevitable outcome of the progress of enlightened thought and civilization. Lyell "forged a masterly propaganda history," showing that religion, philosophy, and anthropomorphic narrow-mindedness "had caused geology to miscarry."[51] The study of the earth needed to be freed from catastrophes and providential miracles. Civilized and progressive minds were reaching the stage of appreciating the uniformity and continuity of nature's economy. "He made uniformitarianism seem natural and inevitable, by assimilating it within a conjectural evolution of the human mind. Lyell offered uniformitarianism not just as a methodology. It was unchallengeable because it was integral to the evolutionary epistemology of the human race."[52]

Lyell's historical introduction showed considerable historiographical sophistication.[53] Darwin's does not. In fact, it is a poorly constructed and rather haphazard enumeration of selected predecessors. Yet the two introductions are alike in that they had similar purposes. Darwin, like Lyell, created a dichotomy of religious obscurantism versus scientific enlightenment, which in his case equaled creation versus evolution. In telling the story of "the progress of opinion on the origin of species," he offered less a record of fact than a self-serving polemic that obscured the scientific alternative to his theory. This was, on the one hand, just an all too commonly used presentation strategy to promote one's own theory; yet on the other hand, Darwin's historiography expressed an essentialist ideology that substituted, with his own truth claims, those made by orthodox religion. To Darwin as to Lyell, as well as to many other Victorian scientists, the scientific study of nature took over from theology as arbiter of absolute truth about the history of the earth and of life. Also, with respect to this quasi-religious characteristic, Darwinism showed its ideological leopard's spots.

What if ... ?

Once again, if the *Origin of Species* had never been written, would that have made a difference to the state of today's evolutionary biology? Ironically, true Darwinians say it wouldn't, because all Darwin did was walk the straight and narrow path of scientific truth, and, if he hadn't led us there, someone else would. After all, was Wallace not waiting in the wings with an identical theory of the origin of species, ready to take Darwin's place if for one reason or another the great prophet of evolution had failed? I argue to the contrary that the biology of origins today would have looked different, possibly fundamentally so. A scientific alternative to Darwinism existed, and it is imaginable that, if this had prevailed in the form of a Bronnian and Owenian program of multiple autogenous origins combined with directional evolution, its characteristic premises might have led to breakthroughs and opened up avenues toward an unfamiliar constellation of theories and practices. The successes and institutional power of these new directions would have been effective self-affirmations. "Darwinism" in the form of Wallace's theory would not have happened. Leaving aside the differences between Darwin and Wallace in their understandings of natural selection—the one focusing on divergence, the other on adaptation—Wallace would never have managed to get natural selection established without Darwin. In Moore's words: "We have every reason to believe that Owen would have upped his own profile, moved to a wider centre-stage, and trumped any move Wallace might have made to usurp him."[54]

So in what ways would the non-Darwinian Owenian program have been different? Let me restrict myself here to a single yet major example of a topic that Darwin ignored but that was high on the autogenists' research agenda, namely the origin of life. A significant feature of autogenous generation was that it made no distinction between the origin of life and the origin of species. The two issues were conflated to form a single scientific question. Darwinism—I contend—set back research into the origin of life by many decades and may well be a principal cause of the embarrassing fact that, until the present day, we have not solved this fundamental problem of organic origins. We neither know how life began nor have we succeeded in reproducing in the laboratory the processes that during primeval times led—one assumes—to abiogenesis.

It is a widely known and often cited fact that Darwin did not include the problem of the origin of life in his magnum opus on the origin of species.

Toward the very end of the book he inserted a couple offhand remarks that seemed more to dismiss than seriously address the question. Darwin wrote: "I should infer from analogy that probably all the organic beings which have ever lived on this earth have descended from some one primordial form, into which life was first breathed," and further he referred to life "having been originally breathed into a few forms or into one."[55]

A long-held view is that Darwin avoided the issue because "[t]his enabled him to work out a theory of descent without bogging down in unanswerable questions."[56] To put it another way: it allowed him to make natural selection seem more efficacious than it otherwise would because natural selection has no obvious grip on the transition from lifeless matter to primitive life (even though more recently attempts have been made to apply natural selection to stages of the primordial abiogenesis process, famously by the Glasgow organic chemist Alexander Graham Cairns-Smith).[57]

Here I have shown that there was more to Darwin's avoidance of the origin of life than the fact that his proposed mode of evolution would fare better without it. His ignoring the issue, unprofessional as this may have been given its fundamental significance, proved tactically clever in that by so doing Darwin averted a confrontation with representatives of the then leading theory of the origin of species—for the most part men who he was anxious to get on his side. In order to succeed, Darwin could ill afford alienating the many powerful and prestigious representatives of the theory of autogenesis. Among its outspoken advocates was an élite group of Continental earth and life scientists; its sympathizers included several eminent men from Darwin's own circle, such as Lyell, Forbes, and J. D. Hooker. Darwin evaded a direct challenge of these men's views, reducing the question of the origin of species to two possible positions: evolution or creation, the one characterized by "impartiality" and the other by a "load of prejudice."[58] Given that the autogenists were not creationists and in fact were severely criticized by creationist colleagues, Darwin's staging of a common creationist enemy facilitated the crossing over by autogenists to the Darwinian camp.

The origin of life was removed from the Darwinian research agenda, even more so after the *Bathybius* debacle, when Huxley, Haeckel, and many other eager converts to Darwinism misidentified albuminous coagulates in ocean floor samples from the *Challenger* expedition as incipient life.[59] It was to no avail that Owen insisted "[t]hat the doctrines of the *generatio spontanea* and of the transmutation of species are intimately connected. Who believes in the one, ought to take the other for granted."[60] The origin of life was dropped as a central concern of the biology of origins, not to be taken up again until approximately a century later in the wake of the work of the American chemist and biologist Stanley Miller. Had the alternative route been followed, it is imaginable that by now the fundamental problem of the origin of life would be solved.

One might counter by arguing that the very tools for creating life in the laboratory, as currently attempted with the promise of success by, for example, the American geneticist J. Craig Ventner and his group,[61] did not exist until the late-twentieth century and that therefore the science of reproducing life could not have developed any faster than it has. Ever since the work by Christian Gottfried Ehrenberg (1795–1876) on foraminifera and radiolaria we have known that even the smallest forms of microbial life are highly complex; and it would appear that we require the technological capabilities of today's genetic engineering to reproduce such microbes in the laboratory. Yet what the autogenists attempted did not necessarily presuppose the instrumental and theoretical know-how of today's molecular biology. They were interested in retracing the historical steps that nature took in creating life, and their aim may well have been accomplished with the time-excluding or at least time-reducing techniques that physiological chemistry set-ups from the early part of the nineteenth century onward possessed. It was entirely possible to produce a range of conditions of temperature, pressure, humidity, composition of inorganic and organic molecules, electrical discharges, etc. that—one assumes—conditioned the emergence of simple life on earth. The fact that we have not yet succeeded in reproducing these historical stages is due—I repeat—to the limitations of the Darwinian program that for many decades discouraged turning the problem of the origin of life into a major interdisciplinary and international research endeavor.

If the alternative to Darwinism had prevailed and managed to trace nature's steps in creating life, today we would be celebrating the accomplishments of the non-Darwinian life sciences as proof that these represented the one and only highway of scientific progress and objective reality. That hypothetical biological world would have been considered the best of all possible ones, just as at present many of us can conceive of no better world than the Darwinian one we have come to inhabit. Let us, in conclusion, remind ourselves of the 1932 address to the Prussian Academy of Sciences by Erwin Schrödinger, Nobel laureate and one of the founders of quantum physics, when he discussed the question "Ist die Naturwissenschaft milieubedingt?" [Is science conditioned by context?]. Stressing the importance of *Gesamtkultur* [collective culture] for our scientific orientation and the sort of questions we ask, he stated:

> Rarely will a thought more powerfully have influenced the direction of our interests in almost all areas of science and life as has the concept of evolution, and that in its general form as well as in the particular way expressed by Charles Darwin.... Just think of how much this thought has taken possession of us, so that we can no longer think without it and hardly notice when we apply it as a matter of course to everything.[62]

This chapter makes a case for the non-inevitability of Darwin's theory and, by so doing, endeavors to remove from Darwinism its protective armor of essentialist claims. The contention by Bob Young, John Greene, Jim Moore, and others, as well as Schrödinger all those years ago, is thus reinforced, namely that sociopolitical concepts, values—ideologies—and material conditions were and are constitutive of Darwin's theory, and vice versa.

CHAPTER SEVEN

Biology and the emergence of the Anglo-American eugenics movement
Edward J. Larson

Francis Galton (1822–1911) believed that some people are born smart, others are born stupid, and the roots of all such hereditary traits reach deep into one's ancestry. He came to these beliefs from personal experiences; his class and racial biases; the inspiration of the evolutionary theories of his first cousin, Charles Darwin (1809–1882); and what he viewed as his own hereditary genius. One of the last significant English gentleman scientists without institutional affiliation, Galton developed his hereditarian notions to such a degree that they helped to lay the foundation for both genetics and eugenics, which in their formative period were allied, overlapping disciplines.

During the late 1800s, most naturalists (including Charles Darwin) accepted a blending view of inheritance whereby offspring manifest a middling mix of their parents' traits. In the tradition of the French evolutionist, the chevalier de Lamarck (1744–1829), many of them also maintained that at least some characteristics acquired by individuals during their lives passed to their descendants. Darwin came to regard acquired characteristics and other environmentally induced alterations in a parent's hereditary material (which he called "gemmules") as the major source for the inheritable variations that fueled the evolutionary process. Galton instinctively rejected this later concept, and planted the seeds for its eventual overthrow in science.

Freed from birth of any obligation to earn a living, after receiving an enormous financial inheritance upon his father's death in 1844, Galton first gained fame during the 1850s as a traveler and explorer. Adventures with Cambridge college chums in the Middle East gave way to serious expeditions through regions of southwest Africa never before traversed by Europeans. His technical and popular accounts of his explorations coupled with his writings about the art of travel made him well known in Britain by the time that his cousin published the *Origin of Species* in 1859. That book, Galton later claimed, changed his life. "I used to be wretched under the weight of the old fashioned 'arguments from design,'" Galton wrote to Darwin. "Your book drove away the constraint of my old superstition as if it had been a nightmare and was the first to give me freedom of thought."[1] Liberated from the perceived limits of Christian mores and psychologically driven to promote the progress of civilization as he saw it, Galton refocused his considerable abilities toward improving humanity by championing a purportedly Darwinian process of selective human reproduction, which he called "eugenics."

Something of a child prodigy, Galton remained inventive throughout his long life. He learned to read at age two, but partied more than he studied during his college years at Cambridge—so that he graduated without honors. Then came his enormous inheritance and African adventures. During the mid-1850s, when in his thirties, Galton settled into the life of a wealthy, well-connected gentleman scholar, dividing his time between his elegant London townhouse and excursions on the European Continent. In addition to his foundational contributions to genetics and statistics, he conducted significant research in cartography, geography, meteorology, psychology, and sociology. Except for his work in eugenics, which was sustained, Galton typically offered brilliant insights into a particular field (such as recognizing the phenomena of anticyclones in meteorology) but showed little follow-through, leading some commentators to dismiss him unfairly as a dilettante. He also invented a host of clever but ultimately worthless gadgets, from underwater reading glasses and high-pitched whistles for testing hearing to a sand-glass speedometer for bicyclists and a self-tipping top hat. Of more lasting impact, he pioneered the use of fingerprinting for personal identification.

Based in part on his experiences in Africa, Galton believed that (on average) blacks are naturally inferior to whites in intelligence and other hereditary traits fitted to civilized life. Other non-white races fared little better in his estimation, and some even worse. Environmental factors can not account for these racial differences, Galton asserted. Blacks in Europe or raised in white families remain much like their savage ancestors while whites visiting or living in Africa maintain their civilized superiority, Galton explained, citing as authority the accounts of whites.[2] These observations convinced him that individuals cannot acquire inheritable attributes through nurture or other environmental factors.

Galton applied similar hereditarian reasoning to those he viewed as superior and inferior within a race, which fit his preexisting dismissive attitude toward welfare programs designed to uplift the underclass. Characteristics acquired by one generation simply do not pass to the next, he believed even before he began his research on heredity. His subsequent scientific findings confirmed these socially constructed opinions.[3] Reading (or misreading) *Origin of Species* apparently supplied him with a seemingly scientific basis for his societal views, and launched him on a midlife career in science to establish their validity.

Galton's key insight, which he belabored in many of his books and articles, was the concept of "hard" heredity, or heredity that cannot change during one's lifetime, as opposed to the "soft" heredity of inheritable acquired characteristics. In his earliest published work on heredity, the 1865 magazine article "Hereditary Character and Talent," Galton rhetorically asked, "Will our children be born with more virtuous dispositions, if we ourselves have acquired virtuous habits? Or are we no more than passive transmitters of a nature we have received, and which we have no power to modify?" "No" and "Yes" were the correct answers, respectively, in Galton's mind. "We shall therefore take an approximately correct view of the origin of our life, if we consider our own embryos to have sprung immediately from those embryos whence our parents were developed, and these from the embryos of *their* parents, and so on for ever," he concluded.[4] For Galton, the body serves essentially as a passive receptacle and replicator for transmitting hereditary information across the generations rather than as an interactive participant in the process.

Seeking a material basis for this hereditary information, Galton commandeered Darwin's theory of gemmules and turned it to his own ends. For Darwin, gemmules represented invisibly small bits of hereditary information collected from every part of both parents and transmitted during reproduction to their offspring. New hereditary variations came from environmentally caused changes in gemmules as thrown off by a parent or received by the offspring. Galton accepted the basic idea of particulate gemmules, which he came to call "germs," but maintained that they remained distinct within the body, guiding its development yet never impacted by it. In reproduction, unmodified germs from both parents combine to compose the germs for their offspring, he asserted (see also Roe, Chapter 2, this volume).

Although all individuals (except identical twins) therefore carried their own unique mix of "germs," Galton viewed such variations as merely a normal aspect of heredity as expressed in a population. Indeed, a born quantifier, he offered mathematical evidence suggesting that the distribution within a population of any given hereditary trait (which for him included complex characteristics such as intelligence as well as such simple ones as height) fits a statistical bell curve, with most individuals falling at or near the species or racial mean and ever fewer toward the limits of variation in either direction. Most people are near the mean height for their race and gender, for example, but only a few greatly shorter or taller than it.

As long as no new hereditary information enters the equation, such as by acquired characteristics (which he rejected) or inborn mutations (which he did not), Galton asserted that normal, continuous variations in the population cannot lead to the evolution of new species. Developing new techniques in statistics to prove his point, he calculated that successive generations descended from even the most gifted individual will naturally regress to its species or racial norm. Sustained selection for the same characteristic in both parents (such as can result from environmental pressures or eugenic mating) can slow the regression, but Galton purported to demonstrate that even it cannot permanently alter species or racial norms. Only discontinuous variations in the form of gross inborn mutations can overcome the limits of ancestral inheritance and spawn the evolution of new species or races, he concluded.[5]

Setting aside his pessimism about the cumulative effect of sustained selection, Galton had hit upon many basic elements in the future synthesis of modern genetics and Darwinian theory—particularly the statistical interpretation of hard heredity. Because of his peculiar concerns, however, he conceived of these elements in a fundamentally different fashion from how later evolutionary geneticists would view them. Nevertheless, Galton's idiosyncratic jumble of revolutionary insights and reactionary ideology inspired a generation of geneticists and eugenicists.

During the late 1800s and early 1900s, different scientists pursued Galton's insights in different directions. In Germany, cytologist August Weismann soon took up the cause of hard heredity—at first independently from

Galton's work but increasingly with reference to it. Weismann called the heredity material "germ plasm," which quickly became its favored name, and localized it on chromosomes in the cell nucleus—but it was conceptually similar to Galton's germs and just as immutable during an individual's life.[6]

In Britain, mathematician Karl Pearson and marine zoologist W. F. R. Weldon carried on Galton's statistical analysis of the cumulative impact over time of normal, continuous variations within populations—a field of study they called biometry. Although Pearson and Weldon did not win many converts, by 1900 they had demonstrated to their own satisfaction that Galton was wrong on one point: sustained selection of normal, continuous variations (such as larger crabs over smaller crabs in Weldon's experiment) could permanently shift the species norm in the selected-for direction. Over generations, they argued, this process could generate new species without dramatic discontinuous variations.[7]

Concurrently, Galton's other principal British disciple, morphologist William Bateson (1861–1926), championed the role of gross mutations in fueling the evolutionary process and defended the position that normal, continuous variations lead nowhere. Although Pearson and Bateson clashed mightily over whether continuous or discontinuous variations fed evolution, both upheld Galton's position on hard heredity and were committed eugenicists. By 1900, Bateson was far from alone in accepting a mutation theory of evolution. Disillusioned by the lack of solid experimental evidence for either normal, continuous variations spawning new species or the inheritance of acquired characteristics, an increasing number of late nineteenth-century biologists turned to mutation theory for answers, with Dutch botanist Hugo de Vries finding some of the most encouraging results in his study of the evening primrose, which appeared to mutate randomly and regularly. American geneticist Thomas Hunt Morgan soon joined the search for mutations in his legendary work with fruit flies.

Despite their disagreements on the means of evolution, all these evolutionary geneticists—Weismann, Pearson, Weldon, Bateson, de Vries, and Morgan—explicitly or implicitly endorsed eugenics. This gave it enormous scientific credibility in Europe and America.

Galton's dream

Galton derived the word "eugenics" from the Greek for "well born," signaling his fascination with the sources of natural ability rather than a preoccupation with the causes of human disability. His 1869 book, *Hereditary Genius*, analyzed the pedigrees of eminent men, whose names Galton culled from biographical reference books and lists of high achievers, leading to the conclusion that genius (in all its forms) runs in families. Galton later pioneered the comparative study of twins raised apart. "Everywhere is the enormous power of hereditary influence forced on our attention," he wrote, "proving the vast preponderating effects of nature over nurture."[8] On the basis of these findings, Galton proposed that society should encourage men and women of hereditary fitness to marry each other and to bear many children—propositions that became known as positive eugenics. "What an extraordinary effect might be produced on our race, if its object was to unite in marriage those who possessed the finest and most suitable natures, mental, moral, and physical!" he exclaimed.[9] Toward this end, and reflecting his belief that physical beauty served as a marker for eugenic fitness, Galton collected data for what he called a "Beauty Map of the British Isles," which graphically presented his assessment of the geographical distribution of attractive female mating stock. "I found London to rank highest for beauty; Aberdeen lowest," the English scholar observed.[10]

Although Galton urged from the outset that society also discourage "weakly and incapable" persons from breeding, and later recommended their compulsory segregation into single-sex institutions, such efforts of so-called negative eugenics never became as much the focus of attention for him as they did for so many later eugenicists.[11] Combining his belief in hard heredity transmitting eugenic and dysgenic traits with his faith in the cumulative impact of marginally raising the reproductive rate for the fit and marginally lowering it for the unfit, Galton affirmed "that the improvement of the breed of mankind is no insuperable difficulty."[12] Indeed, he predicted, "If a twentieth part of the cost and pains were spent in measures for the improvement of the human race that is spent on the improvement of the breed of horses and cattle, what a galaxy of genius might we not create! We might introduce prophets and high priests of civilization into the world, as surely as we can propagate idiots by mating *crétins*."[13]

For Galton, the selective breeding of the human population represented the greatest promise of applied biology and the highest mission of a scientific society. "I take Eugenics very seriously, feeling that its principles ought to become one of the dominant motives in a civilized nation, much as if they were one of its religious tenets," he wrote in the concluding passage of his autobiography. "Individuals appear to me as partial detachments from the infinite ocean of Being, and this world as a stage on which Evolution takes place." In the past, Galton explained, human evolution proceeded fitfully and with great individual pain through natural selection, but with the developed mind of modern man, "I conceive it to fall well within his province to replace Natural Selection by other processes that are more merciful and not less effective. This is precisely the aim of Eugenics."[14]

As Galton conceived it, eugenics would operate solely within a race, and not across racial groups. Like Charles Darwin and German evolutionist Ernst Haeckel, Galton believed in a strict hierarchy of racial types with some subset of Northern Europeans at the apex of the evolutionary pyramid. All three of them assumed that, where different races came into extended contact, superior ones would inevitably supplant inferior ones through natural selection. Galton never suggested that some sort of genocidal artificial selection of races either should be employed to hasten this process or could alter the ultimate outcome. "There exists a sentiment, for the most part quite unreasonable, against the gradual extinction of an inferior race," Galton explained. "It rests on some confusion between the race and the individual, as if the destruction of a race was equivalent to the destruction of a large number of men. It is nothing of the kind when the process of extinction works silently and slowly through" a process of natural selection.[15] This comment reflects the mainstream tenets of late nineteenth-century scientific race theory. Darwin made precisely the same arguments in *Descent of Man*.[16] Only in the twentieth century would some of Haeckel's followers attempt to assist the process through genocide in the Nazi death camps.

Where Galton concentrated his research on blood lines of distinction, pioneering the genealogy of degeneracy fell to a New York social reformer, Richard Louis Dugdale (1841–83). He became interested in the issue when, during an 1874 inspection of conditions at a jail in rural New York State for a prison reform organization, he learned that six of the prisoners were related. Struck by this observation, Dugdale launched an investigation into the lineage of their family, which he called the Jukes, in an effort to uncover the causes of crime (Figure 7.1). Using prison records, relief rolls, and court files, he traced the Jukes's family tree through five generations, back to six sisters, two of whom had married the sons of a Dutch colonist named Max. Dugdale found that, over the years, more than half of the 709 people related to this family by blood or marriage were criminals, prostitutes, or destitute.[17] The publication of Dugdale's findings in 1877 caused a sensation and spawned a small genre of similar studies by

FIGURE 7.1 A member of the "Jukes" family. (Arthur Estabrook Papers, M. E. Grenander Department of Special Collections and Archives, University at Albany Libraries.)

other researchers—all reaching essentially the same conclusion: degeneracy, like genius, runs in families.

Galton's publications also stirred considerable interest. *Hereditary Genius* appeared just as Darwin was finishing the *Descent of Man*, and the latter included his younger cousin's findings within that new book. "We now know through the admirable labors of Mr. Galton that genius," Darwin wove into his text, "tends to be inherited; and, on the other hand, it is too certain that insanity and deteriorated mental powers likewise run in the same families."[18] Like the work of many eccentric geniuses, however, Galton's eugenics was somewhat ahead of its time. For his part, Dugdale simply never anticipated how eugenicists would later use his findings because (unlike Galton) he was not one of them. Neither Galton's 1869 *Hereditary Genius* nor Dugdale's 1877 *The Jukes: A Story in Crime, Pauperism, Disease, and Heredity* provoked a rush to administer eugenic remedies for perceived social diseases. The science of the day did not yet justify them.

The impact of Mendel's laws

Most late nineteenth-century biological and social scientists, including Galton and Dugdale, subscribed to a blending view of heredity under which children inherited a blend of their parents' or other ancestors' traits. Under this view, as Galton demonstrated with regression analysis, the challenge lay not so much in pulling inferior families up to the species or racial norm (because that should happen naturally through crossbreeding) but rather in keeping superior families from falling back toward it. Thus Galton stressed positive eugenics. Discouraging reproduction by the unfit, which later became the central thrust of the eugenics movement, did not appear as critical to early eugenicists as encouraging the fit to intermarry and breed.

A second scientific factor limiting the appeal of eugenics was the widespread acceptance by late nineteenth-century evolutionists (though not by Galton) of Lamarckian notions of soft heredity (see Krementsov, Chapter 9, this volume). It was not that Lamarckians never embraced eugenic remedies—many of them did—it was simply that a belief in the inheritance of acquired characteristics undercut the urgency of such prescriptions. This was apparent in Dugdale's recommendation of how to deal with the Jukes. "Environment tends to produce habits which may become hereditary, especially so in pauperism and licentiousness," he explained in patent Lamarckian terms. "The licentious parent makes an example which greatly aids in fixing habits of debauchery in the child. The correction is change of the environment." In particular, Dugdale argued in *The Jukes*, the way to break a chain of hereditary degeneracy in a family is to remove the children of social misfits from the home and raise them in a healthy environment. Eugenic restrictions on reproduction were unnecessary under this perspective because, as Dugdale asserted, "Where the environment changes in youth, the characteristics of heredity may be measurably altered."[19]

By the beginning of the twentieth century, the scientific theories that held back eugenics began to crumble. Galton, Weismann, de Vries, and others pressed the case for hard heredity. Then, in 1900, the rediscovery of Mendel's laws suggested that parental and other ancestral traits reappear in children and more remote descendants without blending. From its rediscovery, mutation theorists embraced Mendelism as showing how discontinuous variations are propagated. During the 1920s and early 1930s, geneticists R. A. Fisher and J. B. S. Haldane in Britain and Sewall Wright in the USA demonstrated mathematically how Mendelian processes could also account for the evolution of new species through normal, seemingly continuous, inborn variations. Through their work, biometricians and mutation theorists united under the banner of population genetics to rout

Lamarckian notions of soft heredity from biology. For many, including Fisher, Haldane, and Wright, this made the case for eugenics. If there are superior and inferior hereditary traits, they reasoned, and if the impact of these traits on succeeding generations is unalterable by environmental influences or blending, then eugenics must be true—at least if the genes carrying those traits operate as simple Mendelian factors. Indeed, Fisher bent his mind to genetics to prove this very point and thereby encourage the propagation of a better breed of Britons.

"More children from the fit, less from the unfit," became the motto for a new age of eugenicists.[20] Of course, the triumph of eugenics built on a history of greater public acceptance both of hereditarianism in general and of a competitive struggle for existence as the driving force for social and economic progress. It only took a slight twist of reasoning to transpose accepting the natural selection of the fit into encouraging the intentional elimination of the unfit.

The newly energized, early twentieth-century eugenics movement built on the pioneering efforts of Galton and Dugdale. For example, the Eugenics Education Society, which championed negative eugenics in Britain, recruited an elderly Galton to serve as its honorary president. The Galton Laboratory for National Eugenics at University College, London, funded by Galton and directed by his protégé Karl Pearson (1857–1936), acted as the society's research arm. Galton's first cousin once removed, Charles Darwin's middle son Leonard, served as the society's president from 1911 to 1925. In 1933, Fisher assumed the professorship in eugenics endowed by Galton at University College, London. Also drawing on earlier work in the field, the leading American eugenics organization, the Eugenics Record Office (founded in 1910 and operated in conjunction with the Carnegie Institution's Cold Spring Harbor station), distributed and revised Dugdale's classic study of hereditary degeneracy.

The revised study of the Jukes revealed much about the transformation in social-scientific thought brought about by the rise of Mendelism. Where Dugdale's original book categorized family members by their social behaviors, a 1915 revised edition did so by their supposed mental abilities. The revised edition, which included 2,000 more family members, concluded that "over half the Jukes were or are feebleminded."[21] This finding was significant because eugenicists typically viewed "feeblemindedness," which covered various levels of mental deficiency, as an inherited Mendelian trait. "If the Jukes family were of normal intelligence, a change of environment would have worked wonders," a prominent American eugenicist explained at the time. "But if they were feeble-minded, then no amount of good environment could have made them anything than feeble-minded."[22] Low intelligence breeds antisocial behavior, eugenicists assumed, and its propagation weakened the race. Accordingly, the revised study recommended permanent sexual segregation or sterilization for all the Jukes, where Dugdale had prescribed a change of environment for the younger Jukes. In doing so, the 1915 study reflected the means and objectives of the mature eugenics movement.

Putting eugenics into practice

Even though eugenics was born in Britain, methods for its coercive application principally developed first in America.[23] Eugenicists, who were instinctively élitist, never placed as much hope in education to discourage dysgenic unions as they did in it to encourage eugenic ones. Instead, for negative eugenics, they turned to four different methods of legal restriction: marriage laws, sexual segregation, involuntary sterilization, and limits on foreign immigration. In the USA, the first three of these methods required the enactment and enforcement of legislative programs by the various state governments, which became the primary legal objective of the American eugenics movement during the period from 1895 to 1945. Typically, these three methods built upon each other as a series of overlapping and increasingly intrusive steps toward achieving national eugenic improvement.[24]

To the extent that laws restricting dysgenic marriages represented a first step in eugenics reform, it was one that many eugenicists jumped over. States had regulated marriage long before the eugenics movement came into existence, such as by laws forbidding marriage by persons who were underage, closely related by blood, or of different races. Many states also maintained that a marriage involving a mentally ill or retarded partner was either automatically void or voidable by either spouse on the traditional contract-law grounds that the mentally disabled partner lacked the legal capacity to enter into a binding contract. By the early twentieth century, some sexual-hygiene and public-health reformers were calling for laws to limit or prohibit marriage by persons infected with venereal disease. Although eugenicists generally agreed that all of these laws helped their cause, none of them were strictly eugenic in their intent or impact. "When the law prevents marriage on account of insanity, feeble-mindedness, or other hereditary defects, it obviously has a eugenic value," two prominent American eugenicists, Paul Popenoe and Roswell Hill Johnson, explained, "but in so far as it concerns itself with venereal diseases, which are not hereditary, it is only of indirect interest to eugenics."[25]

Building on this long tradition of state oversight of marriage, some eugenicists sought strictly eugenic restrictions over who could marry. In 1896, Connecticut became the first state to comply when it decreed that "no man and woman either of whom is epileptic, imbecile, or feeble-minded" could "inter-marry, or live together as husband and wife, when the woman is under forty-five years of age."[26] Most Northern and Western states quickly followed suit, so that by 1914, over half of the states had imposed new restrictions on the marriage of persons afflicted with mental defects. Only South Dakota and Nebraska went as far as Popenoe wanted, however, by establishing central registries for all mentally defective residents of their states and precluding any listed individual from obtaining a marriage license unless one of the wedding partners was sterile.[27]

As a region within the USA, only the Deep South, which had typically mandated fewer restrictions on marriage than other regions, failed to impose any such limits other than conventional ones based on the ability to enter into a contract.[28] Of course, every state in the Deep South maintained strict antimiscegenation statutes throughout the period; that is, laws banning interracial marriage and sometimes interracial sex between whites and blacks. These laws pre-dated eugenics and were of only incidental interest to eugenicists because social barriers effectively limited interracial wedlock in the region.[29]

Despite the inevitable popular interest in eugenic marriage restrictions, many leading eugenicists dismissed them as hopelessly ineffective. For example, Goddard observed that the feeble-minded so lacked any sense of morality that marriage was not a prerequisite for procreation.[30] A St. Louis psychologist did not include marriage restrictions in the comprehensive package of eugenic reforms that he drafted for the Missouri legislature because, as he explained to a local women's group demanding such restrictions, "the enactment of such a bill would not prevent reproduction on the part of the unmarried feeble-minded girls and many similar statutes in other states were almost dead letters because they were not generally enforced."[31]

After warning "that the denial of a marriage license will by no means prevent reproduction among the anti-social classes of the community," Popenoe and Johnson concluded, "it is a mistake for eugen[ic]ists to let legislation of this sort be anything but a minor achievement, to be followed up by more efficient legislation."[32] The universally acknowledged spokesperson for the American eugenics movement, geneticist Charles B. Davenport, left no doubt about what that more efficient legislation would entail. "No cheap devise of a *law* against marriage will take the place of compulsory segregation of gross defectives," he asserted in 1913, nearly a decade after he founded the Carnegie Institution's Cold Spring Harbor genetics research station, which served as the fountainhead for eugenics in America for nearly four decades.[33]

When eugenicists spoke of segregation, they had something quite specific in mind. "Segregation," Popenoe and Johnson explained, "as used in eugenics means the policy of isolating feebleminded and other anti-social individuals from the normal population into institutions, colonies, etc., where the two sexes are kept apart."[34] Davenport put it more bluntly when he advised that "the proper action in the case of imbeciles or the gross epileptics who wish to marry is not to decline to give them a marriage license, but to place them in an institution under State care during at least the entire reproductive period."[35] This advice, of course, was not limited to those who wanted to marry, but included any such individuals who might reproduce their kind with or without the benefit of a formal marriage.

The rudimentary infrastructure for implementing such a policy existed in many parts of the country long before eugenicists came on the scene. As a humanitarian reform to improve the care and treatment of those suffering from severe mental disease, the movement to build state mental-health hospitals was well under way by the mid-1800s. "By the 1870s," medical historian Gerald N. Grob noted, "mental hospitals had assumed the form that they would retain in succeeding decades," which included the entire eugenics era.[36] This form typically involved a massive remote building housing hundreds or thousands of male and female patients in an authoritarian, custodial environment.

The mid-nineteenth-century movement to build mental health hospitals in the USA was followed by ones to build custodial training schools for the mentally retarded and asylums for epileptics. These developments pre-dated the rise of eugenics and the resulting institutions initially stressed moral treatment, not eugenic control. In his 1877 study of the Jukes family, for example, Dugdale saw schools for the mentally retarded as a means to permanently break the cycle of hereditary degeneracy though environmental reform. By the early 1900s, eugenicists saw them as offering a different type of permanent solution for the same problem.[37]

State hospitals for the mentally ill and schools or colonies for the mentally retarded offered ready-made institutions for sexual segregation once fixed hereditarian ideas of mental disease and retardation took hold. "At first," the early twentieth-century social-work advocate Albert Deutsch noted, "the ideal underlying institutional custody of the feebleminded was to protect the inmates from the dangers of society. The turn of the century witnessed a complete revolution in the custodial ideal; the main object now was to protect society from the menace of mental defectives."[38] The new objective generated increased public support, leading to the construction or enlargement of institutions for the mentally retarded throughout the nation. By 1906, eugenicist Alexander Johnson, a top official in the U.S. National Conference of Charities and Corrections, could report regarding his influential association of social workers, "I believe that every member will agree that the segregation and even permanent detention of at least a great majority, if not all of the feebleminded, is the proper procedure."[39]

Alexander Johnson's brother-in-law, Henry H. Goddard (1866–1957), who as research director for the prestigious Vineland Training School for Feeble-Minded Children in New Jersey became the leading architect of eugenic remedies, described the ideal plan for segregation. "Determine the fact of their defectiveness as early as possible, and place them in colonies under the care and management of intelligent people who understand the problem," Goddard wrote in 1913, "train them, make them happy; make them as useful as possible, but above all, bring them up with good habits and keep them from ever marrying or becoming parents."[40] As Goddard saw it, implementing his plan faced three hurdles, all of which were surmountable.

First, the mentally retarded must be identified, which he proposed doing through the public schools.[41] To facilitate this process, Goddard introduced the Binet–Simon tests for what soon became known as a person's "intelligence quotient," or "IQ," into the USA from France in 1908. Using these tests, Goddard claimed that he could scientifically quantify the mental capacity of any individual. Further, he attempted to grade the mentally retarded into three levels of increasing mental deficiency: high-grade "morons," middle-grade "imbeciles," and low-grade "idiots."[42]

Once identified, mentally retarded children needed to be institutionalized. This posed a problem because, according to Goddard, most such children "are not of so low intelligence that every one, the parents included, is convinced of their defect and is willing to have them placed in separate colonies." In fact, Goddard found that most parents refused to relinquish their children due to "their parental love" and the fact that many of the higher-grade defectives "are trainable to do errands and simple work which brings in a few pennies to the family treasury." As a result, he concluded in a classic testimony to the conflict between governmental versus parental control over the family, "Until we come to the point where society is driven to the extreme of making laws requiring the forcible taking away of these children from their homes and placing them in the colonies, the matter will be an insurmountable difficulty."[43]

Lastly, the states needed to build enough institutions to house all the mentally retarded. Even as early as 1913, Goddard recognized that this represented an enormous and costly undertaking. He calculated that simply housing the estimated 15,000 feebleminded schoolchildren of New York City would require thirty large institutions, which was twenty-six more than then existed in the entire state. "As is often true, however, the difficulty is not really as great as it appears," Goddard optimistically proposed. "It is only necessary to show . . . that the increased cost will be largely offset by the saving" in reduced welfare and prison expenses.[44] Further, Goddard believed that these costs would quickly diminish as the supply of new retarded children was cut off by preventing procreation by known defectives.[45] Finally, he hoped that these institutions could "become partly self-supporting or even completely self-supporting" through the labor of their residents.[46] Vineland's superintendent, E. R. Johnstone, elaborated on how this could happen. "The colony should be located on rough uncleared land," he advised. "Here the unskilled fellows find happy and useful occupation, waste humanity taking waste land and thus not only contributing toward their own support, but also making over land that would otherwise be useless." Once the land became productive, it could be sold for a profit, and the colony moved on to a new tract. In the meantime, it could largely live off its own farming and handicrafts.[47]

Of course, it never worked out as Goddard and Johnstone hoped. The estimated number of mentally retarded children in every state increased with every new count far beyond the capacity of available facilities, and none of these institutions ever became self-supporting.[48] By 1918, Popenoe and Johnson acknowledged that the objection "against segregation is that of expense."[49] A decade later, when the American Eugenics Society tried to justify such expenditures by calculating that $25,000 spent on segregating the original Jukes for life would have saved the state over $2 million in later costs, it immediately added that sterilizing the initial Jukes would have cost less than $150.[50] Even though eugenicists continued to support segregation, its inherent limitations inevitably drove many of them to favor sterilization as well.[51]

Early eugenicists appreciated the efficiency of surgically preventing reproduction by persons suffering from hereditary defects, but they lacked an acceptable means to carry out this remedy. Both euthanasia and castration were suggested, but neither gained much popular support.[52] At most, the emergence of hereditarian explanations for criminal behavior revived interest in castrating rapists and other violent criminals, though the primary arguments for such a penalty still focused on its direct deterrent effects rather than its indirect eugenic benefits. These deterrent effects included both inhibiting initial criminal activity through fear of losing further sexual capacity and pleasure plus preventing subsequent criminal acts by removing sexual passion and desire.[53] Indeed, at the time, some proponents viewed this approach as an enlightened alternative to the then common practice in the American South of lynching black men accused of assaulting white women.[54] The physical and psychological impact of castration on its victims, however, made its use on anyone other than marginalized, individualized or social outcasts unacceptable to the general public. For example, popular outcry ended an experiment with eugenic castration at the Kansas State Home for the Feeble-Minded in the 1890s after forty-four boys and fourteen girls were mutilated. Similarly, an early proposal to asexualize the mentally retarded in Michigan was defeated by the state legislature in 1897.[55]

Fortuitously for the eugenics movement, new surgical techniques for sexual sterilization were then being developed. Vasectomy offered a safe and simple means of terminating a man's ability to reproduce without inhibiting his sexual desire or pleasure. Salpingectomy (surgical removal of the Fallopian tubes) provided a similarly effective but somewhat more dangerous sterilization procedure for women.[56] The significance of these developments for eugenics was immediately recognized. In an 1899 article announcing his pioneering research with vasectomy, Chicago surgeon A. J. Ochsner wrote, "It has been demonstrated beyond a doubt that a very large proportion of all criminals, degenerates, and perverts have come from parents similarly afflicted." After observing that the castration of criminals "has met with the strongest possible opposition because it practically destroys the possibility of the future enjoyment of life," Ochsner offered sterilization as a socially acceptable alternative. "This method would protect the community at large without harming the criminal," he claimed. "The same treatment could reasonably be suggested for chronic inebriates, imbeciles, perverts and paupers."[57]

Eugenicists needed little encouragement. An Indiana reformatory promptly launched an eugenic sterilization effort that began in 1899 as a voluntary experiment with prisoners, but which rapidly expanded until 1907, when the state enacted America's first compulsory sterilization statute.[58] That law mandated the sterilization of certain "confirmed criminals, idiots, rapists and imbeciles" whose condition was pronounced incurable by a committee of three physicians.[59] Fifteen other states followed suit over the next decade, including such progressive trendsetters as California, New Jersey, New York, Michigan, and Wisconsin. "The general advantage claimed for sterilization," Popenoe and Roswell Johnson explained in 1918, "is the accomplishment of the end in view without much expense to the state, and without interference with the 'liberty and pursuit of happiness' of the individual."[60] At least for these avid eugenicists, depriving individuals of the ability to procreate a mentally ill or retarded child did not infringe on their liberty or happiness.

Despite the confident assurance of eugenicists that sterilization was a harmless procedure that did not deprive degenerates of anything worth having, constitutional challenges dogged the early sterilization laws. Seven of the first sixteen statutes were struck down by state or federal courts for violating individual rights of equal protection, due process, or freedom from cruel and unusual punishment.[61] These decisions so hobbled the enforcement of all the early sterilization laws that only the California program, which was repeatedly revised and reenacted, lived up to eugenicists' expectations.[62] Even Indiana's pioneering program was shut down, first by a hostile governor in 1913 and finally by court order in 1920.[63]

The situation changed dramatically in 1927, however, when the USA Supreme Court upheld Virginia's model eugenic sterilization statute in the landmark case of *Buck v. Bell*. "We have seen more than once that the public welfare may call upon the best citizens for their lives," wrote the high court's only surviving Civil War veteran, Oliver Wendell Holmes. "It would be strange if it could not call upon those who already sap the strength of the State for these lesser sacrifices, often not felt to be such by those concerned, in order to prevent our being swamped with incompetence." Writing for a majority of the court that ranged from conservative Chief Justice William Howard Taft to liberal Associate Justice Louis D. Brandeis, the progressive Holmes concluded, "Three generations of imbeciles are enough." Only arch-conservative Roman Catholic jurist Pierce Butler dissented, and he did so without stating his reasons.[64]

The decision in *Buck v. Bell* fueled a resurgence of eugenic sterilization lawmaking that led to the enactment of new statutes in sixteen states during the late 1920s and early 1930s.[65] Confidence in their constitutionality also contributed to the increased utilization of such laws. The annual average number of operations performed under compulsory sterilization statutes in the USA jumped tenfold, from 230 during the 1910s to 2,273 during the 1930s. Although the number of operations gradually declined thereafter, it did not become insignificant until the 1960s, by which time the total exceeded 63,000.[66]

Due to various local factors, California became the center for involuntarily sterilization in America during the first half of the twentieth century. In 1909, it passed the nation's second compulsory sterilization law and, in 1917, revised it to cover any patient at a state mental-health institution "who is afflicted with mental disease that may have been inherited and is likely to be transmitted to descendants [or who suffers from one of] the various grades of feeble-mindedness." Another provision reached "all those suffering from perversion or marked departures from normal mentality or from disease of a syphilitic nature." Under the law, the procedure occurred prior to a patient's release from custody and did not require consent. Reaching beyond institutional walls, the law also directed state mental-health hospitals to perform outpatient sterilizations without charge on "any idiot" upon the written request of a parent or guardian. The law did not require that the operation benefit the patient; societal benefit alone sufficed.[67] Sponsored by the state's Commission in Lunacy under the leadership of Superintendent F. W. Hatch and endorsed by the influential *Los Angeles Times*, these laws passed easily and were vigorously enforced into the 1940s. By 1950, approximately 20,000 institutionalized persons had been sterilized under California law, which represented three times the number of such persons sterilized in any other American state and nearly one-third of the national total.

Even as California wound down its eugenic sterilization program in the 1940s, Virginia and North Carolina expanded their practice of sterilizing patients at state mental-health institutions. The Virginia program had been expressly upheld as constitutional by the U.S. Supreme Court in 1927, which gave it a boost, and it served as a model that the Eugenics Record Office promoted widely. In North Carolina, the physician-philanthropist Clarence Gamble of the Procter & Gamble soap dynasties initiated and funded a series of studies and demonstration projects designed to clean up the state's gene pool through eugenic sterilization programs implemented by state mental-health institutions and county welfare agencies. The Virginia and North Carolina sterilization efforts continued into the 1960s, by which time approximately 7,000 persons had been sterilized in each state. Taken together, California, Virginia, and North Carolina accounted for over one-half of the forced sterilizations in the USA.[68]

Not all eugenicists favored sterilization. It never became popular among British eugenicists, for example, and Davenport remained lukewarm to the idea in the USA. For them and others, segregation represented a more acceptable policy.[69] An article in 1913 by Goddard suggested the reasoning that led many American eugenicists from initially accepting segregation to ultimately advocating sterilization. "If the individuals that are selected for the operation are never to go out into the world, the operation will be of no very great benefit to society," Goddard observed. "But it is true that many institutions for the feeble-minded have inmates that could go to their homes . . . if they were safe from the danger of reproduction." Then, he reasoned, "others could take their places, could be trained to work, sterilized and sent to their homes to be fairly comfortable in those homes. In this way . . . the burden of caring for so many people for their entire lives in colonies would be, to a certain extent, relieved." Following this logic, Goddard and many fellow eugenicists concluded, "it is not a question of segregation *or* sterilization, but segregation *and* sterilization."[70]

Davenport's chief assistant at the Eugenics Record Office, H. H. Laughlin, emerged as the chief national proponent of sterilization laws. From 1910 to 1939, he tirelessly compiled, analyzed, and distributed material on judicial, legislative, and administrative aspects of eugenic sterilization.[71] As early as 1914, he outlined a "conservative sterilization program" for America that aimed at cutting off "the lowest one-tenth of the total population." If the program began promptly, Laughlin estimated that this would "require the sterilization of approximately fifteen million (15,000,000) persons during [the first] interval." After the initial bottom 10 percent was sterilized during the first interval, the effort would focus on the next lowest group. "The infinite tangle of germ-plasm continually making new combinations will make such a policy of decimal elimination perpetually of value," he concluded.[72]

Even this, however, was not sufficient for Laughlin and other radical eugenicists. "As a final factor, the federal government must cooperate with the states to the extent of excluding from America immigrants who are potentially parents and who are by nature endowed with traits of less value than the better ninety percent of our existing breeding stock," Laughlin wrote. "Every nation has its own eugenical problems, and in the absence of a world-wide eugenical policy, every nation must protect its own innate capacities."[73] During the early 1920s, while state sterilization laws hung in limbo due to constitutional concerns, Laughlin concentrated on this issue as "Expert Eugenical Agent" for a congressional committee that drafted new national legislation restricting the entry of non-Nordic immigrants into the USA. It completed the panoply of eugenic methods.

Despite its grandiose aspirations and notable achievements, the American eugenics movement never attracted a mass popular following. A few very active experts, such as Davenport, Laughlin, Popenoe, and Goddard, provided a remarkably large proportion of its apparent scientific substance. A somewhat larger circle of distinguished professionals, including renown zoologist and Stanford University founding president David Starr Jordan, Harvard University biologists Edward M. East and William E. Castle, New York author Madison Grant, American Museum of Natural History President Henry F. Osborn, sociologist Edward A. Ross, birth-control advocate Margaret Sanger, geologist Roswell Johnson, inventor Alexander Graham Bell, and the legendary plant breeder Luther Burbank added their vigorous support. A wide range of American political leaders, including Presidents Theodore Roosevelt, Woodrow Wilson, and Calvin Coolidge, endorsed eugenic measures.[74]

Leading American biologists generally supported, or at least did not publicly oppose, the eugenics movement during its formative years, as did many prominent American psychologists, sociologists, and physicians. Their professional associations generally followed suit, most notably the American Breeders Association, which became the American Genetics Association in 1913, and virtually all state medical societies. A handful of élite philanthropists and foundations provided a disproportionate share of the movement's financial backing, particularly the Carnegie Institute of Washington, the widow of railroad magnate E. H. Harriman, California financier E. S. Gosney, and the Rockefeller Foundation.[75]

Worldwide efforts

In some ways, American eugenics became a model for other countries. Most Canadian provinces followed their counterparts in the USA in enacting eugenic restrictions on reproduction, as did some Australian states. Beginning with the passage of Germany's Law for the Prevention of Genetically Diseased Progeny in 1933, modeled in part on California's compulsory sterilization law, every Nordic nation in Europe adopted stern eugenics legislation. A scattering of Eastern European, Asian, and Latin American nations acted as well, as did some British colonies.

Germany's law, which was the most far-reaching such statute enacted anywhere, mandated the sterilization of persons determined by genetic health courts to suffer from congenital feeblemindedness, schizophrenia, manic depression, severe physical deformity, hereditary epilepsy, Huntington's chorea, hereditary blindness or deafness, or severe alcoholism. "We must see to it that these inferior people do not procreate," the noted German biologist Erwin Baur asserted at the time. "No one approves of the new sterilization laws more than I do, but I must repeat over and over that they *constitute only a beginning*."[76] Some 300,000 persons were sterilized under this law between 1933 and 1939, when it was replaced by a euthanasia program designed to rid the Fatherland (then engaged in a costly world war) of its mentally handicapped "children" (see also Weindling, Chapter 8, this volume).

Despite the enthusiastic support of eugenics by the Liberal Asquith government, including Home Secretary Winston Churchill, and many Fabian Society socialists, Great Britain trailed behind northern Europe in enacting coercive legislation, and ultimately only passed a limited provision for segregating so-called mental defectives. "Modern eugenics," one leading British advocate complained in 1912, "are the work of an Englishman, but the German, with his accustomed painstaking capacity, will probably be the first to turn them to advantage. He shows the wisdom. He looks to other people to pioneer the way, and once the path is open he gallops merrily along the road."[77]

These compulsory programs and proposals represented merely the most notorious product of the eugenics movement. Eugenicists also attempted to educate the public about their theories—not simply so that people would support compulsory laws for others, but primarily so that they voluntarily would adopt such practices themselves. These efforts included public lectures, popular articles, educational movies, religious sermons, classroom instruction, and even so-called "Fitter Family" and "Better Baby" contests. For example, the best-selling high-school biology textbook of the era in the USA, *A Civic Biology*, featured a section on eugenics that identifies mentally deficiency, alcoholism, sexually immorality, and criminality as hereditary; offers "the remedy" of sexual segregation and sterilization for these disabilities; and urges students to select eugenically "healthy mates."[78] Some theologically liberal Protestant clerics in Britain and the USA launched an effort to require certifications of eugenic fitness by engaged couples as a prerequisite for church weddings.

The 1917 motion picture *The Black Stork* offers a telling example of eugenics advocacy from the USA (Figure 7.2). This full-length feature movie, produced by the motion-picture arm of William Randolph Hearst's media empire, glorified the acts of a eugenics-minded Chicago obstetrician, Harry Haiselden. During the mid-1910s, Haiselden made national headlines by assisting in the death through non-treatment of handicapped newborns. *The Black Stork*, which featured Haiselden playing himself, explicitly encouraged couples to undergo physical examinations for fitness before marriage and parents to allow their dysgenic newborns to die.

The movie opens with scenes of people who should not breed: a staggering drunk, a street beggar, a boy on crutches, a flirtatious girl in a mental-health institution—all described as "victims of inherited mental or physical disease." These scenes appear intermixed with ones showing happy youths at play displaying their healthy physiques. The film's plot centers on a seemingly healthy man who carries a so-called hereditary "taint." He marries without informing his bride about the taint and they have a child who displays the defect at birth. The mother's physician urges her to allow the newborn to die, either by withholding treatment or by lethal medication. She agrees, but only after having a vision of the child's future. In his youth, he is taunted for his limp and hunched back. Sinking into crime and despair as an adult, he ultimately sires a brood of similarly disabled children. "God has shown me a vision of what my child's life would be," the mother exclaims at the movie's climax. "Save him from such a fate," she tells the doctor, who answers this mother's plea by euthanizing her baby, whose soul is shown leaping into the arms of Jesus.[79]

As the dramatic appearance of Jesus in *The Black Stork* suggests, religion played a prominent role in the public policy debate over eugenics. In the USA during the 1920s, for example, the American Eugenics Society (AES) formed a committee for cooperation with the clergy that, among other activities, sponsored eugenics sermon contests. These competitions apparently tapped a reservoir of support because they attracted hundreds of sermons from across the country, most by Protestant ministers. The sermons typically proclaimed Christian authority for selective breeding and linked the spiritual advancement of humans to their hereditary endowment. Some traced the eugenic attributes of Christ's pedigree while others foresaw the millennium arriving through the genetic improvement

FIGURE 7.2 Publicity poster for *The Black Stork*. (From *Chicago Tribune*, April 2, 1917, © 2010 Chicago Tribune.)

of humanity. Harry Emerson Fosdick of New York's prestigious Riverside Church, University of Chicago theologian Charles W. Gilkey, Federal Council of Churches president F. J. McConnell, and other prominent American church leaders served on the AES committee and embraced a eugenic gospel of social reform.[80] Fosdick, Gilkey, and McConnell were liberal Protestants, but at least one Roman Catholic cleric served on the AES committee and some theologically conservative Protestant ministers endorsed eugenics.

The pattern followed in northern Europe. In England, for example, the Convocation of the Church of England and several key Methodist church leaders endorsed the 1912 Mental Deficiency Bill, a highly controversial eugenics proposal. Leading British eugenicists regularly appealed for support from religious audiences, and Anglican cleric William Ralph Inge assumed a prominent role in the national eugenics movement.[81] Ministers in state-supported churches provided similar support for eugenic lawmaking in Scandinavia and other Protestant regions of Europe.

Despite this visible Protestant support for eugenic doctrines, Galton, Davenport, and other leading eugenicists broadly denounced Christians for opposing scientific progress. In part, these attacks reflected the timing of the eugenics movement, which occurred at the height of the antievolution crusade in America and at a time when secular scholars generally perceived Christians as historically hostile to science. Eugenicists felt this hostility primarily in the form of vigorous Roman Catholic opposition to sexual sterilization.

Based on its religious commitment to the sanctity of all human life regardless of biological fitness, the Roman Catholic Church emerged as the first major organization to challenge eugenic doctrines. Catholic resistance stiffened as the eugenics movement began advocating sterilization, which the church denounced as violating the so-called natural law linking sexual activity with procreation. The pope formally condemned eugenics in a sweeping 1930 encyclical on marriage.[82] Biting essays from the pen of popular British author G. K. Chesterton, a convert to Catholicism, did much to undermine support for eugenics among the educated classes throughout the English-speaking world. Church publication and pronouncements picked up on many of Chesterton's themes, and carried them to a wider audience. Wherever bills to legalize sterilization surfaced, local Catholic clergy, lay leaders, and physicians took the lead in opposing them. This opposition proved decisive in defeating eugenic sterilization proposals in such heavily Catholic states as Pennsylvania, New York, and

Louisiana; helped to stall eugenics legislation in Britain; and prevented the spread of eugenic restrictions to the Catholic regions of Europe and the Americas.[83]

Some Protestants also voiced religious objections to eugenic legislation, especially in the American South, where fundamentalism was strong. Protestant opponents of eugenics generally did not articulate their position as clearly as either Protestant supporters of the cause or Catholic opponents of it. Nevertheless, the legislative record is littered with comments by individual Protestants denouncing eugenics as unchristian. Taken as a whole, these comments reflect a view that God controls human reproduction, and that neither science nor the state should interfere.

End of an era

Eugenics lost favor almost as quickly as it had gained favor. In reality, it never attracted a widespread public following. Even in its heyday, Chesterton glibly dismissed eugenics as a mean-spirited, élitist "joke" that briefly "turned from a fad into a fashion."[84] Yet it did not encounter much organized opposition, so that a highly motivated eugenics clique could carry the day in some places for a while. Only the Roman Catholic Church offered sustained, organized opposition to eugenics—doing so largely on religious rather than scientific grounds—and no eugenic sterilization laws ever passed in any jurisdiction where Catholics wielded significant political influence. Labor unions, civil-rights organizations, civil libertarians, working-class populists, and some Protestant denominations occasionally resisted eugenics legislation that directly affected their interests.

Except in Germany, eugenics lacked sufficient depth of support to overcome committed opponents. Within the British Parliament, for example, one highly critical backbencher, Leonard Darwin's libertarian cousin Josiah Wedgwood, almost single-handedly stalled passage of eugenics legislation sponsored by his own Liberal Party government in 1912 and 1913.[85] His efforts within Parliament were effective in part because many prominent British eugenicists, including Pearson, Bateson, and Fabian leader Havelock Ellis, openly questioned the wisdom of coercive sterilization programs.[86]

The tide of popular opinion began turning against eugenics and social Darwinism by the 1930s. Intelligence tests administered to U.S. Army personnel during World War I showed alarmingly low scores, especially among immigrants from southern Europe. Initially a clarion call for eugenic exclusion, contributing to passage of the restrictive 1924 Immigration Act, in time these results helped to break the perceived link between low IQ scores and degeneracy. Many of the soldiers scoring low on these tests nevertheless served their country well in war and, presumably, in peace. A new breed of social scientists led by American anthropologists Franz Boas and Margaret Mead challenged the hereditarian assumptions underlying eugenics. The worldwide economic depression that started during the late 1920s and persisted through the 1930s caused many people to question the simple equation of hereditary fitness with social and economic success.

In the face of heightened public skepticism of eugenic claims, some left-leaning biologists in the USA and Britain took this opportunity to try to purge the movement of what they perceived as racial and class bias. "That imbeciles should be sterilized is of course unquestionable," the Marxist geneticist Herman J. Muller (1890–1967) told the Third International Congress of Eugenics in 1932, but warned that economic inequality could mask genetic qualities.[87] At the time, Muller was one of America's leading geneticists and a future Nobel laureate. By 1935, he had moved to the Soviet Union and was urging its government to implement a eugenics program that would inseminate superior women with sperm from men of extreme intelligence, talent, and social feeling. Within a century, he envisioned that most people could have "the innate qualities of such men as Lenin, Newton, Leonardo, Pasteur, Beethoven, Omar Khayyam, Pushkin, Sun Yat Sen, Marx, or even to possess their varied faculties combined."[88] Here was Galton's dream, with the class and race prejudices attributed to so-called mainline eugenicists replaced with Muller's own cultural and ideological ones (see also Krementsov, Chapter 9, this volume).

Few could match Muller's ardor for the cause, but many progressively minded biologists and physicians took a similar tack. In Britain, such prominent geneticists as Julian Huxley, J. B. S. Haldane, and Lancelot Hogben joined in the call for purging eugenics of its reactionary biases to save it from its own excesses. For them, voluntary sterilization offered a better promise than any compulsory scheme. Back in the USA, the reform position appeared in a 1936 report from a blue-ribbon committee of the American Neurological Association, specially chosen for their expertise, chaired by Boston psychiatrist Abraham Myerson. While the report roundly condemned state-ordered sterilization and found "no sound scientific basis for sterilization on account of immorality or character defect," it nevertheless "recommended sterilization in selected cases of certain diseases and with the consent of the patient or those responsible for him." Those diseases included mental hereditary retardation, schizophrenia, manic depression, and epilepsy.[89]

Comparing the programs offered by the two groups, political scientist Diane Paul concludes, "The distinction between reform and mainline eugenicists should not be overdrawn."[90] The public seemed to agree. Reform or not, eugenics steadily lost credibility during the late 1930s and early 1940s. The exposé of Nazi practices stirred and stiffened public opposition to forced sterilization programs. A chastised Carnegie Institution finally closed the Eugenics Record Office in 1940. Following World War II, American philanthropists still interested in such matters, such as Clarence Gamble and the Rockefeller Foundation, shifted their support from eugenic sterilization programs at home to population-control efforts in developing countries. Somewhere in the process, most geneticists stopped publicly espousing eugenic doctrines.[91] Those who continued to do so, like R. A. Fisher, lost public prestige and professional posts. By midcentury, nurture had supplanted nature as the accepted explanation for how people act.

Although geneticists may have repudiated compulsory sterilization and stopped using the term "eugenics," they did not necessarily abandon the hope for improving humanity through genetics. Within the upper echelon of biologists, William Hamilton and James Watson spoke openly of achieving such ends though individual reproductive choices made in light of expanded genetic knowledge. By the dawn of the twenty-first century, the mapping of the human genome, the advent of genetic screening, the increased use of in vitro fertilization, and the promise of germ-line gene therapy gave substance to their hopes. With prenatal genetic testing, parents could begin to select the genotype of their children, at least in cases of in vitro fertilization. For some, this has become "a new eugenics" and its prospects continue to grow.[92] Galton would have been pleased.

CHAPTER EIGHT

Genetics, eugenics, and the Holocaust

Paul Weindling

The period of National Socialism in Germany lasted only a brief twelve years, and yet, over sixty years on from the defeat and death of Adolf Hitler in 1945, we are still coming to terms with the aftermath. The Nazis made frenetic efforts to place one of the world's great scientific cultures on a new basis of race, unleashing a dynamic of aggressive war and genocide in which science took a central role. The endeavor left immense consequences that are still with us. Much effort has been expended on assessing the cultural, ethical, and spiritual fallout from the Holocaust, not least for human genetics and evolution. Survivors of Nazi medical atrocities are still alive; these include victims of sterilization, and child and adolescent victims of human experiments, as with the twins experimented on by Josef Mengele (1911–79). We still have no clear idea of the overall number of victims of Nazi human experiments; the identities of many experimental victims remain unclear, and other victims of racial persecution—so-called "anti-socials," homosexuals, the "workshy," and "hereditarily criminal"—all raise problems of their numbers and life histories.

In turn, unresolved issues have confronted German science. The Max Planck Society, a conglomerate of prestigious German research institutes, made a brave apology to survivors of Mengele's twin camp in 2004.[1] Many institutions for brain research and anthropology held body parts of euthanasia victims until the early 1990s. Because the institutes rapidly divested themselves of these human remains, only exceptionally with an effort to establish the history of the body parts, to identify the person killed, or to trace relatives, questions remain concerning the identification of these persons. Most German scientific institutions used slave labor to sustain scientific research during the war.[2] Issues to do with forced abortion, fertility research, and forced adoption are still being unraveled. The implications of Nazi biology and eugenics then have living consequences now.

Viewed from the twenty-first century, questions surrounding Nazi ideology in evolution and eugenics have taken on a general relevance. One essential issue is whether irrational elements in Nazi science meant that it lay outside the normal framework of scientific endeavor, or whether it took to an extreme the destructive potential of science. Careful historical research has uncovered how biology lay at the core of the Nazi racial culture. Both in terms of ideology and application, biology fueled the racial war to exterminate, kill, and resettle total populations. Only in the past decade have historians realized how important Darwinism and eugenics were for the whole structuring of Nazi society and its racial policies. On the one hand, historians have underestimated the role of biological experts in the realization of Nazi policies. On the other (and this point needs to be stressed), we should not extrapolate from certain nazified and racialized variants of selection and survival of the fittest to the totality of evolutionary theory, Darwinism, and genetics. Thus, a historically distinctive variant of biology should not be confused with evolutionary theory in general. Failure to identify the specific variants of nazified biology and the unwarranted leap to the totality of Darwinism have meant that the implications of these Nazi atrocities for issues of reproduction and population control, as well as for biological research, are still largely unresolved.

One recent shift of interpretation is the realization that Nazi science was innovative in such areas as evolutionary synthesis and medical research. During the 1930s, and in the aftermath of World War II, Nazi science was often dismissed as pseudo-science and characterized as crude and irrational. As long as American and British scientists could regard Nazi biology as fundamentally irrational and unscientific, uncomfortable questions did not need to be addressed about scientific values. Nazis were associated with much vulgar, rabid theorizing about race—for example, how intercourse with a Jew could infect and pollute an Aryan, and once impregnated a person was irremediably "stained" (this is known as telogeny). The 1935 Nuremberg laws that instituted racial segregation were supported by a range of scientific researchers on race and the so-called Jewish problem. What we find is not a monolithic orthodoxy on race, evolution, and society, but a multiplicity of different positions within the Nazi camp. Competition between these positions unleashed a radicalizing dynamic.

The persecution, dismissal, expulsion, and killing of Jewish biologists from 1933 onward was substantial: biochemists, geneticists, and medical researchers were targets. Hans Przibram (1874–1944) of the Prater Vivarium, or Institute for Experimental Biology, was among the biologists killed (in his case at Theresienstadt camp). The Vivarium was important for laying the foundations for ethology, which then flourished under Nazism.[3] Certainly, some German biologists tried to develop a nazified biology. The idea was that the German biologists had a special affinity with nature, and that the German nation had a calling for developing biology and racial studies.[4] Other scientists continued to pursue normal science such as biochemistry or embryology, accepting the absence of Jewish colleagues and Nazi administration. Only rarely would a colleague decline a post left vacant by a Jewish scientist (as did the pharmacologist Otto Krayer [1899–1982], who was later appointed to Harvard). Most German academics thus benefited from the vacancies left by expelled colleagues, and the Nazi authorities continued to make funding available for scientific research, which was seen as important in realizing policies of territorial expansion and racial regeneration. German biologists continued to participate in international scientific circles, and indeed were awarded Nobel Prizes, as for example the biochemists Adolf Butenandt and Richard Kuhn (albeit prevented by Hitler from accepting these honors). During the 1930s and World War II, scientists challenged nazified notions of race and élitism, as opposed to biology per se, which was international, humane, and democratic.

Not only was there reputable and innovative science, but scientists enjoyed substantive support from the Nazi state and party political organizations. Here, we must draw distinctions: the biologist Karl von Frisch (1886–1982)—the discoverer of the dance of the bees as a means of communication—remained detached from Nazi scientific circles. Even so, he published a popularization of evolutionary biology that benefited Goebbels's military fund.[5] By way of contrast, the animal behavior expert Konrad Lorenz (1903–1989) made a career within Nazi scientific institutions and endorsed Nazi values.[6] There was a dense network of state, military, and party funding agencies that provided young scientists with a range of opportunities and older scientists with the chance to develop research groups and apply their research. But to achieve this meant sustaining normal scientific networks in genetics and virology. The nazified organizations and the military expected useful results, as a functioning biology was required for the Nazi vision of modernity and efficiency. At the same time *völkisch* [populist] ideologues and Nazi racial theorists espoused an ideological style of biology with the aims of protecting the healthy German family and strengthening the community.

Darwin's putative guilt

Charles Darwin (1809–82) and Adolf Hitler have been often linked. The assumption made is that Nazi ideologues used Darwin's natural selection as an exterminatory ideology for the Holocaust. Moreover, the evidence for eugenics having supplied the professionally administered techniques of extermination is compelling. Some speak of the thread connecting Darwin to Auschwitz.[7] The idea is that "the survival of the fittest" and the "struggle for existence" (*Kampf ums Dasein*) were used by Hitler in his programmatic tract *Mein Kampf* (1925). Nazi propaganda did come to draw on Darwinian ideas of natural selection to reinforce extermination of the mentally ill and disabled. Hitler read the Ariosophy (a mixture of racial mythology and pantheism) of Lanz von Liebenfels, and anti-Semitic newspapers, but did not study biology or manuals on animal breeding in any serious way.[8] Hitler's library contained tracts on race purity and health that drew on Darwinian language. Hans F. K. Günther's works on the racial composition of the German people found favor among the Nazis. Hitler's library contained several of Günther's works.[9] Heinrich Himmler (1900–45), head of the Schutzstaffel (or SS; the Nazi party's élite group), studied agriculture at the Technical University in Munich, where he encountered the genetics of animal breeding.

Even so, the assumed linkage between a reclusive Victorian naturalist and twentieth-century demagogues and Nazi perpetrators of genocide is curious, strained, and—when historically scrutinized—tenuous. A link would still mean that Darwinian natural selection could be applied in ways inconsistent with Darwin's own rather tentative views on human race and variation. Not all references to struggle were necessarily Darwinian, as there were all sorts of ideas of struggle in German culture: other sources were the ideas of religious enthusiast Jakob Boehme (1575–1624) and the associated pantheistic nature philosophy, and those of the nineteenth-century philosopher Arthur Schopenhauer (1788–1860) on irrational instincts. These ideas of struggle were then vulgarized by anti-Semitic racial propagandists.

Hitler absorbed the idea of natural selection and struggle in a derivative and diluted form. Certainly, Hitler, like many Nazis, regarded nature as a religion, and he appears to have had more of an intuitive, self-educated regard for matters biological. These religious dimensions allow for a philosophical and religious reading of Hitler's *Mein Kampf*. Although his messianic tract has long been assumed by historians to be Darwinist or social Darwinist, the text has rarely been critically examined from the point of view of the coherence of its views on race. Hitler referred to "God," "Goddess," or "the Gods" no less than sixty-four times, suggesting a belief in divine intervention in human history. Hitler thanks God for incidents favoring the German race, especially the demise of the Habsburg rule of Austria–Hungary. There are certainly many references to German racial superiority and Jews as parasites.[10]

Another approach has been to see Darwinism as morally corrosive. Here, the argument is that an ethic of destruction linked Darwin to the Holocaust.[11] The presumed idea is that Hitler and the Nazi leadership espoused a logic of extermination, driving their selective measures to weed out the unfit—with the imposing of selective welfare policies for the racial élite, sterilization, the killing of the mentally ill, and the unleashing of the Holocaust. From this perspective, Darwin's ideas are portrayed as an inhumane, anti-Christian force, undermining moral scruples and inhibitions. Yet this reading of Darwin's biology as a gospel of destruction has little basis in Darwin's actual writings and opinions.

Although Darwin's magnum opus *Origin of Species* left a troubling moral legacy (not least for Darwin), whether it has the destructive moral force claimed for it is dubious. For belief in Christianity was undermined by a range of values. Philologists viewed the Bible as a historical document, and there were widespread currents of popular materialism. The SS was steeped in pagan Nordic ritual; yet SS members had their hereditary health meticulously screened. Hitler's own religious beliefs are a matter of controversy. Originally Catholic, he remained circumspect regarding the Vatican, and no papal condemnation of Nazi anti-Semitism and racial war was issued. Hitler, like other Nazis, was a racial pantheist, believing in nature as a religion as well as a racial force.

What could be simpler than a link between Darwin and the satanic Hitler? The critical view among anti-Nazi scientists was that Nazism spawned debased "pseudo-science." This meant that academic biologists could plead that they had nothing to do with Nazi racism. We might ask, why were humane ethical constraints or commitments removed from Darwinism and eugenics in Germany, when other countries (such as Britain and the USA) also had influential eugenics movements (see Larson, Chapter 7, this volume)? The problems become even more complex if we examine links between evolutionary biology and welfare institutions, and how race became a central value under Nazism.

Darwinism followed a period of scientific materialism in nineteenth-century Germany; evolutionists maintained a variety of positions on science and Christianity. The problem is that, when it comes to ideas of struggle and extermination, thinkers such as Schopenhauer and Friedrich Nietzsche (1844–1900) were influenced by biology (although not necessarily Darwinian evolution), and themselves exerted influence on biology—as, for example, on the defining of instinct. Even though Darwin was a central figure in evolutionary thinking, putting Darwin on an iconic pedestal is out of place here. Evolution in German biology was a composite involving a range of theoretical contributions.

Turning to the values of Hitler's *Mein Kampf*, we find that these also reflect a range of ideas, but an unequivocal Darwinism is not amongst them. We find Hitler's ardent anti-Semitism and anti-Bolshevism, and also a desire to expand German "living space" (*Lebensraum*). Hitler did have a sense of eugenics: "that the most beautiful bodies find one another and thus help in giving the nation a new beauty." This was an aesthetic physical form of eugenics. He denounced biological inferiors and "ingenious weaklings," as well as those with sexually transmitted diseases.[12] He subsequently, in 1929, demanded euthanasia for the mentally ill. A Nazi may have been a Darwinian, but it is a false logic to ascribe Nazism to Darwinism. Darwinism and Nazism had quite separate historical roots. The problem is rather to discover how these two quite separate traditions fused in Nazi Germany than to rush to the assumption that Darwin begat Hitler.

In seeking to link Darwinism to the Holocaust, there is a sense of expectation that by doing so one might discredit Darwinian theory as a whole. A hard and fast link between Darwin's ideas and the Holocaust is tenuous; Darwin's own views of "lower" and "higher" races varied, not least as Darwin argued for a constant process of transmutation (see Rupke, Chapter 6, this volume). Importantly, we need to take account of sexual selection as a significant factor in evolution. Darwin's and Hitler's actual views on evolution and race were poles apart. Darwin had variable views on race, the human intellect, and mutualism, and did not focus exclusively on the annihilatory struggle in nature.[13] Alfred Russel Wallace (1823–1913), the codiscoverer of natural selection, was far stricter than Darwin in adhering to natural selection, and yet his dualism meant that his social values were resolutely humane. Herbert Spencer's (1820–1903) evolutionary ethics of "survival of the fittest" were interpreted individualistically in the USA, whereas in continental Europe many appreciated how Spencer also demonstrated the need for organic integration. Spencerians raised a cautionary voice against atomistic individualist struggle.[14] Darwinists were conspicuous in their lack of anti-Semitism and racial prejudice (but see also Sivasundaram, Chapter 5, this volume).

Darwinismus

We understand German Darwinism, or *Darwinismus*, to be culturally and scientifically distinctive. The Germans focused on "the organism"—a multicellular entity. We find in the biologist Ernst Haeckel (1834–1919) not only a talented but a strongly anti-clerical and yet pantheist ideology (Figure 8.1). For him biology as a unified science of all sentient life from plants to people meant a challenge to the ideas of a divine or autonomous intellect. Seen in this crusading role, Haeckel supported the Monist League, which was joined by some Protestant pastors. He was strongly anti-Catholic. Haeckel's semi-mystic tone controversially established biology as a surrogate for revealed religion.[15]

Haeckel is customarily seen as the leading prophet of German Darwinism. He is associated with ideas of applying natural selection to human society, and one might add of applying evolution to cell biology and embryology, areas in which German biologists excelled. Yet Haeckel was no orthodox Darwinist. On the one hand, he was a Lamarckian. On the other, Haeckel drew on Darwin's observations on sexual selection in explaining "the differences between the races of men." Haeckel observed how the ancient Spartans killed the weak, sick, and deformed newborn. Historians of Nazi euthanasia cite this text to show Haeckel as a forerunner of Nazi extermination programs. But Haeckel saw the practice as an antithesis of modern civilized society, rather than as a prescriptive model.[16]

Darwinian texts by Thomas Henry Huxley (1825–95), Spencer, and the physiologist John Berry Haycraft (1857–1922) had a strong impact in Germany between 1870 and 1914. The German authors Alexander Tille (1866–1912) and Alfred Ploetz (1860–1940), who first elaborated "racial

GENETICS, EUGENICS, AND THE HOLOCAUST 199

FIGURE 8.1 Ernst Haeckel (1834–1919), leading German biologist and Darwin popularizer whose biology became a surrogate for religion and who considered that the Indogermanic race had far outstripped all other races of men in mental development. (Courtesy of the National Library of Medicine.)

hygiene" in 1895, and the biologist August Weismann (1834–1914) drew on British social Darwinists (see Larson, Chapter 7, this volume). All these texts interpreted Darwin's ideas in divergent ways. Weismann rejected the Lamarckism of Herbert Spencer, and in developing his hereditarian theory of an immutable germ plasm, he introduced works based on strict Darwinian natural selection by the political commentator Benjamin Kidd (1858–1916) and Haycraft. Kidd visited Weismann in 1890 to prepare a major review of his theories.[17] Kidd used Weismann's principle of immutability of the germ plasm to refute socialism.[18] Haycraft's *Naturliche Auslese und Rassenverbesserung* (original published as *Darwinism and Race Progress*) and Ploetz's treatise *Grundlinien der Rassen-Hygiene* [Fundamentals of racial hygiene], published in 1895, made Weismann's determinist *Keimplasma* [germ plasm] central to social Darwinism. Weismann's hereditary determinism displaced the belief in environmentalist progress of Haeckel and Spencer, whose ideas of social altruism were underpinned by those of Jean-Baptiste Lamarck (1744–1829) (see Krementsov, Chapter 9, and Ruse, Chapter 10, this volume). Others, such as the zoologist Richard Semon (1859–1918) continued to defend Lamarckism, as in Semon's mneme theory of inherited cultural memory.[19] Despite the rediscovery of Mendel's genetics in 1900, and the interest in Weismann's theory, there continued to be a strong theoretical divergence in German biology. Most biologists recognized the coordinating role of the state as a biologically evolved factor. We might see various branches of science—genetics, bacteriology, anthropology—as shaping what Zygmunt Bauman has called the "gardening state" mentality (with echoes of T. H. Huxley's prologemena to *Evolution and Ethics* of 1893).[20]

By the turn of the century, social scientists such as Friedrich Hertz (1857–94) in Vienna were warning against race theory as a tool of political oppression. Hertz presciently wrote on "modern race theories" before the founding of any eugenics society, but he constantly revised his analysis to take account of eugenics and racial hygiene.[21] In his final post as political scientist at Halle from 1930 to 1933, he directly attacked the Nordic racial theories of the populist race researcher Hans Günther (1891–1968).[22] Racial biology developed not in a linear path but in response to radical critiques on scientific fallacies and inhumane political consequences.

One broader inference is that if only Christian ethics had remained firm, the biologists and eugenicists would never have committed inhumane atrocities. Yet the practice of Christian ethics did not always remain firm. Under Nazism the Protestant welfare organization, Innere Mission [Inner mission], favored coercive sterilization. The eugenicist Otmar von Verschuer (1896–1969) was a disciple of the dissident Confessing Church of Pastor Martin Niemoeller, and yet also supported Mengele in Auschwitz concentration camp. Catholic eugenicists, notably the Jesuit biologist Hermann Muckermann (1877–1962), were active in establishing the Kaiser Wilhelm Institute for Anthropology (KWIA). Despite preparation of an encyclical against anti-Semitism, *Humani Generis Unitas* [The unity of the human race], under Pope Pius XI, his death in 1939 gave way to a more conciliatory policy, initiated by his successor Pius XII, toward the rising onslaught of Nazism.[23] Protestantism raises a number of paradoxes. On the one hand, there was a vigorous dissident group of German

Protestant pastors who led their churches in the so-called Confessing Church, refusing to submit their congregations to Nazi control.[24] Many pastors lost their lives in the Nazi camps, the best known being Dietrich Bonhoeffer, who was implicated in the Abwehr conspiracy to assassinate the Führer, and executed just days before the Allies liberated his camp. On the other hand, Protestants often supported Nazi racial biology and such measures as compulsory sterilization. Otmar von Verschuer, whose assistant was Mengele, was a member of the Confessing Church. As Widukind Lenz (the son of eugenicist Fritz Lenz remarked), Verschuer worshipped God and Adolf Hitler.[25] In reality, Nazism split the German churches, eventually leading to a remarkable movement of reconciliation between those who had, or had not, collaborated with the Nazis once the war was over.

Natural and/or sexual selection

One crucial issue is that Darwinian evolution involved both natural and sexual selection, the latter often overlooked. The race hygienist Ploetz argued that a humane eugenics policy should consist of reproductive controls, focusing on chromosomal engineering. The physician Wilhelm Schallmayer (1857–1919) produced the scheme of the physician as an official of the state. People should have an annual medical examination and only the fit should be allowed to reproduce. Control on marriage was an answer to so-called "racial poisons" such as tuberculosis, venereal disease, and alcoholism. Schallmayer articulated a clear model of eugenic medicine.[26] He advocated a shift from the physician taking care of the sick individual to becoming a state official, evaluating medical conditions in terms of degenerative diseases from a social and racial perspective. Inhabitants should be subjected to annual screening and issued with health passports. Sanitary inspection on borders and immigration controls were the corollary. He saw the possibility of a United States of Europe running collective institutions. Schallmayer's biological model was one of professional control that could be inserted in any welfare state system, liberal-democratic, socialist—or Nazi.

By the Weimar period, Germany's first democratic republic, which lasted from 1919 to 1933, the social hygienist Alfred Grotjahn's "reproductive ethics" or *Fortpflanzungshygiene*, advocated controls on reproduction as being at the center of social hygiene. The concern was to weed out supposedly inferior persons who could cause the spread of psychological and physical disabilities. Within the framework of a democratic state, professional experts sought to identify those of "lesser value"—the so-called *Minderwertigen* [inferior ones]. On the other hand, those of high eugenic worth were to be encouraged to have large families, and there were special incentives for the *Kinderreichen*—the "child rich." Eugenics was encouraged to be seen as part of a human economy—or *Menschenökonomie*—promoting a fit and strong population. Ethnic minorities such as Jews, and Sinti and Roma (Gypsies) were not the targets of Weimar eugenic welfare schemes, although psychopaths, vagrants, the mentally ill, and the disabled were. Hereditary health indices—*Erbkarteien*—were compiled in prisons and welfare offices. Population was the "organic capital of the state." These concepts allowed for a biologization of public health and social welfare. Grotjahn saw the registration of pregnancies as a way of preventing high rates of abortion. On the other hand, where pregnancies were deemed likely to result in defects, there was support for eugenic abortion. The clinics distributed many thousands of pamphlets on eugenics.[27] We can see the welfare of Weimar Germany or "Red Vienna" (the socialist municipality from 1918 to 1934) more as developments of Darwin's sexual selection. The idea of a struggle for survival in society was attenuated by the welfare state providing marriage allowances and maternity benefits, as well as sickness insurance for the family.

Nazism represented a cataclysmic social upheaval, and a new set of rationales for violence and social action. Whereas the welfare eugenics of Weimar and Austria was inclusive, the Nazi paradigm was selective. A shift from sexual selection—the Darwinian and eugenic framework for Weimar welfare—to natural selection reinforced the Nazi ideology of a racial struggle. Nazi race theory used depictions of the struggle for survival in nature to justify such exterminatory measures as the killing of psychiatric patients. The Nazis attempted to revive the primitive vigor of the German race. Rebirth and national rejuvenation would come through, recovering this primitive vigor. Darwin's natural selection now became a racial slogan for a war against the weak as a racial burden. This marked a profound contrast to Weimar welfare, where sexual selection was the overarching paradigm.

Historians have overlooked the rationales for this shift, preferring instead to draw lines of direct continuity without examining the transformation of biological rationales. The objects of attention shifted from social problem groups to race. There were continuities, as when the Nazis seized *Erbkarteien* to identify those deemed to have inherited illnesses such as schizophrenia, the supposedly "asocial," hereditary criminals, and the sexually perverse. Compulsory sterilization might be seen as a rapidly introduced transitional measure that was not explicitly racial according to the original formulation of the law, which focused on genetically inherited disorders such as schizophrenia and Huntington's chorea (now known as Huntington's disease), but was imposed within a racial context with the idea of strengthening the German race. Attitudes shifted from population and social welfare to race. We see this with the Nuremberg laws for segregating Jews, the so-called Law for the Protection of German Blood and Honor, of 14 November 1935. This provided a context for the Marriage Health Law, imposed at the same time, which required hereditary health certification prior to marriage.

Race theory

Nazi values of race purity and psychological types sit uneasily with Darwinism: first, race was not a fixed value with Darwin. Second, additional notions like blood purity were prominent, with blood being the carrier of racial and psychological traits. The Nuremberg laws were intended to segregate Jews from contact, especially sexual contact, with Germans. The Blood Protection Law forbade all sexual relations between Germans of the Aryan race and Jews. The law was extended by decree to outlaw marriages and sexual relations between Aryans on the one hand and Jews and non-whites on the other.

In addressing the problem of Nazi eugenics we need to move away from ill-defined notions of racial utopias, which have slight explanatory value only as long as the utopian blueprint is not subjected to scrutiny. Policies of racial rejuvenation were based on the idea of recovering the primitive vigor of the putative Aryan, Germanic, Nordic, or Teutonic race, defined in a multiplicity of ways. The converse was indeed the destruction of races and pathogenic human types who posed a threat to racial regeneration; among these were the physically disabled, the mentally ill, homosexuals, asocials, hereditary criminals, and racial undesirables such as Jews and Roma. If we are to see biology as central to Nazi Germany and the genesis, planning, and implementation of the Holocaust, we need to reconstruct distinctively nazified forms of biology over which there was never consensus. But retrospectively ascribing the guilt of genocide to Darwin deviates from evolutionary notions of civilization, altruism, and the similarity of mentality whatever the physical differences among humankind. These evolutionary views were widespread among biologists and allowed the formulations of anti-Nazi positions.

The situation becomes more complex once Mendelian genetics is added to the methodology and theory of racial classification. We know from the work of the geneticist Benno Müller-Hill in the 1980s and from subsequent archival studies in institutions like the KWIA that geneticists were active in establishing racial categories and in adjudicating on individual cases.[28] The KWIA's first director, Eugen Fischer, conducted early research into "degeneracy" in mixed races in the wake of the genocide of the Herero, a tribe of the Bantu group in German South West Africa (today Namibia). The shift of anthropology away from the liberal Rudolf Virchow's advocacy of cranial measurements to biological classification rendered anthropology a useful tool for advocates of a racial state.

Yet the environmentalist critique of Franz Boas, the cautionary position of genetic anthropologists in the widely disseminated *We Europeans* by the demographer A. M. Carr-Saunders, Julian Huxley, and the anthropologist A. C. Haddon, and the role of anthropologists in attacking Nazi race theory as pseudo-science and racist (a novel term introduced in 1922 as a critique of the Nordic racist Hans Günther) indicates that anthropology as a discipline maintained a critical vigilance against unscientific and inhumane appropriation of its theories.[29] Races et Racisme [Races and racism], a group of French scholars who issued a monthly bulletin of the same name, clashed directly with Nazi anthropologists, notably Robert Ritter (1901–51), and with the psychiatric advocate of coerced sterilization Ernst Rüdin (1874–1952), at the Paris population congress of 1937.[30] While the critique—sustained during World War II by anthropologists in exile (such as Paul Rivet)—did not halt the Holocaust, it does establish that biological anthropology was not in itself genocidal.

There had been, since the late nineteenth century, support for Greater Germany's responsibility for German populations beyond the territorial borders of the Reich, and scattered German communities in Eastern Europe and overseas. These ideas were taken up by the Nazis. But contrary to what one might expect, there was no standard Nazi orthodoxy on race purity, no defining treatise, and no conclave or panel of Nazi biologists or ideologists who defined the biological traits of the German race once and for all. While biology was held to be fundamental for the Nazi social order, there was never an absolute standard or authority. While the Final Solution was predicated on racial classification, we find that Nazi race theory was an ideological battleground of academic and party political factions.

This situation can be taken as being in accord with the view of Nazi historians that the anarchy of competing and conflicting "polycratic" institutions radicalized the Nazi state. The assumption has been that the fault lay with the Nazi state interventions in science, rather than the enthusiastic mobilization of many scientists in the service of Nazism.[31] This requires a shift of priorities from that prevailing in the current historical orthodoxy. Nazi racial science remains scarcely examined in its theoretical or institutional contexts. While one might have expected rigorous historical scrutiny of such organizations as the Racial Political Office of the Nazi Party, the Race and Settlement Office of the SS, or such vanguard organizations as the SS Ahnenerbe (the disarmingly named Ancestral Heritage Society) to determine their role in the dynamic of destruction, historians have been loath to engage with the biological sciences. Historians of Nazi Germany have curiously not seen race within a scientific framework. They have variously characterized the mythical and pseudo-scientific aspects, or the racial utopian views of racial science as a virtual religion. There has been less interest in defining the form that the utopia was to take, its scientific rationales, and methods of realization.[32] The biology of race remains relatively unexamined.

We are hardly better served on the history of science side. Historians of "deadly medicine" and Nazi biology have vacillated over how to conceptualize the links between Nazi race policy and science. Opinions differ over whether genetics and human genetics were inextricably entwined with eugenics and race. Some regard genetics as strictly distinct, even dividing plant and animal genetics from human genetics.[33] Others see these technical specialties as broadly linked to a nazified state and society. These issues come to be of importance as a geneticist—for example Hans Nachtsheim at the KWIA—could deny any links to Nazi research.

Biochemistry flourished under Nazism. One important case study has been made of the theory of defense enzymes proposed by Emil Abderhalden (1877–1950). Müller-Hill and Ute Deichmann showed how this was the basis of racial research in Auschwitz associated with exterminatory practices.[34] Organizations like the Deutsche Forschungsgemeinschaft, under Rudolf Mentzel (1900–87), and the Kaiser Wilhelm Gesellschaft (KWG), under Albert Vögler (1877–1945), both organizations that funded academic research, were subject to manipulation by leading Nazis and the SS. When German research was placed on a war footing, racial research was also prioritized.

Genetics had an important role in stimulating the development of demands for sterilization and biological databanks. Hitler took up various themes such as sterilization and the damage to the nation's hereditary stock through sexually transmitted disease. But he preferred the nationalist mythology of blood purity—he warned how German blood could be corrupted by mixing with Jews (Figure 8.2). Once human genetics and serology were brought to bear on the issue of "half Jews" and "quarter Jews," a wide spectrum of positions was possible. In part, these tensions represent a power struggle between Nazi ideologists and scientists. The expectation that Hitler would assist in implementing eugenics arose in July 1933, when the Nazis rapidly passed a sterilization law. The psychiatric geneticist Ernst Rüdin, director of one of the first eugenics research institutes, took a lead in drawing up the Sterilization Law. This was targeted at a range of clinical conditions, notably schizophrenia, muscular dystrophy, Huntington's chorea, epilepsy, severe mental defect, inherited deafness, and chronic alcoholism. Sexual and mental abnormalities attracted especial interest. Otmar von Verschuer led the way in twin studies using the Frankfurt public health clinic as a basis. Nazi public health administration was centralized, so as to identify the "genetically defective" and issue orders for sterilizations. Sterilization was authorized by tribunals of two doctors and a lawyer. At least 375,000 individuals were sterilized by the German authorities (including after 1938 in annexed Austria), and there were an estimated 5,000 deaths from complications. The KWIA provided training courses for SS doctors. The law marked a starting point for the sterilization of the physically disabled (such as the deaf) and epileptics, not least because disabilities were taken as markers of mental defects. Moreover, the compulsory sterilization law marked a starting point for the onslaught on homosexuals, and persons deemed asocial, that is predisposed to criminality, who were incarcerated in concentration camps. The extension of sterilization to euthanasia marked a direct link to the Holocaust, and at times camps were deemed to be reservoirs of experimental subjects for a variety of often gratuitously cruel and fatal experiments. The social historian Karl Heinz Roth coined the chilling phrase "the Final Solution to the Social Problem" to condemn how social policy was permeated by an exterminatory eugenics.

But what was missing from all these biologically "degenerate" categories were the "racial parasites" denounced by Hitler in *Mein Kampf*. Leading Nazi doctors like Gerhard Wagner (the first Reich doctors' leader or *Reichsärzteführer*) attacked the sterilization law as being insufficiently

FIGURE 8.2 Propaganda slide entitled "The Jew a Bastard," illustrating different racial types, and characterizing Jews as a "bastard" race. (Courtesy of the U.S. Holocaust Memorial Museum.)

racial. After all, if you were diagnosed as a genetically predisposed alcoholic you could be sterilized, even if you were a member of the Nazi Party. Moreover, as the Nazis had scant respect for civil law or scientific theory, if you fell into one of the categories for sterilization, and were also a Jew, or of mixed race, you might be more likely to be sterilized. Thus 385 "mixed race" children were forcibly—and (even by German laws at the time) illegally—sterilized in 1937 for racial reasons. First, they were subjected to extensive evaluations from a psychological, anthropological, and genetic point of view. They would have been aged between thirteen and sixteen. These were cruelly referred to as the "Rhineland bastards," mixed-race children fathered by black French troops who occupied the Rhineland after World War I.

On 15 September 1935, the German government passed the Reich Citizenship Law, which effectively limited citizenship of Germany to only those of "German and related blood who through their behavior make it evident that they are willing and able faithfully to serve the German people and nation." Jews and other racial non-Germans were subjected to a range of discriminatory measures.

The Blood Protection Law, proclaimed on the same day, forbade all sexual relations between Germans and non-Germans, the difference being decided by citizenship. This effectively forbade marriages and sexual relations between Germans, and Jews and non-whites. At the same time, the Marriage Law was passed. This demanded hereditary health examinations prior to marriage. Nazi health propaganda encouraged people of good eugenic breeding stock to have at least three children. The idea was that health offices would register the birth of the unfit. Marriage certificates involved tests to make sure that no one married who was suffering from a sexually transmitted disease or carrying a genetic disease. These laws can be understood within a framework of sexual selection.

In June 1936, a Central Office to "Combat the Gypsy Nuisance" opened in Munich. This office became the headquarters of a national databank on so-called Gypsies. Robert Ritter, a medical anthropologist at the Reich Health Office, concluded that 90 percent of the Gypsies native to Germany were "of mixed blood." He described them as "the products of matings with the German criminal asocial sub-proletariat" and as "primitive" people "incapable of real social adaptation."

Hitler's biology

Biology fit the Nazi worldview and its economic and military priorities. Nazism represented an alternative form of modernity, and ideas of rebirth were central to this futurist vision.[35] Ploetz was among the early "racial hygienists" who spoke of recovering a primitive racial vigor. This would boost immunity to disease, national intelligence, and the quality of future generations. Figures like the right-wing medical publisher Julius Lehmann, when in the Landsberg prison in 1924, drew Hitler's attention to the eugenic handbook by the botanist Erwin Baur (1875–1933), the anthropologist Eugen Fischer (1874–1967), and the racial hygienist and human geneticist Fritz Lenz (1887–1976).[36] Heinrich Himmler's interest in plant and animal breeding, and the Nazi agriculturalist Richard Walter Darré (1895–1953), the ideologist of Blut und Boden [blood and soil], the ideology of racial descent rooted in a homeland, both prioritized racial eugenics. Lenz obligingly advised the SS on marriage suitability and racial fitness, and counseled fellow racial hygienists in 1930 that National Socialism offered the best hope for realizing eugenics.

Yet the fit between human genetics and National Socialism is less close than one might expect. Nazi leaders were not willing to defer to scientific experts. If we look at the biographical details of Lenz, we can see his promotion to the KWIA and as holder of a university professorial chair of racial hygiene in Berlin in 1933 as a reward for having aligned eugenics with Nazism. After 1941 the nationally oriented Lenz extricated himself from the Nazi web. He clashed with Himmler over the biological value of the illegitimate children, and increasingly kept aloof from Nazi racial measures once the Holocaust was under way. By contrast there was Otmar von Verschuer, KWIA director from 1941, the human geneticist who actively sought to promote Nazi racial agendas and collaborated with his assistant Mengele in obtaining body parts and blood samples from Auschwitz. The psychiatrist Ernst Rüdin and an early member of the German Society for Racial Hygiene, Paul Nitsche (1876–1948), became involved in Nazi euthanasia of the so-called incurably mentally ill or defective. Whether one should perform racial adjudications was a key issue. Certifying racial descent and fitness was lucrative for the expert, and could mean exclusion from civil society and persecution. A few racial hygienists, for example Wilhelm Weitz, professor of internal medicine, declined to carry out such certifications. We find then that racial hygienists were divided over whether to follow the ever more radical Nazi exterminatory agenda, or whether to attempt to insulate themselves from a destruction that would ultimately destroy its Nazi progenitors.

Mendelian genetics enjoyed a high status in Nazi Germany. The German ancestry and identity of botanist and monk Gregor Mendel were stressed. Although there were advocates of Lamarckism and non-Mendelian patterns of heredity (such as cytoplasmic inheritance), Mendelian biologists enjoyed considerable status. *Die Evolution der Organismen* [The evolution of organisms], published in 1943, developed the evolutionary synthesis in Germany.[37] Certainly mainstream eugenics was Mendelian, but there were enough non-Mendelian biologists oriented to environmental factors. Within the SS there were Mendelians and an antithetical group of environmentalists who stressed the Umwelt [environment] and "geo-political" factors. Others focus on "applied biology" in such areas as understanding the behavior of parasites such as lice.[38] Extermination was a major interest. We see how divergent and opposing groups of scientists and doctors aligned their research in various ways with Nazism.

Even so, Nazi race ideologists were impatient with academics, and, sensing this, many academics mobilized their researches and adapted their mental attitudes to Nazism. Himmler expressed contempt for academic scientists. The SS had its own phalanx of academic members, such as the anatomist August Hirt (1898–1945) at the nazified Reich University of Strasburg, where he assembled a Jewish skeleton collection from seventy-eight victims selected in Auschwitz. The SS was less sure of its control over medical faculties and universities. For example, Himmler was unable to find a university willing to award a higher doctorate to the air force doctor and human vivisector Sigmund Rascher (1909–45). The SS had its Ahnenerbe research organization, which sponsored archaeology and prehistory, as well as medical research, including experiments on human beings.

Human biology

The KWIA under Fischer, and from 1941 Verschuer, supported Nazi racial policy. Institute members adjudicated on questions of racial identity and sterilization. Verschuer cultivated connections to the Racial Political Office and gave energetic support to a nazified eugenics. He had two assistants in Auschwitz, Siegfried Liebau and Mengele. Here the connection between biological anthropology and the Holocaust is direct. A comparison with Lenz's situation suggests a lack of coercion and central state direction, and instead a degree of self-mobilization of the scientists who were willing to energetically advance National Socialist ends through biology—and enough were willing.

The psychiatrists Ernst Rüdin and Carl Schneider (1891–1946) supported the radicalization of euthanasia, as did those scientists who saw euthanasia as an opportunity to cut up brains to find an abnormal structure. Schneider, a professor of psychiatry, was not only an adjudicator for euthanasia, he also saw this as an opportunity for histopathological research. He wanted to determine the difference between inherited and acquired mental deficiency. Fifty-two children were examined, each for six weeks in the clinic. They were subjected to many forms of physical force and terror as part of psychological and psychiatric tests, including a painful X-ray of one of the brain ventricles. The children were also held under cold and warm water to test their reactions. Schneider intended to correlate the results with anatomical lesions. In the event, twenty-one of the children were killed deliberately, so as to compare the diagnosis, made when they were alive, with the post-mortem pathological evidence.

The Reich Health Office imposed severe measures against the Roma, or "Gypsies." Robert Ritter directed measures of registration and psychological evaluation of the Roma. He was supported in this work by psychologists and racial anthropologists. Their observations were followed by incarceration of Roma in concentration camps, notably Auschwitz. Here, they became victim to Mengele's research, and ultimately to the exterminatory agendas in Nazi biological racism.

Mengele

Dr. Dr. Josef Mengele stands as one of most celebrated examples of a genetic vivisector and racial exterminator. In 1935 he took a doctorate in anthropology in Munich before qualifying in medicine in 1938 at Frankfurt with an M.D. thesis on the genetics of cleft palate. At the Institute for Hereditary Biology and Racial Hygiene, he worked as assistant to Otmar von Verschuer, who was interested in the genetics of twins. He joined the Nazi Party in May 1938, the SS in September 1938, and the Waffen-SS (the true military wing of the SS) in July 1940. In November 1940, Mengele was assigned to the SS Race and Settlement Office dealing with ethnic German returnees. In June 1941, he joined a combat unit as a medical officer and received the Iron Cross. From January to July 1942 he was a member of the SS's international Viking Division, and was again decorated for his frontline service. After a further period with the Race and Settlement Office, and visits to Verschuer, he was sent to Auschwitz as a camp doctor in April 1943. He combined sanitary responsibilities—supervising the "Gypsy Camp," protecting the camp staff from infection—with undertaking racial selections of the newly arrived, sending to their deaths in the gas chambers those who were deemed "unfit" or simply not needed for forced labor at the time. Mengele's scientific research facilities were extensive. He used his position to carry out genocidal selections to find twins and other persons of medical interest, such as those with growth or other anomalies.

Mengele exemplified the scientific drive to produce outstanding data. Coercive human experimentation increased in terms of numbers of experiments as the war progressed, reaching a high point in 1944. By this time, most experiments were for racial and medical research, suggesting that scientists saw a unique opportunity for research not otherwise possible on human subjects. About 900 children endured Mengele's twin camp, and he scoured the victims being selected for the gas chambers or for forced labor for additional subjects. Most but not all of his subjects were twins; children announced they were twins in the hope of surviving. They came from throughout Central and Eastern Europe: Romania, Hungary, Czechoslovakia, and Poland. Most were Jewish, although some were Sinti (also a Gypsy community) and Roma who were killed when the Auschwitz "Gypsy Camp" was liquidated, and their bodies were then dissected.[39]

Verschuer obtained a grant from the German Research Fund (DFG) for research on hereditary pathology, focusing on blood proteins, and introduced Mengele to the KWIA. Mengele injected infective agents to compare their effects, and cross-injected spinal fluid. He would sometimes order the killing of a victim so that internal organs could be analyzed. He assisted in obtaining blood and body parts for Berlin colleagues who would use them for research. Under this DFG project, Mengele assisted in supplying the heterochromic eyes of a Sinto family to Karin Magnussen (1908–97), a geneticist and Nazi activist in Nachtsheim's Department of Hereditary Pathology in Berlin. Magnussen conducted serial research on iris structure in schoolchildren. When anomalies in the iris of the family of Otto Mechau from Oldenburg came to light, she examined the family members in August 1943 before their deportation to Auschwitz. She then assisted the SS anthropologist Liebau in Auschwitz, and made strenuous efforts to secure the Mechaus' eyes through Mengele. This grim example shows how a network of human geneticists stood behind a racial atrocity.

Aftermath

Biology was multifaceted in Nazi Germany and cannot be reduced to Darwinian natural selection. Along with racial extermination came plans to regenerate select racial groups (*Aufartung*), and a range of initiatives in Mendelian population screening. Destruction was hailed as regenerative, restoring primitive racial vigor. The "Generalplan Ost," Hitler's plan for a new empire in the east, showed how the conquest of Eastern Europe went hand in hand with plans for population and racial engineering; inadequate German stock was transferred back to Germany for reconditioning, and vigorous German populations were transferred to the east. The horticultural researcher Konrad Meyer took a key role in drafting the plan, a transfer of ideas of plant breeding to notions of strengthening the German race while eradicating "degenerates." This plan was central to the agenda of a racial war in the east as the Nazis planned for an eventual victory. Overall, the dimensions of destruction were quite staggering. Science was a factor in the vast system of population clearance and destruction, but had to be blended with devotion to Nazi values that are difficult to derive from evolutionary biology. The Anti-Defamation League has judiciously commented: "Hitler did not need Darwin to devise his heinous plan to exterminate the Jewish people and Darwin and evolutionary theory cannot explain Hitler's genocidal madness."[40]

What we have is not a consequence of Darwinism, but the confluence of two streams of ideas and ideology. The fusion was accomplished by a generation of capable but opportunistic and also fervently Nazi scientists. How this fusion occurred is a matter for historical reconstruction rather than the theoretical deduction of ethically inhumane consequences.

The question arises whether there was an adequate response by the international scientific community to the perpetrators of atrocities both during and after the war. The difficulty was that too radical a condemnation could affect all studies of heredity. One stance was again to condemn the racial atrocities as pseudo-science. The argument was also used by the psychiatrist C. P. Blacker to distance British eugenics from German racial measures.[41] Figures such as Lenz under Allied occupation and then Verschuer were reappointed in the postwar Federal Republic, although others such as Rüdin were held in punitive detention but never tried. Yet we also find Allied directives for the occupation forces to seize records concerned with race and genealogy, and to detain and interrogate involved scientists. Eugenics was a concern at the Nuremberg Trials, even though a planned trial of geneticists was never realized. Raphael Lemkin (1900–1959), who coined the term "genocide" to cover physical and cultural destruction, spoke of sterilization as being implemented by "professional genocidists."[42] The Cold War reduced the Allies' will to prosecute academics whose expertise might be usefully exploited by the occupying powers.

At the Nuremberg Trials, we find that there was a partial condemnation, and some safeguards against the worst abuses, not least the principle of informed consent in medical research. The onward dynamic in scientific research, and limitations on legal processes, meant that more was not possible. Important safeguards were placed in the form of the Genocide Declaration and the Declaration on Human Rights by the United

Nations. Other schemes included the Nuremberg Code requiring an enlightened (and later) informed consent, and the British writer Aldous Huxley, author of *Brave New World*, mooted the idea of a Hippocratic Oath for Scientists. UNESCO vacillated over the admissibility of "race" as a scientific concept. It is true that the biologists Julian Huxley and Ronald Fisher were prepared to turn a blind eye to the record of such scientists as Verschuer and Konrad Lorenz, and that a thorough evaluation of the culpability of biologists in the implementation of the Holocaust was never fully undertaken. A trial of geneticists, as proposed at Nuremberg in 1947, would have been difficult at the time of the Stalinist persecution of genetics in the Soviet Union, while the atom bomb created new possibilities for radiation genetics. Not only did genetics have a strategic niche, but a postwar trend among scientists was to seek resources while resisting accountability or responsibility, calling it state interference with liberty.

From the critique of racial science in the early twentieth century to post–World War II calls for protecting the rights of individuals and minorities, there are abundant indicators that scientists have long been faced with options regarding race and its political implications. Yet calls for race to be condemned as a dangerous myth were never as influential as they might have been expected to be. Instead, we find that biology under Nazism was being written off as pseudo-science. By this means, scientists distanced themselves from Nazi racial atrocities. Issues to do with the role of evolutionary biologists in their affiliations to Nazi Party and SS organizations, and in the waging of the racial war in the east, were never resolved. The idea of a non-eugenic human genetics underlined the break with an uncomfortable past. To understand the Nazi biological maelstrom, context and precise historical reconstruction remain crucial.

CHAPTER NINE

Darwinism, Marxism, and genetics in the Soviet Union
Nikolai Krementsov

In August 1948, the Soviet government banned teaching and research in genetics. Within a year, the doctrine of agronomist Trofim Lysenko—dubbed "Soviet Creative Darwinism"—replaced genetics in curricula and research plans of biological, medical, veterinary, and agricultural institutions not only in the Soviet Union but also in practically every country of the newborn "socialist camp." These events sent shock waves through the international biological community. Many Western observers became convinced that the banishment of genetics was somehow rooted in Soviet official ideology—Marxism: "Obviously, one of the reasons why the Communists destroyed genetics, if not the chief reason, was the fact that it was incompatible with the biological doctrines of Marx and Engels."[1] At the same time, they saw Lysenko's doctrine as "Marxist genetics."[2]

Ironically, the Western observers simply took for granted Lysenko's own version of the events. It was Lysenko himself who had declared that his doctrine was Marxist and Darwinist, while portraying genetics as anti-Marxist and anti-Darwinist. Western commentators even used the same quotations from the works of Karl Marx and Friedrich Engels as did Lysenko to prove the "unbreakable links" between Marxism and Soviet Creative Darwinism. Lysenko's opponents in the Soviet Union, however, claimed exactly the opposite: it was Lysenko's views that were incompatible with both Marxism and Darwinism. Soviet geneticists had also referred to Marx and Engels, though with a quite different goal—to prove that genetics, not Lysenko's doctrine, was the real example of "Marxist biology."

The "death" of Soviet genetics ignited a long debate about the interrelations among genetics, Darwinism, and Marxism. Some observers had suggested,[3] and historians later tried to prove, that in the Lysenko affair, "farming was the basic problem, not theoretical ideology,"[4] and that the Soviet genetics controversy bore no relation to Marxist philosophy.[5] Many Western scholars have since maintained that Marxism had nothing in common with Lysenkoism, though it had and has great significance for understanding genetic and evolutionary controversies.[6] Others have insisted that Marxism did play a crucial role in the rise of Lysenko and his doctrine.[7] Indeed, the latest reexamination of the 1948 events has claimed that Marxist ideology lay at the heart of Joseph Stalin's personal support of Lysenko.[8]

In this chapter I address this ongoing dispute from a somewhat different perspective. Instead of seeking to determine whether Lysenko's doctrine or Soviet genetics was truly "Marxist" and "Darwinist," I examine how various links among Darwinism, genetics, and Marxism were constructed, developed, and exploited by historical actors—evolutionists, geneticists, and Marxists—during the first decade of the Soviet régime. I argue that, as a result of massive campaigns to popularize both Marxism and Darwinism, coupled with a fierce debate over the inheritance of acquired characteristics initiated by Soviet geneticists in the 1920s, Darwinism was transformed from a specialized, often esoteric knowledge of the laws and principles of evolution, heredity, and variability into a public cultural resource—"Marxist-Darwinism"—readily available not only to professional biologists and professional Marxists, but also to any graduate of the Soviet educational system, from secondary schools to universities. The public nature of Marxist-Darwinism made its meanings highly dependent on particular institutional, intellectual, and ideological contexts, within which various interested parties deployed this cultural resource. Although the 1920s discussions of evolutionary questions in terms of Marxism represented only a small fraction of research and publications on evolution, heredity, and variability in the country, they set definitive parameters for subsequent debates on these issues in Soviet Russia, including the range of participants and the type of arguments used to defend/attack particular intellectual positions and institutional actions. Struggles for control over Marxist-Darwinism—competition over which among interest groups set the terms, forms, and norms of its meaning—characterized much of the subsequent development of biology in the Soviet Union. As the debates over "issues in genetics" from the 1930s through the 1960s demonstrate, the meanings of Marxist-Darwinism were created and re-created according to changes in its constituent components—both Marxism and Darwinism—and according to particular uses interested parties found for this cultural resource.

Darwinism and Marxism in Imperial Russia

Let me begin with Darwinism. In Russia as elsewhere in Europe, Charles Darwin's *Origin of Species* fueled a lively debate that spread far beyond the biological community, drawing in philosophers, theologians, political commentators and ideologues, sociologists, physicians, writers, and economists.[9] The polemics between proponents and critics of "Darwinism"—a convenient label covering all the various aspects of Darwin's concept of biological evolution[10]—unfolded in academic journals, novels, and popular magazines, and raged at scholarly conferences, proceedings of learned societies, and dinner parties in private homes.

Although the first Russian translation of the *Origin of Species* appeared only in January 1864,[11] more than seventy articles in Russian popular and professional periodicals had addressed various issues raised by Darwin's work during the four years since its appearance in England. In the 1870s, Darwin's original works, including *The Descent of Man*, *Variation of Animals and Plants*, *The Expression of Emotions*, and *The Voyage of the Beagle*, all appeared in Russian, translated and edited by such prominent biologists as physiologist Il'ia Sechenov, paleontologist Aleksandr Kovalevskii, and botanist Andrei Beketov.

A plant physiologist, Kliment Timiriazev, became one of the most active propagandists of Darwinism in Russia.[12] As early as 1864, the twenty-one-year-old student at Moscow University published a series of articles on Darwin's theory and its critics in Russia's leading monthly, which in the next year appeared as a book, entitled *A Short Outline of Darwin's Theory*. In 1877, during a trip to England, Timiriazev paid Darwin a visit and had a long conversation with his hero. Later, as a professor at the Moscow Agricultural Academy and Moscow University, Timiriazev covered Darwin's theory in his courses. In 1883, he expanded his "outline" into a volume of over 400 pages, entitled *Charles Darwin and His Theory*, which went through six editions and became a standard text on Darwinism in

Russian. In the early 1890s, Timiriazev delivered a series of public lectures on Darwinism, publishing several of them under the general title "Historical method in biology" in the influential monthly *Russian Thought*. Two of the four different editions of Darwin's collected works that appeared in Russia between 1896 and 1910 came out under Timiriazev's editorship. Timiriazev's untiring "struggle for Darwinism" even earned him a nickname "Darwin's Russian bulldog."

Many Russian biologists took an active part in the concurrent debates over the issues of heredity and variability in relation to the problems of biological evolution.[13] Every single concept of heredity and variability advanced during the late nineteenth and early twentieth centuries— Lamarckian "inheritance of acquired characteristics," Darwin's "pangenesis," Gregor Mendel's "hybridization laws," Francis Galton's "ancestral law," August Weismann's "germ plasm," Hugo de Vries's "*Mutationstheorie* [mutation theory]," and William Bateson's "genetics"—received ample attention from, and found proponents and critics among, Russian biologists.

The Russian biological community did not simply absorb the latest advancements of their Western colleagues but actively contributed to the ongoing debates, developing original concepts of heredity and evolution. In 1880, zoologist Karl Kessler suggested that mutual aid, not Darwinian struggle for existence, was the leading factor in biological evolution. The prominent theoretician of anarchism Prince Petr Kropotkin developed this idea further in a series of publications in the 1900s.[14] In 1899, botanist Sergei Korzhinskii proposed a concept of "heterogenesis," which in certain ways prefigured and influenced de Vries's "*Mutationstheorie*." A few years later, two botanists, Konstantin Merezhkovskii and Andrei Famintsyn, advanced a theory of "symbiogenesis," which placed symbiosis at the center of evolutionary processes.[15] In Russia as elsewhere in the world, the rise of experimental biology fostered both critiques of Darwin's evolutionary concept as a purely observational generalization and attempts to investigate evolutionary processes through rigorous experimentation.

By the turn of the twentieth century, the Russian biological community had recognized virtually unanimously the importance of Darwinism, even though many biologists expressed reservations about some of its components, particularly the concept of the struggle for existence.[16] By 1909, as the public celebrations of Darwin's centennial coupled with the fiftieth anniversary of the first publication of the *Origin of Species* demonstrated, many leaders of Russian biology considered themselves Darwinists. Russia's two Nobel Prize winners—zoologist Il'ia Mechnikov and physiologist

Ivan Pavlov—each contributed an article to a special jubilee volume, *In Memory of Darwin*, while Timiriazev supplied five different pieces.[17]

Meanwhile, what was happening with Marxism? In contrast to the rest of Europe, in Russia Karl Marx's *Das Kapital* won quick, if short-lived, recognition.[18] Indeed, Russian was the first foreign language into which the book was translated in the spring of 1872, and reviewers in Russian periodicals greeted it with much enthusiasm. As compared to the wide debate over Darwinism, which involved practically all segments of the Russian intelligentsia, the reaction to Marxism was limited mostly to economists, political commentators and ideologues, and sociologists. Furthermore, after the initial wave of acclaim, interest in Marxism in Russia waned.[19] With the advent of industrialization the second wave of interest rose towards the end of the nineteenth century, as many Russian intellectuals turned to Marxism in an attempt to understand the political and economic development of the country and to envision its future: the second Russian edition of *Das Kapital* appeared in 1898. The defeat of the first Russian revolution of 1905, however, led to widespread disillusionment with Marxism. At the time of Darwin's centennial, few Russian intellectuals would have called themselves Marxists, and most of those that did resided outside of Russia, in political exile in Western Europe.

The decades before the Bolshevik Revolution of 1917 saw very limited effort to establish connections between Darwinism and Marxism. In general, Russian biologists showed little interest in Marxism. Similarly, Russian Marxists—busy with political struggles and internal infighting—showed little interest in biology. Following the lead of the founders of Marxism—Karl Marx and Friedrich Engels themselves—Russian Marxists acknowledged the importance of Darwinism as a "materialist concept" of biological evolution. Although they criticized social Darwinism as an unjustified expansion of Darwin's views on biological evolution to human society, they generally abstained from entering the debate over the exact contents of Darwin's evolutionary concept or its relation to ideas about heredity and variability. Instead, the links between Marxism and Darwinism were created on a different plane. Following the lead of Engels, who in his eulogy at Marx's grave had equated Darwin and Marx as scientific geniuses each of whom had accomplished a revolution by bringing "historical method" to his field, Russia's leading Marxist, Georgii Plekhanov, stated:

> Darwin succeeded in solving the problem of the origins of plant and animal species in the struggle for survival. Marx succeeded in solving the problem of the

emergence of different types of social organization in the struggle of men for their existence. Logically, Marx's investigation begins precisely where Darwin's ends.... The spirit of research is the same in both thinkers.[20]

Similarly, in his analysis of Darwin's influence on the development of sociological thought, one of the founders of Russian sociology and Marx's personal acquaintance, Maksim Kovalevskii, drew parallels between Darwin's struggle for existence and Marx's class struggle.[21] Perhaps the only Russian Marxist who attempted to go beyond the simple parallelism of Darwinism and Marxism was Aleksandr Bogdanov. In his 1899 book, tellingly entitled *Major Elements of a Historical Outlook on Nature: Nature, Life, Psyche, Society*, Bogdanov proposed a "synthesis" of the two doctrines through the application of Darwin's main analytical categories of "selection" and "adaptation" to a Marxist analysis of social evolution, suggesting that selection and adaptation work the same way in nature and in society.[22]

At that time, however, none of these attempts to develop connections between Darwinism and Marxism attracted much attention, either among Darwinists or among Marxists. The situation changed dramatically after the Bolshevik Revolution, as a result of the largely independent but converging efforts by both Russian biologists and Russian Marxists.

Darwinism as Marxism

In October 1917, a radical faction of the Russian Social Democratic Labor Party, the Bolsheviks, carried out a coup d'état in Petrograd (renamed Leningrad in 1924) and declared the establishment of a socialist republic. In March 1918, the Bolsheviks concluded a peace treaty with Germany, ending Russia's participation in World War I. But within a few weeks, a civil war erupted, forcing the Bolshevik government to move the capital of the country to Moscow. The Bolsheviks' Red Army finally triumphed in 1921, and in 1922 the Union of Soviet Socialist Republics was established, engulfing almost all the territories of the former Russian Empire.

From the faction's very birth in 1903, the Bolsheviks claimed Marxism as the theoretical foundation for all their actions and policies. When the Bolsheviks took power, Marxism became official state ideology, and its propaganda became a main focus of the new state's educational efforts. State and party agencies, particularly the People's Commissariat of Enlightenment (Narkompros), waged a massive popularization campaign.

During the civil war, Narkompros had moved to reform existing curricula to include the teaching of Marxism in every educational institution, from primary schools to universities. Narkompros established a number of specialized institutions of "Communist learning"—Communist universities and "institutes of red professors" to prepare the cadres for disseminating Marxism.[23] Narkompros also sponsored the creation of specialized periodicals, such as *Communist Thought* and *The Communist's Companion*, as well as the publication of countless booklets, pamphlets, leaflets, and books popularizing Marxism, including separate editions, excerpts from, and compilations of its classic works—by Marx, Engels, and the leader of the Bolsheviks, Vladimir Illych Lenin. As part of this concerted effort, several works by Engels exploring the issues of biological evolution appeared in Russian translation for the first time, including "The Role of Labor in the Origin of Man from Ape" in 1922 and "The Dialectics of Nature" in 1925.[24]

It was within this huge educational/propagandistic campaign that the first systematic attempts to forge links between Darwinism and Marxism took place. In 1923, in a series entitled "Problems of Marxism," Bolshevik historian Moisei Ravich-Cherkasskii published a special volume under the title *Darwinism and Marxism*. In the preface, he explained the necessity of studying Darwinism for any "conscientious Marxist": "It is hardly possible to consider the learning of Marxism to be not only completed, but even generally serious and sufficiently deep, if one did not learn—at least on an elementary level—the historical and philosophical relation of Marxism to Darwinism."[25] The volume illustrated this "relation" with excerpts from the works of prominent Western Marxists and their critics, including Engels, Ludwig Buchner, Enrico Ferri, Karl Kautsky, and Ludwig Woltmann. The compiler saw his work as "a modest beginning that would serve as a stimulus to create something bigger, more elaborate for the [use of] wide masses of workers' intelligentsia."[26] It was indeed just the beginning. In 1925, Ravich-Cherkasskii published a second, much expanded edition of the volume. During intervening and subsequent years, a number of publications with similar titles appeared in Russia.[27]

Many of these publications continued the earlier emphasis on parallels between Darwinism and Marxism. As Leon Trotsky, one of the leaders of the Bolsheviks, reiterated in 1923, "in relation to social phenomena, Marxism occupies the same position as Darwinism does in relation to the plant and animal worlds."[28] Certain propagandists, however, moved beyond simple parallels aiming to incorporate Darwinism into Marxism.

They began to hail Marxism as a comprehensive "materialistic worldview" (dialectical materialism) that explained not only economic and political developments, but any and every kind of development, including biological evolution. Following the lead of Marx and Engels, who had portrayed Darwin's works as the "materialistic explanation" of biological evolution, they began to present Darwinism as an important component of Marxism itself. As a result, Darwinism was included in the courses on the history of philosophy and on Marxism at various educational institutions and became an important part of the huge antireligious campaign mounted by the Bolsheviks in the early years of their rule.[29]

The campaign for the dissemination of Marxism among the Russian population targeted the Russian scientific community in particular. A special place among the institutions devoted to the propaganda of Marxism was occupied by the Socialist Academy, organized in 1918 by several prominent Bolsheviks and renamed the Communist Academy in 1923.[30] The Communist Academy had initially focused its activities on social sciences and the humanities, but by the mid-1920s, it included natural sciences under its purview. In 1924, the academy created a "section of natural sciences," which fostered the emergence of numerous Marxist societies such as those for "Marxist biologists," "Marxist mathematicians," "Marxist statisticians," and the like. The academy also founded two periodicals (*Herald of the Communist Academy* and *Under the Banner of Marxism*) and undertook the compilation of the first *Great Soviet Encyclopedia*.

The Communist Academy became the main instrument for the introduction of Marxism to the Russian scientific community. In 1922, in the first issue of its mouthpiece, *Under the Banner of Marxism*, the editorial board called upon Marxists to "unfold the banner of militant materialism,"[31] and to launch a broad attack on "idealism" in scientific research. The journal's third issue carried an article entitled "On the Significance of Militant Materialism," written by none other than Lenin himself. Lenin charged scientists with the task of "the struggle against the pressure of bourgeois ideas and the bourgeois worldview." He insisted that every "scientist must be an up-to-date materialist, a deliberate follower of the materialism presented by Marx, that is, he must be a dialectical materialist."[32] Lenin's article became the manifesto of Communist scholars, and some of them took upon themselves the task of rooting out "idealistic" conceptions of biological evolution. In June 1925 a resolution of the Central Committee of the Communist Party recorded: "The infusion of dialectical materialism into entirely new fields (biology, psychology, natural sciences in general) has begun."[33]

This early campaign for the "infusion of dialectical materialism" into biology was carried out primarily by professional Marxists, with one notable exception—Kliment Timiriazev. Timiriazev was the only biologist of note who had accepted the Bolsheviks' coup d'etat without reservation. As early as summer 1918, he joined the Socialist Academy and became a member of the Narkompros Main Scientific Council—the country's leading agency responsible for science policy. In mid 1919, in the journal *Proletarian Culture*, Timiriazev published an article, entitled "Ch. Darwin and K. Marx," introducing a series of links between Darwinism and Marxism that would become a model for many of the subsequent treatments of the issue.[34] Timiriazev was the first to note that Darwin's *Origin of Species* and Marx's *Zur Kritik der politischen Oekonomie* [A contribution to the critique of political economy] appeared in the same year, 1859, and to maintain that "it was more than a simple chronological coincidence." According to Timiriazev, both Darwin and Marx based their theories not on "theology and metaphysics," but on science: on "scientifically explored" and "material" phenomena. Darwin studied "economy of plants and animals" and Marx "economy of human societies." Both explained "evolution"—Darwin biological and Marx social—as a result of "material," "economic" conditions and factors: "Both of them marched under the banner of natural sciences." A year later, preparing a new edition of his 1890s lectures on "historical method in biology," Timiriazev peppered the text with references to Marx and Engels and rewrote the last lecture, expanding on "parallels" between Darwinism and Marxism.[35] He performed the same procedure with a new version of his textbook *Charles Darwin and His Theory*.[36]

Timiriazev died in the spring of 1920, but his works on Darwinism and Marxism lived on. The article "Darwin and Marx" was included in a volume of Timiriazev's collected essays issued the next year, and it appeared in virtually every collection on "Darwinism and Marxism," such as the one prepared by Ravich-Cherkasskii. Timiriazev's *Charles Darwin and His Theory,* issued by the state publishing house in 1921, was reprinted five times in ten years, becoming the standard Soviet text on Darwinism.

Soviet genetics between Darwinism and Marxism: the polemics over Lamarckism

Most Russian biologists greeted the Bolshevik Revolution with distrust, suspicion, and open hostility. Very soon, they developed a functioning symbiosis with the new government, forging good working relations with various agents and agencies of the Bolshevik state.[37] Biologists quickly capitalized on the Bolsheviks' science policy, aimed at the revival and expansion of the Russian science system to increase greatly their institutional base, tapping practically every state agency involved with science, most importantly Narkompros, the People's Commissariat of Public Health (Narkomzdrav), and the People's Commissariat of Agriculture (Narkomzem).

As part of this remarkable institutional growth, a new discipline of genetics emerged in Soviet Russia. Responsibility for the building of the discipline belongs to Iurii Filipchenko, Nikolai Kol'tsov, and Nikolai Vavilov, who mobilized the resources of Narkomzdrav, Narkompros, and Narkomzem to organize research institutes and laboratories, convene conferences, create periodicals, publish textbooks, and train a new professional generation. To legitimize their discipline-building efforts, Russian geneticists conducted a large publicity campaign, popularizing both the actual achievements of genetics in uncovering mechanisms of heredity, including the "laws" of Gregor Mendel and T. H. Morgan's "chromosomal theory," and the promises their discipline held for "mastering control over the nature" of plants, animals, and humans. An important part of this legitimization campaign was geneticists' "contribution to Darwinism."

Across the world, the first two decades of the twentieth century witnessed wide-ranging debates over the validity of Darwin's views on evolution and the proliferation of "alternative" concepts, which historians have described as the "eclipse of Darwinism."[38] The debates largely revolved around two main issues inherent in Darwin's notion of natural selection. Many biologists felt that natural selection as a stochastic process could not explain the apparent regularities and parallelisms of evolutionary trajectories in various taxa. At the same time, many biologists thought that natural selection operated merely as a "sieve"—an eliminating procedure—that by itself could not create either evolutionary novelties or adaptations. The critics proposed a range of alternative concepts to correct the perceived deficiencies of Darwin's natural selection, advocating instead a variety of alternative evolutionary mechanisms. These concepts could be grouped into four major clusters: orthogenesis, that is evolution on the basis of regularities; neo-Lamarckism, that is evolution on the basis of the inheritance of acquired characteristics; mutationism, that is evolution on the basis of mutations; and isolationism, that is evolution on the basis of isolation.

The Russian biological community became actively engaged in these debates, and all four major alternative concepts found adherents and critics among biologists, with geneticists taking a very active part.[39] In 1920, Vavilov published his essay on "the law of homologous series in hereditary variation," which demonstrated that the evolution of plants exhibited certain regularities, implying, though not clearly articulating, a kind of orthogenetic evolutionary change.[40] In 1921, Filipchenko published an expanded version of his 1915 brochure on *Variability and Evolution*, which advocated a mutationist view of evolution combined with orthogenesis.[41] Two years later, he produced a lengthy analysis of "the evolutionary idea in biology," examining various evolutionary concepts and again siding with orthogenesis and mutationism.[42] In 1922, Kol'tsov published an article on the "formation of new species and the number of chromosomes," which also advanced a mutationist concept of speciation.[43]

As part of their legitimization efforts, geneticists also launched a concerted critique of the concept of the inheritance of acquired characteristics or "neo-Lamarckism."[44] The initial spark for the attack likely came from Russia's leading physiologist, Ivan Pavlov. Like many other students of animal behavior early in the twentieth century, Pavlov had endorsed the idea of the evolutionary transformation of acquired behavioral features into hereditary ones. In his report to the 1913 international physiological congress, he suggested that "some conditioned, newly formed reflexes later became transformed into unconditioned [ones]."[45] During 1921–3 one of Pavlov's assistants, Nikolai Studentsov, conducted a series of experiments that seemingly demonstrated such transformation. Studentsov trained a mouse to run to a feeding rack only after a bell had sounded. The formation of this conditioned reflex in the first mouse proved very difficult—Studentsov had to repeat the experiment 298 times before the reflex was established. He then studied the formation of the same reflex in the mouse's offspring. The first generation established the reflex much faster than their parents (after 114 repetitions), the second generation even faster (29 repetitions), the third faster still (11 repetitions), and the fourth almost immediately (6 repetitions). Studentsov concluded that the conditioned reflex of the first generation had become unconditioned by the fourth generation—the acquired reflex was now transmitted through heredity from parents to offspring. In the summer of 1923, on a lecture tour abroad, Pavlov enthusiastically informed his British and American audiences that "the latest experiments (which are not yet finished) show that the conditioned reflexes, i.e., the highest nervous activity, are inherited," which was promptly reported in the pages of *Science*.[46] Studentsov's results appeared in the first 1924 issue of the *Russian Physiological Journal* edited by Pavlov. Furthermore, in April 1924, Pavlov himself delivered a public lecture in Leningrad on "the newest research" conducted in his laboratories, devoting a considerable part of his report to Studentsov's experiments, which was highlighted in the country-wide press coverage of the event.[47]

Apparently, Pavlov's endorsement of Lamarckian inheritance greatly disturbed Russian geneticists; the opinions of Russia's only living Nobelist carried considerable weight among the Bolshevik leadership and could thwart geneticists' efforts to legitimize their discipline.[48] As Kol'tsov noted, "if the cited article [in *Science*] did not bear the signature of Pavlov, we would have simply ignored it."[49] But it did, and Kol'tsov paid Pavlov a visit to convince him that Studentsov's experiments were flawed. It was actually not the mice, but rather Studentsov, Kol'tsov suggested, whose performance had improved over time. Kol'tsov proposed to repeat the experiments in such a way as to exclude any possible influence of the experimenter on the process of reflex formation. Reportedly, Pavlov agreed with Kol'tsov's objections and assigned another assistant to construct a special apparatus for this purpose.[50]

Pavlov, however, did not publicly retract his earlier statements in support of Lamarckism. Perhaps to ensure that the public (and the patrons) were informed of the incident, in mid-1924, Kol'tsov published an article entitled "The newest attempts to prove the inheritance of acquired characteristics," which advanced his arguments against Studentsov's experiments in particular and Lamarckian inheritance in general.[51] A few months later, Filipchenko joined the battle by publishing a booklet entitled *Are Acquired Characteristics Inheritable?* The booklet included a Russian translation of an article by the leading American geneticist T. H. Morgan (who had attended Pavlov's lecture in Boston) and an article by Filipchenko himself.[52] Both articles answered the question posed in the title of the booklet in the negative.

Although Russian geneticists were adamant in rejecting the inheritance of acquired characteristics, several biologists defended Lamarckism. They launched a counterattack on genetics, which they saw merely as a modern version of preformationism, presenting their own views as a kind of epigenesis.[53] They regularly reviewed and abstracted works by Western proponents of Lamarckism, including the notorious experiments of the Austrian biologist Paul Kammerer on the inheritance of the acquired coloration patterns in salamanders.[54] In 1924, three students at Moscow University, zoologists Evgenii Smirnov, Boris Kuzin, and Iulii Vermel', published a 200-page volume of "essays on evolutionary theory."[55] Dedicated to Kammerer, the volume sharply criticized geneticists' views on the role of mutations in the evolutionary process. The critics focused in particular on the notion of the gene. Defending the leading role of the environment in inducing evolutionary changes, they could not accept geneticists' postulates of the "immutability" of the gene and the "accidental"—independent from environmental influences—nature of mutations. In 1925, Kammerer's voluminous treatise on "general biology" appeared in Russian translation with a sympathetic (though anonymous) foreword.[56] The same year, a prominent botanist, Vladimir Komarov, published a laudatory biography of the founder of the concept of inheritance of acquired characteristics—Jean-Baptiste Lamarck.[57] In 1927, Smirnov published a lengthy survey of "the problem of the inheritance of acquired characteristics," citing Kammerer's and Pavlov's works.[58] The same year, two different publishers almost simultaneously released the Russian translations of Kammerer's book, *The Enigma of Heredity*.[59]

Geneticists responded to the growing popularity of Lamarckism by producing highly critical accounts of their opponents' research and publications. Thus a student of Kol'tsov, Aleksandr Serebrovskii, offered a detailed critique of major publications by Russian Lamarckists, while a student of Filipchenko, Theodosius Dobzhansky, provided a lengthy analysis of the concept of the inheritance of acquired characteristics.[60]

The heated atmosphere of debate over Lamarckism also apparently stimulated Russian geneticists' search for novel ways of reconciling their own views on heredity and variability with Darwin's views on evolution. At the height of the polemic, in 1926, the head of genetics at Kol'tsov's Institute of Experimental Biology, entomologist Sergei Chetverikov, published an article on "certain moments of evolutionary process from the viewpoint of modern genetics."[61] Chetverikov addressed the central subject of Darwin's theory—the origin of species—and laid a solid foundation for a synthesis of genetics and Darwinism. His article shifted the focus of evolutionary thinking from individual to population as the basic unit of evolutionary change. Chetverikov developed an extensive research program on the genetics of natural populations and the role of various factors (such as natural selection, isolation, and mutations) in speciation, which became the major preoccupation for a large group of Moscow geneticists. At the time, however, Chetverikov's concept did not engage the polemic between geneticists and Lamarckists.

Although the debates over "alternative" evolutionary concepts had originated within the biological community, professional Marxists soon joined the fray.[62] In 1923, preparing the publication of a volume of Plekhanov's collected works, prominent Marxist historian Dmitrii Riazanov critically commented on Plekhanov's remarks that were supportive of de Vries' *Mutationstheorie*.[63] Another Marxist, Vladimir Sarab'ianov, disagreed with Riazanov, and, in a long essay entitled "The looming question," which appeared in the *Communist's Companion*, he came to defend not only mutationism, but also orthogenesis and Lamarckism.[64] Several other Marxists soon published arguments for and against mutationism in its relation to Darwinism in *Under the Banner of Marxism*.[65] Some Marxists criticized Darwinism for its neglect of "revolutions," embodied, for them, in sudden "mutations," while others attacked mutationism for its neglect of environmental influences on evolutionary processes.

Geneticists' scathing critique of Lamarckism sparked a sharp polemic among Marxists affiliated with various new "communist" institutions. The Sverdlov Communist University, the Timiriazev Biological Institute (Figure 9.1; this was renamed in 1924 the Timiriazev Scientific Research Institute for the Study and Propaganda of the Scientific Foundations of Dialectical Materialism), and the Communist Academy constituted the stronghold of Lamarckism. Two Marxist societies organized in 1924 under the auspices of the Moscow University medical school, the Circle of Materialist Physicians and Leninism in Medicine, also lent support.[66] Starting in 1925, the Communist Academy section of natural sciences regularly held meetings to discuss the interrelations among Darwinism, Marxism, Lamarckism, and genetics. At one such meeting in late November 1925, a professor of biology at the Sverdlov Communist University, embryologist Boris Zavadovskii, delivered a long report on "Darwinism and Marxism," which advocated a "synthesis" of Lamarckism and genetics.[67] In early 1926, the academy leadership even planned a special laboratory to prove the existence of the inheritance of acquired characteristics and invited Paul Kammerer to head it. Kammerer had accepted the invitation, but on the eve of his departure to Russia, faced with accusations of scientific fraud, he committed suicide, and the plan fell through.[68]

Recognizing the threat, several geneticists with Marxist sympathies, notably Aleksandr Serebrovskii, joined the Communist Academy to lead

FIGURE 9.1 The reading room of the library at the Timiriazev Institute (ca. 1931). The slogan at the top reads: "The front of the struggle against religion"; the banner behind Lenin's bust reads: "Materialism [is] our weapon." (Courtesy of Nikolai Krementsov.)

the criticism of Lamarckism from the inside. On 12 January 1926, Serebrovskii delivered a long talk on "Morgan's and Mendel's theory of heredity and Marxists."[69] His talk advanced the notion that modern genetics represented a "truly materialist," and hence Marxist, view of heredity and variability. In November, Serebrovskii also took an active part in the organization under the academy's auspices of the Society of Marxist Biologists. In subsequent years, geneticists and their critics regularly debated the interrelations among genetics, Lamarckism, Darwinism, and Marxism within the academy, with proceedings appearing on the pages of the academy's oracles.[70]

The polemics over Lamarckism raged from 1924 to 1930. By all counts, geneticists and their Marxist supporters won the debate. In May 1927, an active member of the Communist Academy, Maks Levin, published a critical review of Evgenii Smirnov's latest monograph on the inheritance of acquired characteristics in the Communist Party's mouthpiece, *Pravda*. Included in the review was an excerpt from Pavlov's letter to a colleague dated 1 March 1927, which stated that experiments using the new apparatus did not confirm the prior suggestion about the hereditary transmission of conditioned reflexes.[71] Geneticists succeeded both in publicizing Pavlov's retraction from Lamarckism and in enlisting his authority to support their own views.

The same year, geneticists also countered the major objection of their opponents—the "immutability" of the gene and the "accidental" character of mutations. A student of T. H. Morgan, Herman J. Muller, discovered that mutations could be induced by X-ray irradiation, and thus demonstrated that the gene could be altered by an environmental factor. At the end of July 1927, Muller published his finding in a short article in *Science*.[72] In early September, Serebrovskii published an enthusiastic account of Muller's discovery on the pages of *Pravda* under the title "Four pages that shook the scientific world," literally equating the import of Muller's discovery for genetics to that of the Bolshevik Revolution for the world.[73]

Serebrovskii's untiring crusade against Lamarckism within the Communist Academy "converted" to genetics a number of young Marxists, including Solomon Levit, the founder of the Circle of Materialist Physicians, Vasilii Slepkov, an active proponent of "socialist eugenics," and the rising science administrator Isaak Agol, all of whom had previously subscribed to Lamarckism. All three became ardent supporters and students of genetics. Each of the three converts published extensive critiques of their former comrades "from the viewpoint of dialectical materialism."[74] As a result, the supporters of Lamarckism became marginalized and lost their institutional strongholds, most importantly the Timiriazev Institute (Agol was appointed the director in early 1930) and the Communist Academy, which came to be dominated by geneticists and their Marxist allies.

Darwinism as a cultural resource

The results of the 1920s debates over the inheritance of acquired characteristics transcended the geneticists' victory. In the course of the debates, Darwinism emerged as a powerful cultural resource employed by every interested party to its own advantage. What made Darwinism so influential in various institutional and intellectual struggles of the 1920s was its "hybrid" nature as a biological and a philosophical/ideological doctrine.[75] On the one hand, Darwinism was hailed as the most important theoretical conception in biology, which provided a scientific foundation for *all* biological research. On the other hand, Darwinism was lauded as an important part of Marxism—a "scientific foundation of dialectical materialism," as the full name of the Timiriazev Institute indicated.

This hybrid nature gave much flexibility to the actual contents of polemical and theoretical constructions covered by the label "Darwinism." On one hand, adherents of every competing evolutionary doctrine—be it mutationism, isolationism, orthogenesis, or Lamarckism—claimed that their own views represented "true Darwinism," while those advocated by their critics and opponents exemplified "anti-Darwinism." On the other hand, during the 1920s, the label "Marxism" covered a number of often contradicting statements, declarations, and assertions produced by various individuals and groups, each of which claimed to advance dialectical materialism as originated by Marx, Engels, and Lenin. The fierce polemics over the contents of "Marxism" that raged within the community of Soviet Marxists during the 1920s added much volatility to attempts to construct links between "Darwinism" and "Marxism."

This elasticity of both "Darwinism" and "Marxism" allowed a group of Marxists specializing in "dialectics of nature" to assert their own authority over Darwinism. This group took as its task "to develop the dialectical-materialist methodology" for biology.[76] In the course of the debates over "Darwinism and Marxism," this group introduced a new Marxist lexicon and a new polemical style into discussions of evolutionary issues. Like similar polemics in other fields of knowledge, the debates quickly moved from issues of validity of certain scientific concepts, such as Lamarckism and genetics, to issues regarding which of the competing concepts (and its advocates) was more Marxist—and hence more Darwinist (or vice versa)—than the other.

Within this new lexicon, "Marxism" served as a synonym for monism, materialism, and dialectics, while "Darwinism" became equated with materialism and hence "Marxism." At the same time, "Darwinism" functioned as an antonym to vitalism, spiritualism, theology, metaphysics, idealism, teleology, dualism, and transcendentalism in biology. In the new Marxist vocabulary, "Darwinism" was progressive, while all of its antonyms were characterized as reactionary. As this string of pejorative terms indicates, the actual scientific content of competing biological concepts became subordinate to their perceived philosophical/ideological foundations. As Agol stated in his monograph eloquently entitled *Dialectical Method and Evolutionary Theory*, "Darwinism is an intrinsic materialistic dialectic of biology.... Darwinism as a method for studying the animal and plant kingdoms is one of the areas of the application of the dialectical method."[77]

Emulating the style of inner-party struggles against various "oppositionists," many Marxist polemicists adopted a militant tone and endorsed orthodoxy in scientific debates. Marxist writers were intolerant of criticism of their own pronouncements, even if such criticism was of a purely scientific nature; it was immediately countered by calling its author anti-materialist, anti-Darwinist, or anti-Marxist. One scholar remarked that if scientists were to follow this practice, "it would be logical to label a mathematician who criticized some mistakes and shortcomings in another mathematician's work as an anti-mathematician, and to label a critic of poetry an anti-poet."[78] Labeling opponents and name-calling became a characteristic feature of the Marxist polemical style.

In many polemical writings, reference to "sacral texts" produced by the founders of Marxism—Marx, Engels, and Lenin—replaced references to results of scientific observations and experiments. Professional Marxists with no biological training saw fit to enter the debates and pass judgments on questions of biological evolution with no other reference than Engels' "Dialectics of Nature." For instance, in 1926, the director of the Marx–Engels Institute, Abram Deborin, published a series of articles in *Under the Banner of Marxism* on "Engels and dialectics in biology."[79] Along with Marx, Engels, and Lenin, a number of Marxist polemicists began to hail Timiriazev as a "founding father of Darwinism," using his statements to support their own, often diverging views on evolution, variability, and heredity.[80] Thus debates began to revolve not over scientific facts, hypotheses, and conceptions per se but over which of the debated concepts corresponded more closely to the pronouncements of the "founding fathers" of both Marxism and Darwinism.

Just as the two parallel systems of scientific institutions—academic and communist—coexisted in the 1920s, so too did the traditional scientific and Marxist treatments of evolutionary issues coexist in different settings; the former appeared mostly in professional periodicals, the latter in party publications and the popular press. Toward the end of the decade, polemics spilled onto the public arena.

In 1927, the head of Narkompros, Anatolii Lunacharskii—the official patron of all "pure" science in the country and an active member of the Communist Academy—wrote a film script based loosely on Paul Kammerer's life story. The movie, entitled *Salamander*, hit the screens all over the country in late 1928. It portrayed Kammerer as a progressive scientist, whose research on the inheritance of acquired characteristics in salamanders (Figure 9.2) earned him the hatred of the clergy and the bourgeoisie.

FIGURE 9.2 Students at the Timiriazev Institute working with salamanders. (Courtesy of Nikolai Krementsov.)

His opponents (who apparently supported the views of modern genetics on the impossibility of Lamarckian inheritance) plot to destroy his reputation by tampering with his research. Driven out of his laboratory and haunted by lack of money, the scientist is saved from poverty and disgrace (and his salamanders from inevitable death) by the invitation of the Soviet government to come to Moscow to continue his research. The official press greeted the movie with enthusiasm, while geneticists bombarded Lunacharskii with critical letters and offers of a remedial course in modern genetics.

In contrast to the "communist" and popular press, academic publications maintained neutrality in the Marxist disputes over evolutionary issues. During the 1920s, the majority of Soviet biologists stayed clear of Marxist debates. At the three congresses of Russian zoologists, anatomists, and histologists (1922, 1925, and 1927), the links between Marxism and Darwinism were completely absent from the discussions, even though the questions of evolution, heredity, and variability occupied almost half of their proceedings. Similarly, in a jubilee account of the development of Darwinism during the first decade of the Soviet régime written for the tenth anniversary of the Bolshevik Revolution, Russia's leading evolutionary morphologist Aleksei Severtsov did not even mention Marxism.[81] Furthermore, the first Soviet edition of Darwin's collected works issued in 1928–9 under the editorship of Russia's foremost ornithologist, Mikhail Menzbir, contained not a single reference to the classics of Marxism.[82] Most academic publications on evolution and genetics remained true to their professional standards and maintained the traditional style of scientific criticism, discussing evidence, logic, and the novelty of research methods and theoretical constructions, while ignoring any connection of the debated questions with Marxism. For instance, in his introductory essays to Darwin's collected works, Menzbir offered a detailed analysis of geneticists' views on and contributions to evolutionary issues. Although siding with geneticists in their rejection of Lamarckism, Menzbir criticized them for "locking themselves up in the laboratory," neglecting "the creative role of natural selection in evolution," and "ignoring much of zoological material" in their discussions of evolutionary processes. But he never once referred to Marxism.[83] Some biologists even derided attempts to infuse Marxism in their science, though not publicly. For instance, during a speech on "the tasks of Marxists in the field of natural sciences" in April 1929, the chairman of the Communist Academy section of natural sciences received from the audience an unsigned note in verse:

"Catholics" now have new tasks—
To take over the natural science,
To convert botany and zoology
Into Marxist-theology.[84]

One can see a clear generational divide in the involvement of Soviet biologists in the discussions over Marxism and Darwinism: while older scientists largely abstained from exercising Marxist lexicon and adopting "communist" polemical culture, many of their younger colleagues eagerly entered the debates. It is indicative that no one among the first generation of geneticists, including Kol'tsov, Filipchenko, and Vavilov, utilized the Marxist lexicon or took part in the debates over Darwinism and Marxism. In contrast, many of their students, including Serebrovskii, Boris and Mikhail Zavadovskii, Agol, and Levit became actively engaged in Marxist polemics. For some biologists, such as Serebrovskii, Levit, and Agol, dialectical materialism perhaps indeed became a genuine source of inspiration and influenced their research programs and scientific writings.[85] But at the same time, for them as for many others, Marxism provided a convenient rhetorical cover in distinguishing their personal research interests, institutionalizing their own approaches, or simply advancing their own careers within the Soviet science system. The younger generation of Soviet biologists struggled to establish and maintain their own institutions and careers when available institutional niches within the academic science system had already been occupied by their older colleagues. Many of the younger biologists looked for support within the system of "communist" science or in the institutions altogether removed from academia. Thus Boris Zavadovskii, an active participant in the debates over Marxism and Darwinism, established his laboratory within the Sverdlov Communist University, while his brother Mikhail created a large research department at the Moscow zoo. References to the philosophical/ideological value of their work may have helped younger biologists to bolster their appeals for state support by displaying their loyalty to their employer—the Bolshevik state.

Yet the 1920s debates over Darwinism and Marxism produced more than convenient rhetoric. The fierce polemics over Lamarckism, coupled with the active propaganda of both Darwinism and Marxism during the 1920s, transformed specialized, often esoteric knowledge about the laws and principles of evolution, heredity, and variability into a *public* cultural resource—"Marxist-Darwinism"—readily available not only to professional biologists and professional Marxists, but also to any graduate of the Soviet educational system, most importantly to state and party officials at all levels. The inclusion of Darwinism as a part of Marxism into curricula of educational institutions—from secondary schools to the courses on Marxism taught at the Communist Academy and communist universities—helped to disseminate a simplified, stripped-down knowledge of evolution, heredity, and variability. Similarly, the inclusion of Marxist evaluations of Darwinism into biology curricula helped to spread the particular Marxist lexicon and the militant polemical style that emerged during the 1920s debates over the Lamarckian inheritance of acquired characteristics.

"Soviet creative Darwinism"

In early 1930, it seemed that the geneticists' victory over their Lamarckian opponents established firmly their control over the meanings of "Marxist-Darwinism" by asserting the notion of genetics as both a "truly Marxist" and a "truly Darwinist" concept of heredity and variability, while marginalizing the concept of the inheritance of acquired characteristics as both anti-Marxist and anti-Darwinist. By the end of the year, this control had slipped away from geneticists' hands in a renewed attack on their discipline and its practitioners as a result of a drastic change in the meanings of "Marxism." This change was spurred by the vicious campaign against "mechanisticism and menshevizing idealism," which Stalin and his lieutenants identified respectively as a "left" and a "right" deviation from an orthodox version of Marxism-Leninism.[86]

The new attack on genetics reflected dramatic changes in the institutional, economic, ideological, and political landscapes of Soviet Russia induced by the "Great Break" of 1928–30. These changes included the collectivization of the peasantry, crash industrialization, the introduction of planned economy, the establishment of strict controls by the Bolshevik Party's central organs and Stalin personally over all facets of life in the country. These changes resulted in the rapid "Stalinization" of Soviet science and a complete reorganization of the system of higher education. The arguments advanced against genetics by its various critics at this time involved not so much appeals to Lamarckism as to the newly introduced norms of *partiinost'* (literally, party-ness, which actually meant blind obedience to the ever-changing party line), *practicality*, and *patriotism*.[87]

The majority of critics came from a new generation (born around 1900) of "Marxist biologists" who had graduated from the Soviet educational system, particularly the institutions of communist learning. This generation had imbibed Marxist lexicon and had mastered Marxist polemical style during their student years. In the late 1920s, they eagerly answered Stalin's famous call for "a mass attack of the revolutionary youth on science" in order "to replace the old guard."[88] This generation became the major force in a massive campaign "to mobilize science for the construction of socialism" inaugurated by the Great Break. In the field of biological sciences, particularly active were embryologist Boris Tokin in Moscow and professional Marxist Isaak Prezent in Leningrad. Both rose to prominence and seized important administrative positions in the wake of the campaign "against mechanicists and menshevizing idealists." In 1931, Tokin became the chairman of the Society of Marxist Biologists and replaced Agol as the director of the Timiriazev Institute. The same year, Prezent came to head the society's chapter in Leningrad, as well as the section of biology within the newly created Institute of Natural Sciences under the Communist Academy's Leningrad Branch. Prezent also established the first teaching department of "dialectics of nature and the general theory of biology" at Leningrad University. In the subsequent years, both of them put considerable efforts in defining the contents of Marxist-Darwinism. In 1934, Tokin and M. I. Aizupet produced a collection of quotations from the classics of Marxism, entitled *K. Marx, F. Engels, V. Lenin on Biology*,[89] while Prezent compiled an over-500-page *Reader on Evolutionary Theory*, nearly one half of which was taken up with excerpts from works by the same authors.[90]

The rapid expansion of the science system during the late 1920s to early 1930s accorded Soviet scientists huge amounts of material and structural resources, which of course spurred fierce competition among various disciplinary groups, age cohorts, and individuals. According to the new party policies, the access to these resources depended not so much on the applicant's scientific qualifications as on his/her loyalty to the party line, demonstrated by the applicant's "class origins," mastery of Marxist lexicon, and readiness to fulfill directives of the Bolshevik Party and its leaders. The heavy pressure on the scientific community "to mobilize science to serve socialist construction" prodded Soviet scientists to adapt their professional culture to the new situation. They developed numerous rituals and adopted the now obligatory rhetoric of *partiinost'*, practicality, and patriotism to demonstrate their loyalty to the régime. To give but one example, in contrast to the three previous meetings, the 1930 congress of Soviet zoologists, anatomists, and histologists formulated its main subjects as "tasks of zoology in relation to the [first] five-year plan of the development of the national economy" and "dialectical materialism as the foundation of scientific research."[91] The congress, for the first time, "elected" the Politburo of the Bolshevik Party as its "honorary presidium," formed a "communist section," and heard several reports on Darwinism and Marxism.

These new contexts dramatically influenced the contents of Marxist-Darwinism and fueled competition for control over this powerful cultural resource, aptly manifested during the country-wide celebration of the fiftieth anniversary of Darwin's death in April 1932. In the words of the Communist Academy leadership, the jubilee was to become "a wide political campaign,"[92] and it did. The campaign spanned hundreds of meetings, lectures, exhibitions, radiobroadcasts, and publications in professional and popular periodicals, as well as in the daily press. On 18 and 19 April, respectively, the country's major newspapers, *Izvestiia* and *Pravda*, each devoted nearly half of its contents to the jubilee. *Izvestiia* published ten

(!) articles under the three general slogans/headlines: "Let's mobilize millions for the struggle against 'scientific' obscurantists who use Darwin's doctrine to legitimize the dominance of the bourgeoisie, colonial exploitation, and imperialist wars," "Let's use all the advances of bourgeois science and technology, having critically reworked them on the basis of Marx's theory," and "Armed with the method of dialectical materialism, let's put evolutionary science to the service of socialist plant- and animal-breeding, to the fulfillment of the historical decisions of the 18th party conference."[93] *Pravda* picked two general headlines: "In the countries of dying capitalism and rotting bourgeois culture, Darwinism is on trial as the accused" and "The working class, armed with Marxist-Leninist theory, takes everything truly scientific from Darwinism for the struggle to build socialism."[94] With Tokin at the helm, the Society of Marxist Biologists quickly compiled a collection of articles from the daily press, issued under the title *Darwin's Doctrine and Marxism-Leninism*.[95]

"Marxist biologists" contributed the majority of the jubilee publications. Tokin and M. Mitskevich called for putting "Darwinism to the service of socialist construction,"[96] while Prezent pontificated on "Darwin's theory in light of dialectical materialism."[97] Furthermore, for the first time, a representative of the highest party organs, a member of the Politburo, directly intervened in the discussions over Marxist-Darwinism: on 21 April, a leading party theoretician, Nikolai Bukharin, delivered a report on Darwinism and Marxism to the joint jubilee meeting of the USSR Academy of Sciences and the Communist Academy.[98] The report was immediately published in a new journal, *Socialist Reconstruction and Science*, issued as a booklet, and used as an introduction to a new edition of the *Origin of Species*.

For their part, Soviet biologists did not stand on the sidelines. Botanists and zoologists, physiologists and plant breeders, geneticists and ecologists, embryologists and morphologists maneuvered to demonstrate their devotion to Darwinism and to exert their own control over the contents of Marxist-Darwinism. The titles of articles they published in *Pravda* and *Izvestiia* speak for themselves. The botanist Boris Keller praised the development of "Darwinism in the country of growing socialism." The embryologist Boris Zavadovskii fought "against perversions of Darwin's doctrine." The geneticist Levit stamped out "capitalists' attempts" to construct links between "Darwinism, racial chauvinism, and social-fascism." The plant physiologist Nikolai Kholodnyi insisted that "Darwin's works must be on the reading table" in every home in the country.

The headlines in *Izvestiia*'s and *Pravda*'s jubilee issues, together with the titles (not to mention the substance) of articles appearing on their pages, show that the Marxist lexicon created in the 1920s was now expanded. This vocabulary added to "Darwinism" a new adjective, "Soviet." A major focus of the entire jubilee campaign was the insistent claim that only in the USSR was Darwin's legacy being truly developed, while in the capitalist West, it was being perverted, misused, or outright forbidden. Accordingly, only Soviet scientists and Soviet research could now be truly Darwinist. It was within this context that a popular slogan, "The Soviet Union is the second birthplace of Darwinism," was coined. In this patriotic context, the authority of Timiriazev as the founding father of "Soviet Darwinism" rose substantially, as witnessed by a very unusual jubilee held in May 1935: the fifteenth anniversary of Timiriazev's death.[99]

Another new adjective attached to "Darwinism" was "creative." The central tenet of the jubilee campaign was the "practicality" of Darwinism. Numerous publications proclaimed repeatedly that "Darwinism" was not merely a theoretical concept, explaining evolutionary processes in materialist terms; it was a practical instrument for the transformation of nature, or, as Bukharin put it, "a scientific force of production," "zoo- and phyto-engineering," which "is directly connected to the process of material production" and which "transforms agriculture into a science-based branch of socialist industry."[100] The "transformation of the nature of plants and animals" and the "creation of new varieties of plants and animals for Soviet agricultural production" was now set as the main task of Soviet Darwinists. It was within this vocabulary of practicality that a new term, "agrobiology" (an analog of "agrochemistry," contemporaneously praised as the solution for the régime's agricultural difficulties), was coined, and that the views of Ivan Michurin—an amateur plant breeder hailed as the "Russian Luther Burbank"—became identified with Creative Darwinism.[101]

Genetics and agrobiology

For geneticists, the Great Break brought both losses and gains. In 1929, several geneticists, including Chetverikov, the founder of Russian population genetics, were arrested and exiled from Moscow. The next year, as a result of an attack on eugenics as a "bourgeois perversion" in the popular press, the Russian Eugenics Society was dissolved, and geneticists lost the important source of legitimization for their discipline.[102] At the same time, the Great Break accelerated the institutional expansion of Soviet genetics. From 1930 to 1935, a small laboratory of human genetics directed by Levit grew into a large Institute of Medical Genetics under the Narkomzdrav patronage. During 1930–1, genetics departments and laboratories were established at Moscow, Leningrad, and Kazan universities. After Filipchenko's death in 1930, Vavilov inherited his genetics laboratory, and three years later he converted it into a large Institute of Genetics under the auspices of the USSR Academy of Sciences.

The policy of "mobilizing science to the service of socialist construction" especially stimulated the growth of institutions for agricultural genetics. The severe agricultural crisis set in motion by collectivization led to the creation in 1929 of the Lenin All-Union Academy of Agricultural Sciences (VASKhNIL) under the presidency of Vavilov. On the base of his Institute of Applied Botany, Vavilov created the gigantic All-Union Institute of Plant Breeding, with hundreds of affiliated experimental stations and laboratories scattered across the country. In 1931, Vavilov helped to establish the All-Union Institute of Animal Breeding, where Serebrovskii came to head a large genetics department. With the introduction of obligatory planning in science, geneticists led the way in 1932 by organizing a well-publicized conference "On the Planning of Breeding and Genetics Work." At this conference, under the pretext of "better planning and organization" of research, they discussed and elaborated an agenda aimed at the further institutional expansion of their discipline.[103]

By 1932, agricultural institutions became a bastion of Soviet genetics. The rapid expansion of institutions for genetics fostered the arrival of a new cohort of practitioners and a new interest group of self-defined agrobiologists. In 1932, a young agronomist, Trofim Lysenko, emerged as the spokesman for agrobiology, and his newly created journal, *Bulletin of Vernalization*, became the group's mouthpiece. Two years later, Isaak Prezent joined the agrobiologists' camp and became its chief ideologist and theoretician. Lysenko and Prezent elaborated a theoretical framework for agrobiology, drawing upon various concepts of plant physiology, cytology, genetics, and evolutionary theory. Echoing the main precept of the current version of Marxist-Darwinism—the transformation of nature of plants and animals—they proposed a doctrine of "the transformation of heredity" as the major tenet of agrobiology. They rejected the concept of the gene as a material unit of heredity, denied Mendel's laws, and claimed that external conditions could directly affect the "heredity" of plants. They cast their doctrine as the scientific basis for the whole of

Soviet agriculture and began competing with geneticists for control over agricultural institutions. The so-called Lysenko controversy had begun.

Marxist-Darwinism became a major instrument and a battlefield of the competition. Agrobiologists quickly capitalized on the current rhetoric of practicality, patriotism, and *partiinost'* embedded into the latest version of Marxist-Darwinism, and worked hard to mold it to their own advantage. They accused geneticists of "impracticality," pointing out that much of genetic research dealt with "useless" flies—the fruit fly *Drosophila*—instead of practically important organisms utilized in agrobiologists' studies. The term agrobiologists coined to label their opponents' research—"formal" genetics—was a transparent reference to "formalism," one of the stigmatized "isms" of Soviet political rhetoric, used as an antonym to both "materialism" and "practicality." Playing the patriotic card, agrobiologists appropriated Michurin and Timiriazev as the founding fathers of their doctrine and contrasted its "native roots" with genetics' "foreign origins," inventing a string of pejorative labels—Weismannism, Mendelism, Morganism—to denote the Western, bourgeois, capitalist, and even theist (playing on Mendel being a monk) nature of genetics. They accused geneticists of "deviating from the party directives in agriculture" and of an inability to resolve the current agricultural crisis in the country.[104]

The agrobiologists' attack on genetics proved effective. In 1935, the main spokesman for genetics, Vavilov, was dismissed from the presidency of VASKhNIL, while Lysenko and a number of his supporters were appointed members of the academy. In early 1936, Lysenko assumed the directorship of the Odessa Institute of Genetics and Breeding, replacing one of Vavilov's allies in this post. The same year, the circulation and the volume of his journal, now named *Vernalization*, expanded considerably, while the plans to create a *Soviet Journal of Genetics* were shelved. At the same time, one of Lysenko's disciples became the director of the Institute of Animal Breeding, previously a genetics stronghold.

In an attempt to recapture control over agricultural institutions and to fend off agrobiologists' efforts to employ Marxism-Darwinism to their own advantage, geneticists had initiated a public discussion on "issues in genetics" held in December 1936 at a general meeting of the VASKhNIL.[105] Presided over by several party officials in charge of agriculture and science, the discussion featured personal attacks on rival scientists and revolved around the political, ideological, and practical connotations of genetics and agrobiology. Agrobiologists took on Kol'tsov and Serebrovskii, accusing them of "fascist links," exploiting ties between the early development of Soviet genetics and eugenics and between eugenics and the Nazi concept of a "higher race" to condemn genetics as a "fascist science." They attacked Vavilov, portraying his work "The law of homologous series in variation" as "anti-Darwinist." They criticized geneticists for their "purely theoretical" works and their lack of practicality. Geneticists, in turn, accused Lysenko and Prezent of "Lamarckism" and "perversions of Darwinism." They also defended their science by reference to a wide antifascist campaign launched by several prominent British and American geneticists against the use of genetics to support Nazi racial policies.[106]

Geneticists managed to halt their opponents' takeover of agricultural institutions. The resolution adopted by VASKhNIL's ruling body endorsed the expansion of experimental work on "issues of heredity" and provided additional funding for genetics research. Alas, geneticists did not succeed in recapturing control over Marxist-Darwinism. In April 1937, as part of the public celebrations of the fifty-fifth anniversary of Darwin's death, *Pravda* carried an article, entitled "On Darwinism and certain anti-Darwinists," signed by Iakov Iakovlev, the chief party warden of the VASKhNIL discussion. The former commissar of agriculture and currently head of the party's Agriculture Department, Iakovlev maintained that Lysenko and his allies were advancing "Darwinism," while genetics was "anti-Darwinist."[107] The first round of the Lysenko controversy ended in an uneasy balance, with geneticists holding on to their institutional positions and agrobiologists establishing their authority over the meanings of Marxist-Darwinism.

Dialectics of coevolution: science and cultural resources

As we have seen, during the first decade of the Bolshevik rule, largely independent but converging efforts of certain Marxists and biologists "grafted" Darwinism onto Marxism in Russia. Every interested party employed the resulting hybrid—Marxist-Darwinism—as an influential cultural resource to advance its own agendas. The 1920s discussions of evolutionary questions in terms of Marxism represented only a small fraction of research and publications on evolution, heredity, and variability in the country. Nevertheless, they set definitive parameters for ensuing debates over these issues, including the range of participants and the type of arguments used to defend/attack particular intellectual positions and institutional actions, amply illustrated during the 1920s debates between geneticists and proponents of the inheritance of acquired characteristics.

As evident in the first round of the Lysenko controversy, the public nature of Marxist-Darwinism made its meanings highly dependent on specific institutional, intellectual, and ideological contexts, within which the competing groups exploited this cultural resource. At the second round of the Lysenko controversy in October 1939, the party patrons assigned Marxist philosophers, members of the editorial board of *Under the Banner of Marxism*, to adjudicate a public debate between geneticists and agrobiologists. At this debate, philosophers rebuffed attempts by both geneticists and agrobiologists to exercise control over Marxist-Darwinism in order to appropriate such control for themselves. Yet just a month later, in November, during the country-wide celebration of Darwinism's next jubilee (the eightieth anniversary of the *Origin of Species*) representatives of all three groups—Vavilov, Lysenko, and the editor-in-chief of *Under the Banner of Marxism* Mark Mitin—again laid claims for exclusive control over the meanings of Marxist-Darwinism, delivering keynote addresses to a special jubilee meeting of the USSR Academy of Sciences. At the same time, the sudden change in Soviet foreign policy—the August 1939 Soviet–German pact—profoundly affected the arguments in the debate, forcing agrobiologists to drop their depiction of genetics as a "fascist science" and geneticists to leave out their earlier emphasis on the antifascist attitudes of their British and American colleagues.[108]

In their competition for resources, interested parties deployed Marxist-Darwinism as a convenient language shared by the biological community and its party-state patrons, defining and redefining its vocabulary according to current political, ideological, and economic priorities of the régime. In the immediate post–World War II years, geneticists capitalized on the revival of Soviet–Western relations and on the decline of ideology and ideologists' role in the party apparatus to improve their institutional positions and public image. They orchestrated a broad campaign to publicize the "modern synthesis" of genetics and Darwinism effected by their American and British colleagues in the 1930s and 1940s, preparing Russian translations of major publications on the "synthesis" by Ernest Mayr, Theodosius Dobzhansky, Julian Huxley, and George Simpson. Geneticists formed an alliance with other biologists involved in evolutionary studies and deployed the advances of their foreign colleagues to recapture control over both Marxist-Darwinism and the genetics institutions seized by Lysenko and his supporters during previous years.[109]

In 1946–8, geneticists and their allies even attempted to transform "Soviet Creative Darwinism" from a publicly accessible cultural resource back into a specialized research field controlled by professional biologists by organizing discussions of recent advances in evolutionary synthesis and genetics within the walls of the country's leading scientific institutions—the USSR Academy of Sciences and Moscow University. Lysenko and his disciples countered this move by bringing the discussions back to the pages of the daily press and party periodicals and inviting the lay readership and party officials to adjudicate their disagreements with geneticists. Agrobiologists also revived the old debate over the role of the struggle for existence in biological evolution to portray evolutionary synthesis as an antipatriotic, capitalist, Western perversion of Darwinism.[110]

The growing Cold War gave agrobiologists an opportunity to launch a counteroffensive. With the shift in Soviet foreign policy from collaboration to confrontation with its wartime allies, the elaborate Anglo-American links that had served Soviet geneticists so well became a dangerous liability extensively exploited by agrobiologists to discredit the spokesmen for genetics. The Cold War also elevated the role of ideology in the Soviet polity, and several Marxist philosophers quickly joined in the renewed attack on genetics and geneticists to regain their own positions within the party apparatus. With the Cold War confrontation reaching its apex in the summer of 1948 over the status of Berlin, Lysenko managed to secure Stalin's personal approval of his "Soviet Creative Darwinism" and seized complete control over the country's biological institutions, which resulted in the "death" not only of genetics but also of modern evolutionary studies in the Soviet Union.[111]

Competition over which of the many interest groups would set the terms, forms, and norms of the meaning of Marxist-Darwinism characterized much of the subsequent development of Soviet biology. In 1951, Soviet biologists organized a new counterattack on agrobiology; as one would have expected, the main target of this offensive was the agrobiologists' version of Marxist-Darwinism, particularly Lysenko's concept of the generation of species and his denial of the struggle for existence.[112] By the mid-1960s, Soviet biologists had finally succeeded in overthrowing Lysenko's domination of their field, restoring the institutional basis of genetics, and breaking the "indestructible links" between Marxism and Darwinism. They managed to effect the separation of authority; now professional biologists controlled research on evolutionary questions (Darwinism), while professional philosophers exercised control over the critique of Western "deviations" and ideological issues in contemporary evolutionary studies (Marxism).

As we have seen, the meanings of "Marxist-Darwinism" were created and re-created according to changes in its constituent components—both "Marxism" and "Darwinism"—and according to particular uses interested parties found for this cultural resource. For scientists involved in evolutionary studies, Marxist-Darwinism served as an important tool in competition for resources, offering a convenient lexicon to translate their own, often quite esoteric research interests into the language understandable to their party patrons. They quickly adjusted their vocabulary to incorporate the latest pronouncements by party leaders and their proxies to the scientific community. Marxist-Darwinism became a negotiating language employed by many biologists to legitimize and justify their research agendas—as well as institutional and career ambitions—in the eyes of state officials. The party patrons, in their turn, used Marxism-Darwinism to inform the scientific community of the newest twists in the party line, reflecting their own economic, political, and ideological priorities of the time. For many scientists not directly involved in debates over evolutionary issues, Marxist-Darwinism provided an appropriate rhetoric for demonstrating their loyalty to the régime by joining a condemnatory (or laudatory) chorus singing to the latest party tune in order to protect themselves from possible attacks by competitors, ideologues, and demagogues.

It is tempting to see the above story as an embodiment of the "failed Soviet experiment," as an idiosyncratic tale of the arranged marriage between Soviet ideology and biological knowledge, and many commentators have succumbed to this temptation. Yet despite many unique features, the story of Marxist-Darwinism also illuminates certain general principles involved in the transformation of scientific knowledge into a cultural resource; principles whose actions could be seen in historical and contemporary public debates over evolution, eugenics, infectious diseases, genetic engineering, race, human origins, organ transplantation, and many other "hot" biological subjects.

The migration of scientific knowledge into the public sphere entails much more than the creation of a convenient rhetoric and a negotiating language, which mediate the interactions between scientists and their patrons. It also creates a zone of contest between science and culture writ large, a zone that is opened for use and appropriation by any interested party, be it theologians, public health officials, movie producers, philosophers, business managers, writers, or state bureaucrats. In the process of becoming a cultural resource, particular scientific knowledge about certain natural objects or processes is simplified, fragmented, and reduced to a few, often disjointed, general statements. It loses much of its conventionality, its embeddedness in a particular mode of production, its hypothetical, relative, and provisional character, acquiring instead dogmatic formulations in the form of various "isms" associated with a particular field of knowledge and doctrinaire imagery embodied in its exalted founding fathers and popular cultural icons (from Frankenstein to the mushroom cloud).

A cultural resource generates public debates whose forms, tones, and terms differ considerably from those used by scientists involved in the production of the scientific knowledge that gave birth to this resource. The arguments used in scientific debates (logic, verifiability, controlled experiment, disciplinary consensus, technical apparatus, and so on) lose their convincing power and are replaced by such "social" arguments as economic utility, political expediency, ideological conformity, and cultural affinity. A cultural resource acquires a life of its own, evolving in accord with its own environment (of which science per se constitutes only a small part). The transformation of scientific knowledge into a cultural resource undermines scientists' control over their product and its applications. It opens the scientific field from which this cultural resource emerged to incursion by "outsiders"—from competing disciplinary groups to politicians—eroding its professional autonomy, disciplinary boundaries, and established scientific cultures.

Acknowledgments

I am grateful to Denis R. Alexander and Ronald L. Numbers for their helpful suggestions and criticism during the preparation of the manuscript. I am indebted to Aleksandra Bekasova, Galina Savina, and Marina Sorokina for their help in locating and copying relevant materials in various libraries and archives in Russia. Support for research leading to this article was provided by the Standard Research Grant from the Social Sciences and Humanities Research Council of Canada.

CHAPTER TEN

Evolution and the idea of social Progress
Michael Ruse

Every historian of evolutionary theory knows that there is a connection between the idea of organic evolution and the social doctrine or ideology of Progress. However, the exact connection is far from well known. Once it is revealed, much light is thrown on the history of evolutionary theory—a history that stretches from the beginning, at the start of the eighteenth century, to the present, as we enter the twenty-first century. It is this tale that I tell in this chapter, and, as I set out to do so, I start with three definitions, one generally accepted "fact," and one tempting interpretation.

The definitions are of evolution, of social or cultural Progress, and of Providence. In this chapter, in talking of evolution I am talking of the development of organisms, by natural processes, from much more primitive forms.[1] Generally this is thought to involve a tree of life, that is shared origins, although some early evolutionists (notably Jean-Baptiste Lamarck; see below) did not subscribe to this. By Progress, I mean the idea or hope that humans through their own efforts can make things better—socially, educationally, in health care, in knowledge and understanding.[2] I take this notion of social and cultural Progress to be opposed to the idea of Providence, where I am now using this second term to refer to the belief that we ourselves can do nothing and that hope for the future is dependent completely on God's grace. I realize that there are those who would argue that God works through human-driven Progress and hence Providence and

Progress should not be opposed. But mine was certainly the common usage in the eighteenth century and there is a strong thread (often evangelical) to those who use the term in this sense today. Billy Graham in his crusades made it very clear that this was his understanding, in the sense that humankind had no real hope of real Progress aside from the grace of God.[3] Note that (in common with other scholars) I capitalize the human notion of "Progress," to distinguish it from the idea that organisms show some kind of upward climb—from primitive to complex, from "monad to man" in the traditional language. This I shall refer to as "organic" or "biological progress," or simply as "progress." I take it that organic progress is not simply a statement about what happens, but in some sense implies that the endpoint is better than the beginning. It is not just a matter of complexity but of complexity representing things that are improvements, like sharp eyes rather than crude light detectors and internal heat mechanisms rather than the warming and cooling contingencies of sun and wind and so forth.

The first part of my generally accepted "fact" (accepted by me) is that early evolutionists were out-and-out biological progressionists. Consider Erasmus Darwin (1731–1802), the grandfather of Charles Darwin.

> Organic Life beneath the shoreless waves
> Was born and nurs'd in Ocean's pearly caves;
> First forms minute, unseen by spheric glass,
> Move on the mud, or pierce the watery mass;
> These, as successive generations bloom,
> New powers acquire, and larger limbs assume;
> Whence countless groups of vegetation spring,
> And breathing realms of fin, and feet, and wing.
> Thus the tall Oak, the giant of the wood,
> Which bears Britannia's thunders on the flood;
> The Whale, unmeasured monster of the main,
> The lordly Lion, monarch of the plain,
> The Eagle soaring in the realms of air,
> Whose eye undazzled drinks the solar glare,
> Imperious man, who rules the bestial crowd,
> Of language, reason, and reflection proud,
> With brow erect who scorns this earthy sod,
> And styles himself the image of his God;
> Arose from rudiments of form and sense,
> An embryon point, or microscopic ens![4]

The second part of my generally accepted "fact" is that if you look at the writings of today's evolutionists in their professional publications—*Evolution* and the *American Naturalist*, for instance—you have to trawl hard to find talk of this biological notion of progress. I am not saying that it never occurs—in fact, I shall be discussing one very prominent living evolutionist who insists on talking about such progress—but generally it comes in op-ed pieces in newspapers or places where one can be more reflective. Overall, no one is setting out to show that guppies are better than fruit flies, and humans better than guppies—to take three much-discussed organisms.

The tempting interpretation gives a reason for this decline of progress-talk in evolutionary discussions.[5] It is simply that, in the early days, evolutionary thinking was basically a function of, an epiphenomenon on, culture—in particular, evolutionists to a person were human-driven Progressionists, and they read their ideology into the living world, coming up with a biological progressionist picture of the past, namely their version of evolution. However, as happens with all successful sciences, as the years went by, more empirical facts came on board. At the same time, evolutionists wanted to be able to predict facts, they wanted to make their science consistent with other sciences, they wanted to unify their various disparate explanations, and so forth. In the language of the philosophers, evolutionists wanted to make their science more epistemically respectable. Hence, cultural values like Progress had to take second place to epistemic values such as predictive fertility, and eventually the cultural values were eliminated from the science.

In the particular case of evolutionary theory, the social and cultural notion of Progress took a heavy knock when, in 1859 in his *Origin of Species*, Charles Darwin (1809–82) came along with his mechanism of natural selection (Figure 10.1).[6] The whole point, as Darwin himself noted in a comment on the flyleaf of a book (the *Vestiges of the Natural History of Creation*), is that selection relativizes everything. "Never use the word higher or lower."[7] Which is better, having a big, strong beak or having a small, delicate beak? As Peter and Rosemary Grant showed, in their brilliant, several-decade study (starting in the 1970s) of finches on an islet in the Galapagos Archipelago, it all depends. Sometimes in times of food abundance, better to be small because with your more refined beaks you can manage to eat a wider range of foods. Other times, when food is scarce, better to be big and strong, because you alone can manage the tough foodstuffs.[8] Which is better, being bright or being stupid? Again it

ON

THE ORIGIN OF SPECIES

BY MEANS OF NATURAL SELECTION,

OR THE

PRESERVATION OF FAVOURED RACES IN THE STRUGGLE
FOR LIFE.

By CHARLES DARWIN, M.A.,

FELLOW OF THE ROYAL, GEOLOGICAL, LINNÆAN, ETC., SOCIETIES;
AUTHOR OF 'JOURNAL OF RESEARCHES DURING H. M. S. BEAGLE'S VOYAGE
ROUND THE WORLD.'

LONDON:
JOHN MURRAY, ALBEMARLE STREET.
1859.

The right of Translation is reserved.

FIGURE 10.1 The title page to the first edition of *The Origin of Species*. (From Charles Darwin, *The Origin of Species*, 1859).

all depends. Brains require protein, and big brains require a lot of protein. As the late Jack Sepkoski, a paleontologist, once said memorably: "I see intelligence as just one of a variety of adaptations among tetrapods for survival. Running fast in a herd while being as dumb as shit, I think, is a very good adaptation for survival."[9]

Then, if the relativism of selection were not enough, at the beginning of the twentieth century came the theory of genetics, Mendelism, on which Darwinian selection was to depend. It is a crucial claim of Mendelism that the raw materials of evolution, the mutations, are random—not in the sense of being uncaused, but in the sense of not occurring to need. A prey may need a new protection against a predator, but there is nothing in theory to suppose that the right mutation will appear. There is certainly nothing in theory to suppose that the mutations will be making organisms ever better on some absolute scale of nature.

Darwinian selection and Mendelian mutation were not accepted by evolutionists because the theories have pretty faces. Through these causal factors, evolutionary thinking became much more epistemically powerful. Hence, as predicted, progress-thinking was expelled from evolution-thinking. That is why today in the professional biological journals we find no progress-talk. The cultural notion of Progress has lost its grip.

So much for definitions, facts, and interpretation. This chapter is revisionist. I accept the definitions and the fact. Biological progress talk of the monad-to-man variety has vanished from professional evolutionary writings. I reject the interpretation and explanation. It is quite wrong. However, this chapter is not an exercise in negativism. The wonderful thing is that there is a true explanation and it makes more sense of everything. We may not have organic progress in biology, but we can have a human notion of Progress in the history of science.

The link between evolution and cultural-social Progress

Let us use the traditional explanation's rough, three-part division of the history of evolutionary theorizing. The first reaches from the beginnings of evolutionary thought at the start of the eighteenth century to the publication of the *Origin of Species* in 1859.[10] The second is from the publication of the *Origin* to the entry of genetics into evolutionary thinking—it is hard to fix an absolute date to this, but let us say 1930, which was when the great population geneticists like Ronald A. Fisher (1890–1962), J. B. S.

Haldane (1892–1964), and Sewall Wright (1889–1988) began publishing their major works. The third extends from 1930 to the present, at the beginning of the twenty-first century. Let us start with the first period.

Although there were proto-evolutionary ideas in antiquity—Empedocles' speculations for instance—the first real evolutionists appeared on the scene at the beginning of what we now call the Enlightenment.[11] The Encyclopedist Denis Diderot (1713–84) is a good example. "Just as in the animal and vegetable kingdoms, an individual begins, so to speak, grows, subsists, decays and passes away, could it not be the same with the whole species?"[12] He made no bones about being a social Progressionist and seeing a link between his social views and his scientific speculations. "The Tahitian is at a primary stage in the development of the world, the European is at its old age. The interval separating us is greater than that between the new-born child and the decrepit old man."[13]

Likewise, Erasmus Darwin, quoted above, was keen on cultural and social Progress. He was a member of the Lunar Society (1765–1813), that group of British industrialists, natural philosophers, and intellectuals who would get together in the Midlands to talk and promote science and technology and industry. These were men committed to inventions, to machines and factories, and to new economic ideas like the division of labor, all making it that much more possible to produce goods cheaply and efficiently. They were also men committed to ancillary ideas and movements, for example in improved and universal education. Erasmus Darwin was always devising new contraptions, for instance a horizontal windmill, an untippable carriage, a lift for barges going through canal locks; he was a physician at the forefront of medical thinking; and he was much interested in education. For two daughters whom he set up as schoolteachers, he wrote a little manual on female education.[14] For Darwin *grandpère* there was a direct and explicit link between his enthusiasm for human-driven Progress—he was a big chum of Benjamin Franklin (1706–90), a Founding Father of the USA, and all for the American Revolution—and his thinking (in prose and verse) about evolution, something for him that was synonymous with biological progress. "This idea [that the organic world had a natural origin] is analogous to the improving excellence observable in every part of the creation; ... such as in the progressive increase of the wisdom and happiness of its inhabitants."[15]

The best known and most detailed of the pre-Charles Darwin evolutionists was the French biologist Jean-Baptiste Lamarck (1744–1829). His version of evolution was progressionist through and through.

Ascend from the simplest to the most complex; leave from the simplest animalcule and go up along the scale to the animal richest in organization and facilities; conserve everywhere the order of relation in the masses; then you will have hold of the true thread that ties together all of nature's productions, you will have a just idea of her pace, and you will be convinced that the simplest of her living productions have successively given rise to all the others.[16]

Especially in his major evolutionary work, *Philosophie Zoologique*, published in two volumes, he endorsed the old idea of a Great Chain of Being—organisms can be ordered from the primitive to the complex (with humans at the end)—and then simply argued that this is the history of life also.[17] This progressionist picture reflects Lamarck's strong commitment to cultural Progress. He was a friend of the *philosophes*, those French thinkers who made human-driven Progress their intellectual backbone, and he himself wrote approvingly of such ideas (see Roe, Chapter 2, this volume).

Science, as you might expect, figures high in this business. "In considering each society with respect to its degree of civilization, one might say that there is a direct proportion between the sophistication of science required for its members' well-being and the needs that they express."[18] Although if truth be known and proof is needed of Lamarck's (known) sympathy with Progress, ask yourself how it was that Lamarck—a minor aristocrat—did not merely survive the Revolution but thrived, getting a good job in the new science establishment. This was not a man preaching the virtues of Providence.

Finally, to complete the picture, jump forward to the 1840s and to the publication of the scandalous, anonymously authored evolutionary tome, the above-mentioned *Vestiges of the Natural History of Creation* (now known to have come from the pen of the Scottish publisher Robert Chambers; see Harrison, Chapter 1, this volume). The links between evolution and biological progress, and then on to cultural and social Progress are spelled out explicitly.

A progression resembling development may be traced in human nature, both in the individual and in large groups of men ... Now all of this is in conformity with what we have seen of the progress of organic creation. It seems but the minute hand of a watch, of which the hour hand is the transition from species to species. Knowing what we do of that latter transition, the possibility of a decided and general retrogression of the highest species towards a meaner type is scarce admissible, but a forward movement seems anything but unlikely.[19]

The status of early evolutionary thinking

Take now as established the link between evolution and Progress in the century and a half before the *Origin of Species*. Let us dig a little more into the status of evolutionary thinking in this period, particularly in the light of the link with such human-driven Progress. I do not think that anyone could or would claim that this measured up to the standards of mature science—what I will call "professional science"—by the standards of the time back then. If we think of the work of Antoine Lavoisier in chemistry or of Augustin-Jean Fresnel in the theory of light, the speculations of the *Vestiges* are simply not in the same category. This is not to say that there were no solid facts as we would judge them, or as they should be judged back then. The fossil record is really not the best candidate for anything very much. Until well into the nineteenth century it was very spotty indeed. Certainly, a progressive fossil record was not the evidence sought by Erasmus Darwin. But there were some other facts that we would judge pertinent, most particularly homologies—the isomorphisms between organisms of very different species. The French evolutionist Étienne Geoffroy Saint-Hilaire (1805–61) thought these very important.[20] And embryological findings were also coming in and suggesting that there are interesting links between organisms.[21]

So in a sense, the easiest thing would be to label pre–*Origin of Species* evolution as "pre-scientific" or "proto-scientific." But this really does not capture the way that people then generally regarded evolutionary speculations. People, those for evolution and those against evolution, knew full well that it was the human notion of Progress that counted. This and this alone was what mattered. Most particularly, when the critics attacked evolutionary ideas, it was the ideology that they despised and wanted to counter. Take Erasmus Darwin. He was not only in favor of the American Revolution, he was in favor of the French Revolution—at least until things got out of hand, or off with heads. His conservative critics could see how the doctrine of Progress had to be countered and how the purveyor of Progress had to be destroyed. They did not go after his science. They went after his Progressionist poetry, writing a devastating parody. Darwin had written a work, "The Love of the Plants," so the conservatives—led by the politician George Canning—wrote "The Loves of the Triangles."[22] In case anyone was in any doubt as to their intention—to destroy human-driven Progress as a viable idea—in the introduction to the poem (published in their magazine the *Anti-Jacobin*—the title reflecting their opposition to the English supporters of the French Revolutionaries), they firmly linked Darwin's thinking to other Progressionists whose ideas they parodied in another poem, "The Progress of Man."

> **Sample: Having killed a pig, the savage turns to bigger game:**
> Ah, hapless porker! what can now avail
> Thy back's stiff bristles, or thy curly tail?
> Ah! what avail those eyes so small and round,
> Long pendant ears, and snout that loves the ground?
> Not unreveng'd thou diest!—in after times
> From thy spilt blood shall spring unnumber'd crimes.
> Soon shall the slaught'rous arms that wrought thy woe,
> Improved by malice, deal a deadlier blow;
> When social man shall pant for nobler game,
> And 'gainst his fellow man the vengeful weapon aim.[23]

Not much social or cultural Progress here. In this poem also, by reference to his major evolutionary work *Zoonomia*, Erasmus Darwin is picked out explicitly as a target of their scorn.

The same story continues in the nineteenth century. Georges Cuvier (1769–1832), the French naturalist and zoologist, was no evolutionist, not really because of the empirical evidence (although he did note that the "mummified organisms" of Egypt seem very similar to living organisms) but primarily because of his commitment to Kantian thinking about final causes—the teleology of organisms precludes significant changes.[24] However, the chief factor in what became a driving hostility was his loathing of the idea of human, cultural Progress. He too saw it as the cause of the French Revolution, and later—under Napoleon and then under the Restoration—he became an ultra-bureaucrat and defender of stability and the status quo. He went after Lamarck and Geoffroy Saint-Hilaire at least as much for their politics as for their science (see Roe, Chapter 2, this volume).[25] And finally consider Adam Sedgwick, professor of geology at Cambridge University and Anglican parson, mentor of Charles Darwin and the most vicious opponent of *Vestiges*. (Having opined that a work so dreadful must have been written by a woman, he then pulled back declaring that no woman could be so degraded as to pen so evil a work: "the ascent up the hill of science is rugged and thorny, and ill suited for the drapery of the petticoat.")[26] He was explicit in his detestation of Progress.

What wonderful days we live in! Parsons turning blind Papists; ... men and women talking to one another at 100 miles distance by galvanized wires; Old England lighted with burning air; the land cut through by rails until it becomes a great gridiron; men and women doing every day what once was thought no better than a crazed dream, doubling up space and time and putting them in their side-pockets; new planets found as thick as peas; nerves laughed at, and pain driven out of the operating room; some sleeping comfortably, some cutting jokes as you are lithotomizing them or chopping off their limbs (this I have not seen, but D.V. I hope to see it soon) in short 'tis a strange time we live in! But is there no reverse to this picture? Yes! a sad and sorrowful reverse! our friends are dying around us; famine is stalking around the land; peace is but a calm before a tempest; sin and misery are doing their work of mischief; by God's judgement the same kind of disease which has destroyed the daily bread of our Irish brethren, may, for aught we can tell, next year consume our daily bread by attacking the grain on which we live. And then what becomes of art and science and civilization? Gold will not feed us; the heart of man will not beat by steam.[27]

Moreover, Sedgwick linked cultural Progress, biological progress, and evolution as an unholy triad. He may have been sarcastic about the Oxford Movement (when people like the theologian John Henry Newman left the Church of England and became Roman Catholics). He was incandescent about evolution. "I am no believer either in organic or social perfectibility and I believe all sober experience teaches us that there are conditions both moral and physical which must entail physical and moral pain so long as the world lasts."[28]

Given this sense of opprobrium felt toward evolutionary thinking, the best term to use to refer to such thinking before Darwin is "pseudo-science." This captures both the odor of fanaticism about the supporters and about the critics, and the stench of non-respectability, relished by supporters and hated by critics. Evolution certainly had all of that. I stress that in thus characterizing evolutionary thought, one is not being anachronistic, bringing today's judgments to bear on the past. The term "pseudo-science" was in operation before the middle of the nineteenth century (it was used by the physiologist François Magendie in 1843); more importantly the idea of pseudo-science was much alive at that time. In 1784, Louis XVI had convened a committee (led by Benjamin Franklin and including Antoine Lavoisier) to consider mesmerism (animal magnetism), which was occasioning scandal as impressionable people (mainly women) got together in what resulted in mass hysteria and fainting fits. The committee made short order of the practice and its theory. It failed every epistemic test—the ideas were contrary to what we know about the nature of fluids, it covers up anomalies, and it fails to make any predictions except where the results are known already. "Since the imagination is a sufficient cause, the supposition of the magnetic fluid is useless."[29] This precisely characterized other areas of enthusiasm, like phrenology, the science of brain bumps, something in Britain notoriously connected with radical, materialist (or at least deist) views about human psychology and nature (see Sivasundaram, Chapter 5, this volume). (It was phrenology of which Magendie was speaking.)[30] And the same applied to evolution. I am not saying that every evolutionist was a charlatan. Erasmus Darwin was deservedly famous for his medical skills. Lamarck was rightly respected as a taxonomist—although interestingly, he was also derided for his ludicrous and very expensive claims about meteorology. Robert Chambers (1802–71) was a top-flight businessman. But none of this denies the status of evolution. Interestingly and pertinently, Chambers set out to write a book on phrenology and then changed topics, to evolution, when he was halfway through.

The conclusion is that, before Charles Darwin, evolution was an epiphenomenon of the ideology of Progress, a pseudo-science and seen as such. Liked by some for that very reason, despised by others for that very reason.

The *Origin of Species*

Charles Darwin set out to change all of this. Let no one underestimate his intelligence and professionalism, understanding this latter term not so much in the sense of someone who makes a living from science—Darwin was always supported by the vast family fortune, powered mainly by the pottery works started by his grandfather Josiah Wedgwood (1730–95)—but someone who has training and energy and above all a respect for the proper standards of science, and who is in turn respected by the scientific community for his or her attitudes and achievements. For a start, do not think that it was by chance that Darwin stumbled over the evolution issue, and made it his life's work. It is true that, in a way, the idea of evolution as such was rather thrust on Charles Darwin—he was after all the grandson of Erasmus. And he did go off on *HMS Beagle* and visit the Galapagos Islands in 1835, the peculiar distributions of whose denizens called out for a transformist origin. But Darwin knew that the origin of organisms was the big question offering the biggest prize, of fame and respect. In 1836, the philosopher-astronomer John F. W. Herschel (1792–1871) referred to it, in a letter to his friend Charles Lyell, as "the mystery of mysteries."[31] It was this very phrase that Darwin used at the opening of the *Origin of Species*.[32]

It is one thing to be an evolutionist. It is quite another to be a Darwinian, referring to the man who discovered natural selection, the force behind evolution. As a graduate of the University of Cambridge, the home of Isaac Newton, Darwin knew he had to find a mechanism for evolution—a cause akin to the Newtonian mechanism of gravitational attraction. Here his training kicked in.[33] Darwin had read the authorities on scientific methodology and excellence. One was the above-mentioned John F. W. Herschel, author of *A Preliminary Discourse on the Study of Natural Philosophy*, a work that Darwin read in 1831, just after it was published—and which he admitted in later life had left a lasting impression.[34] Another was the man shortly to become the master of Trinity College, Cambridge, William Whewell (1794–1866), historian, philosopher, mineralogist, physicist, tidologist, geologist (well, he was president of the society), and general man of science in every way. In 1837, the year of its publication, Darwin read Whewell's *History of the Inductive Sciences* twice; he knew Whewell intimately (he knew him when he was an undergraduate, and then they were on the governing board of the Geological Society together) and constantly discussed with him topics on the nature of science; and then in 1841 Darwin read a detailed review, penned by Herschel, of Whewell's *Philosophy of the Inductive Sciences*.[35]

From Herschel and Whewell, Darwin learned that a good cause is a *vera causa*—a true cause.[36] This is what Newton had insisted upon. Unfortunately Newton was a bit vague about the actual criteria, although it was generally agreed that such causes ought to be embedded in axiomatic law systems, what today's philosophers call "hypothetico-deductive" systems. Herschel favored a kind of empiricist approach, insisting that true causes are either seen directly or inferred by analogy.[37] Whewell favored a kind of rationalist approach, insisting that true causes be at the apex of a unified theory, what he called a "consilience of indications."[38] Darwin, the clever fellow, set about satisfying everyone!

Nothing was very axiomatic, admittedly, but the central arguments of the origin, first to a struggle for existence and then to natural selection, were at least quasi-axiomatic, and made up of general laws about organisms.

> A struggle for existence inevitably follows from the high rate at which all organic beings tend to increase. Every being, which during its natural lifetime produces several eggs or seeds, must suffer destruction during some period of its life, and during some season or occasional year, otherwise, on the principle of geometrical increase, its numbers would quickly become so inordinately great that no country could support the product. Hence, as more individuals are produced than can possibly survive, there must in every case be a struggle for existence, either one individual with another of the same species, or with the individuals of distinct species, or with the physical conditions of life. It is the doctrine of Malthus applied with manifold force to the whole animal and vegetable kingdoms; for in this case there can be no artificial increase of food, and no prudential restraint from marriage.[39]

Note that, even more than a struggle for existence, Darwin needed a struggle for reproduction. With the struggle understood in this sort of way, given naturally occurring variation, natural selection follows at once.

> Let it be borne in mind in what an endless number of strange peculiarities our domestic productions, and, in a lesser degree, those under nature, vary; and how strong the hereditary tendency is. Under domestication, it may be truly said that the whole organization becomes in some degree plastic. Let it be borne in mind how infinitely complex and close-fitting are the mutual relations of all organic beings to each other and to their physical conditions of life. Can it, then, be thought improbable, seeing that variations useful to man have undoubtedly occurred, that other variations useful in some way to each being in the great and complex battle of life, should sometimes occur in the course of thousands of generations? If such do occur, can we doubt (remembering that many more individuals are born than can possibly survive) that individuals having any advantage, however slight, over others, would have the best chance of surviving and of procreating their kind? On the other hand we may feel sure that any variation in the least degree injurious would be rigidly destroyed. This preservation of favourable variations and the rejection of injurious variations, I call Natural Selection.[40]

The Herschelian demand for direct or analogical experience was met through artificial selection. It was this that led Darwin in the first place to natural selection, and now it served a justificatory role as well as a heuristic one. Darwin introduced the whole topic of evolution by drawing attention to the successes of breeders in making new forms, and then, again and again in the *Origin of Species*, he referred back to artificial selection to bolster his case for natural selection. A prime example was of his embryological discussion, where he argued that embryos of different species are similar because selection had not ripped them apart. He pointed out by analogy that breeders do not select on juvenile forms and hence the young—puppies, foals—are far more similar than are the adults of different breeds.[41]

The Whewellian demand for a consilience was met through the second half of the *Origin of Species* when, having established natural selection, Darwin then applied it across the biological spectrum—instinct and behavior, paleontology and the fossil record, geographical distributions of organisms, systematics and classification, anatomy and the above-mentioned embryology. Selection explains in all of these areas and conversely these areas make selection plausible.

Darwin knew full well what he was about. The following letter to the English botanist George Bentham (1800–1884; nephew of the philosopher Jeremy Bentham) is typical.

> In fact the belief in natural selection must at present be grounded entirely on general considerations. (1) on its being a vera causa, from the struggle for existence; & the certain geological fact that species do somehow change (2) from the analogy of change under domestication by man's selection. (3) & chiefly from this view connecting under an intelligible view a host of facts.[42]

How does one judge the science of the *Origin of Species*? It was not perfect. Darwin had a lot of problems with heredity—he knew this and others knew it too. Darwin had problems with the fossil record—there were no then-known pre-Cambrian organisms. Again, the physicists (ignorant of the warming effects of radioactive decay) argued that the earth's lifespan is too short for a leisurely process like natural selection. But overall, the *Origin* is a pretty good attempt at producing epistemically satisfying science. Something we much prize today is Darwin's realization that function—the adapted nature of organic features such as eyes and teeth and hands and the rest—is all-important. Darwin had learnt this from his natural theological training. He joked that he could repeat Paley's *Natural Theology* by heart.[43] Darwin had also learnt this from his teachers and mentors—Sedgwick,

John Henslow the botanist (1796–1861), and Whewell especially (who was the author of one of a series of natural theological tracts, the *Bridgewater Treatises*; see Topham, Chapter 4, this volume).[44] Cuvier with his teleological thinking, something he expressed in science as "conditions of existence"—those principles spelling out the adaptive structures that are essential for life and reproduction—was the hero of the hour. This is what made natural selection so convincing to Darwin. It speaks not just to change, but to function, to adaptation, to what today's evolutionists call "organized complexity."

Hence, if one were to judge the *Origin of Species* in isolation, I would not hesitate to speak of it as (judged by the standards of his day) mature science or, in another way, as professional science—meaning science that is turned out by people who have training, and who are passing on work that others can use. Despite the problems, there was work that one could do. For instance, one could test selective pressures and their results on fast-breeding organisms. Shortly after the *Origin* was published, the naturalist Henry Walter Bates (1825–92) did just this with his brilliant observations and experiments on butterflies, establishing the selection behind the mimicry that now bears his name.[45] More than this, one can see how (in the Darwinian scheme of things) biological progress might start to take a back seat. Bates is not saying that one form of mimicry is better in some absolute sense than another, or that the original form is better or worse than the mimicker. Biological progress is really not a relevant item.

This is not to say that Darwin never said anything about this kind of progress. He certainly believed in the cultural-social notion of Progress. After all, he was a beneficiary of the Industrial Revolution. He and his first-cousin wife Emma, likewise a grandchild of Josiah Wedgwood, owed their very financially secure lives to the efficacy of machines and the political-economic ideas that made the revolution so successful. More than this, thanks especially to Charles's close friendship with his older brother, another Erasmus and very much a man around (London) town, ideas about improved societies, better education, more efficient public health schemes, and all such things were part of the whole milieu in which Charles Darwin was immersed, particularly in the creative times after the *Beagle* voyage, when he was working towards his mechanism. There is therefore no great surprise that, at times in the *Origin of Species* (first edition), Darwin acknowledged that the fossil record is upwardly progressive in a sense, nor that there are those famous flowery passages towards the end of the *Origin*, where Darwin talked of the move up to our species. "And as natural selection works solely by and for the good of each being, all corporeal and mental endowments will tend to progress towards perfection."[46] But he knew full well that biological progress-talk was playing with fire and so generally he downplayed it. Unlike his grandfather Erasmus Darwin and the Frenchman Lamarck, it was certainly not the backbone of Charles Darwin's theory, to use an apt metaphor. By the third edition of the *Origin* (1861), Darwin felt sufficiently confident of the success of his evolutionary ideas to float notions of biological progress more explicitly, but even here it had to be tied into natural selection and was not inevitable. It is more a product of what today's evolutionists call arms races, the effects of competition between lines. It was not a value notion in any absolute sense. It kept the relative value sense of selection.

> If we take as the standard of high organisation, the amount of differentiation and specialisation of the several organs in each being when adult (and this will include the advancement of the brain for intellectual purposes), natural selection clearly leads towards this standard: for all physiologists admit that the specialisation of organs, inasmuch as in this state they perform their functions better, is an advantage to each being; and hence the accumulation of variations tending towards specialisation is within the scope of natural selection.[47]

Triumph of a popular science

Darwin produced a theory that (in the language of Thomas Kuhn)[48] he intended as a paradigm. There was to be a total shift to professional science. However, things did not work out quite this way. By now, Darwin was a sick man. He was not about to set up a whole functioning discipline of selection studies—university departments, professors, students, journals, and the like. A discipline where (having now just talked of the social side), the work being done would be (to talk also of the epistemological side) mature science in the sense of aiming to be predictively fertile and so forth (falsifiable if you are a devotee of the philosopher Karl Popper). If this kind of science was to be done, it was to be done by others. As it was. The early 1860s was the time when British science was "professionalized" in a big way by people like Thomas Henry Huxley (1825–1895) and others. They wanted quality science; they wanted functioning science; they wanted supported science. And so, at places like the new college in South Kensington (the Royal College of Science, now part of Imperial College, financed by the profits from the Great Exhibition of 1851), Huxley and his supporters set about making for a new professional science, in his case a new professional biology.[49]

Despite, or in respects because of, all of this activity, Darwin's vision, the theory of the *Origin of Species*, rather slipped through the cracks. On the one hand, it just wasn't the kind of science that people like Huxley were interested in. The 1830s might have seen function and adaptation ride high. Today, in the first decade of the new millennium, function and adaptation ride high. Not so in the decades after the *Origin*. By the 1860s, everyone was much more interested in form—homologies and so forth. Evolution certainly spoke to these—so note that I am not in any sense suggesting that people were unappreciative of evolution or less than thankful to Darwin for making the idea of evolution something that any sensible person should now accept. But in the new biology, selection was, if anything, an irrelevance. Adaptation is just a nuisance when you are trying to discern homology. Huxley even went as far as to deny that there could be any point to the markings of butterfly wings, the very phenomenon on which Bates was doing such fundamental Darwinian work![50]

What happened was that, rather than selection studies, people turned to the fossil record full-time and spent their efforts trying to work out paths of evolution—phylogenies. In this they were aided, at least they thought they were aided, by embryology, which supposedly gives clues as to what happened in the past. It was Huxley's counterpart in Germany, Ernst Haeckel (1834–1919), who formulated and popularized the slogan "ontogeny recapitulates phylogeny," the so-called "biogenetic law"[51] (Figure 10.2).

On the other hand, the trouble with evolution—particularly the evolution of the *Origin*—was that it seemed not to have any cash value. Huxley knew full well that, if you are to start a scientific discipline, you must have support—there must be people who want to employ or otherwise value your students to pay for your research and your fellow teachers. Physiology, Huxley sold to the medical profession. His argument was simple: We will give students three years' grounding in physiology and then you can take them and make them doctors. The medics—energized after disasters in the previous decade like the Crimean War (where more soldiers died of disease than of battleground injuries)—loved this proposition and entered fully into the bargain. Anatomy, Huxley sold to the teaching profession. Another argument: Latin and Greek are outdated training in this new industrial age—far better that students have hands-on experience of cutting up dogfish and the like. And the teachers loved this too. Huxley himself

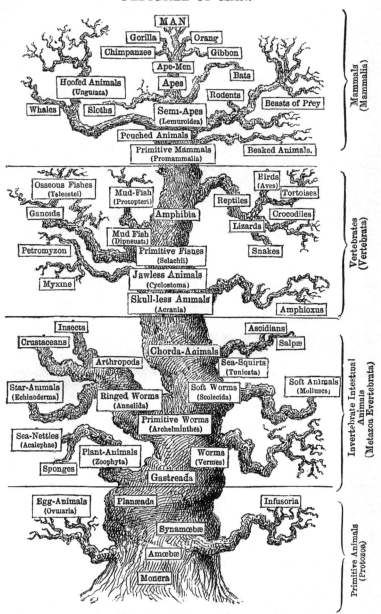

FIGURE 10.2 The tree of life as drawn by Darwin's great German supporter, Ernst Haeckel. Note how strongly progress is built into the picture. (From Ernst Haeckel, *The Evolution of Man*, 1896.)

sat on the first London School Board and he set up teaching sessions in the summer for teachers. His most famous graduate was sometime teacher and then novelist H. G. Wells.[52]

Evolution did not fit into this picture. It does not cure a pain in the belly and it seems a bit risqué for the school classroom. So Huxley simply did not teach evolution to his students. He really did not. I have examined three sets of lecture notes, kept by students. In 165 lectures, you get less than half an hour on evolution and but a passing mention of selection.[53] Note again, however, I am not saying that evolution got passed by. Anything but! T. H. Huxley and everyone else were ardent evolutionists. Just not in the classroom, please. Other than in phylogeny tracing, it simply got shunted aside as professional science. It was certainly shunted aside as professional science in any significantly causal sense.

So what did happen to evolution? It got used in a different way, as a kind of counter to organized religion—to the Anglican religion in particular. T. H. Huxley (I write of him, but he is a place holder for many) saw that, if Britain was to be reformed—better education, good drains, proper local government, doctors curing not killing, soldiers and civil servants appointed on merit and not through money and connections—then the ideology behind the opposition must be crushed. And Huxley, with reason (although not always as much reason as he supposed), saw Anglican theology and practice as supporting the status quo. "The Conservative Party at prayer," as people joked.[54] Justifying inequality.

> The rich man in his castle,
> The poor man at his gate.
> God made them high and lowly,
> He ordered their estate.[55]

Huxley needed his own ideology—he always thought in religious terms (note how a collection of his essays were called *Lay Sermons*, for instance)[56]—and evolution would provide it. But not just any evolution. An evolution of biological progress, pointing to human-driven Progress. An evolution against Providence, which put everything in the hands of God, and justified the status quo. So from lectern to lectern, from workingman's club to workingman's club, from presidential speech to presidential speech, from essay and book and letter to the London *Times*, Huxley preached evolution—progressive evolution.[57] Far from biological progress vanishing after the *Origin of Species*, it took on renewed life!

Behave like the Victorians and leave Darwin qua scientist on one side, although I think that by the time he came to write the *Descent of Man* he had rather given up and gone with the tide, giving a much more progressivist picture of human evolution (up to Englishmen) than you would have expected from the author of the *Origin of Species*.[58] Ask how post-*Origin* evolution generally could be so progress impregnated? How did people justify biological progress? In the simplest possible way. They just took it for granted! Like preachers of religion who beg you to believe, who appeal to your hearts rather than your heads, there were the preachers, the prophets, the gurus, of progressive evolution who likewise appealed to hearts rather than heads. In particular, Herbert Spencer (1820–1903)— man of letters, amateur biologist, self-appointed philosopher, founder of sociology, friend of Huxley and the man who turned him to evolution— preached Progress, progress, and evolution.[59] Far more read than Darwin, Spencer was the public face of evolution and for him progress of all and any kind—something which comes about through a weird combination of thermodynamics, British non-conformist liberalism, and a kind of mystical world spirit moving things upward (the Unknown)—was absolutely basic. His adoring public—in Britain, in America, in India, in the Far East (both China and Japan)—literally took his writings as gospel.

Simply put therefore, after the *Origin of Species*, evolution did change its status. It was no longer accepted simply because of an enthusiasm for human-driven Progress—nor was it rejected for this reason, although fewer and fewer as the years went by felt the need for rejection anyway. (The American religious fundamentalist story is noted as an exception.) Darwin did change things, and his arguments, skillfully constructed epistemically and backed by facts, made evolution as such reasonable. Yet, through and through this, biological progress implying cultural and social Progress was the backbone of the system. The cultural notion of Progress may not have been virtually the sole epistemological support of evolution, but it was still a driving factor in what made evolution an attractive idea. And for this reason, evolution did not enter the portals of respectable, professional science. It became a kind of popular science—certainly a science of the public domain—a kind of secular religion. The evolutionists even built cathedrals to their religion, except that they were not called cathedrals. Instead of going to church on Sunday morning, the young, forward-looking family of the late nineteenth century would set off on Sunday afternoon to the great new museums of natural history. There they could marvel at the new fossils being dug up in the American west—

Stegosaurus, Triceratops, Megalosaurus, Brontosaurus, Tyrannosaurus—and at the same time imbibe the new ideology. Just as the museums showed the wonders of technology, the testaments to the human Progressive spirit, so the museums showed the wonders of evolution, as life moved progressively from the blob to the new Victorian man and woman. For all of its powers as a scientific mechanism, for all of its pointing the way to a science shorn of talk of progress (where this implies some kind of absolute value), natural selection was ineffective against this.

The coming of genetics

In 1920s Britain, J. B. S. Haldane started publishing a formidable series of papers on selection and the flow of genes in populations, capping this in 1932 with his book *The Causes of Evolution*. In 1930, Ronald A. Fisher, who had been working and publishing on these things for nearly twenty years, brought out *The Genetical Theory of Natural Selection*. And across the Atlantic, the former worker for the U.S. Department of Agriculture and now faculty member at Chicago, Sewall Wright, announced his "shifting balance theory of evolution," publishing it in his "Evolution in Mendelian Populations," in 1931, followed by a short conference paper—introducing his famous metaphor of an "adaptive landscape"—in 1932.[60] "Population genetics"—the application of natural selection to genes in groups—had arrived and with it the foundation, the core, for the reemergence of Darwin's ideas and the development of the modern theory of evolution, generally known in Britain as "neo-Darwinism" and in America as "the synthetic theory of evolution."

Quickly, after the theoreticians had done their work, the empiricists moved in. Britain's most important hands-on evolutionist was the Oxford biologist E. B. Ford (1901–88), who was to found the school known as "ecological genetics"—members of which did hugely successful studies of snails and butterflies and other fast-breeding organisms.[61] In America, the most important was the Russian-born geneticist Theodosius Dobzhansky (1900–75). In 1937, he published his major work, *Genetics and the Origin of Species*. In relatively short order following Dobzhansky came *Systematics and the Origin of Species* (1942) by the German-born, American-residing ornithologist Ernst Mayr; *Tempo and Mode in Evolution* (1944), by the paleontologist George Gaylord Simpson; and finally in 1950 the botanical contribution, *Variation and Evolution in Plants*, by G. Ledyard Stebbins.[62]

By any standards this body of work was mature, epistemic, professional. This was not apparent overnight, of course, but from the first determined to be so, and increasingly successful in its aims. Evolutionary studies became all for which Darwin hoped. And with this went the decline of the monad-to-man type of progress in evolutionary discussions—in professional evolutionary discussions, that is. Search as you may through *Genetics and the Origin of Species* or in *Tempo and Mode and Evolution*, and you simply don't find it. The same is true in Britain—with a point-proving exception to be noted below. If you look at the work of those in Ford's group—A. J. Cain, P. M. Sheppard, H. B. D. Kettlewell—you search in vain for talk of progress.[63] Finally, the link between evolution and cultural Progress seems to be severed—and what more reasonable interpretation than that, with epistemic success, came the expulsion of the cultural? As evolutionists used their new theory to predict and to explain, Progress became increasingly otiose, and indeed unlikely and unwelcome given the new mechanisms.

A reasonable interpretation, but quite wrong! Not one—perhaps Ford being the only exception, and he was a very odd man in every respect—of the new evolutionists had given up on human-driven Progress or on biological progress. Indeed, for many if not most the very reason they had got into evolutionary studies was because they were ardent Progressionists, and the proof of this ideology was their lifelong aim. Fisher, for instance, thought that God had created organisms progressively through natural selection and that our task here on earth is to promote eugenics to prevent the decline of the race. The *Genetical Theory of Natural Selection* is designed explicitly to prove this point. Wright, thanks to the influence of his father and of his chemistry teacher at Harvard, L. J. Henderson, was an ardent Spencerian, and his theory of evolution was modeled explicitly to lead to Spencerian conclusions about upward rise.[64] Dobzhansky never concealed his enthusiasm for biological progress. Later in life, he became the American president of the Teilhard de Chardin Society—Teilhard being a French Jesuit paleontologist who saw the whole of life leading up the Omega Point, which he identified with Jesus Christ. And so the story goes. George Gaylord Simpson was a fanatic about biological progress—and cultural Progress, which (in good comic-book fashion) he identified as the "American way" against those "dreadful Ruskies."[65] And Mayr to the end of his very long life (at the age of 100) constantly chided those who doubted Progress in society and progress in evolution. Of an earlier discussion of this topic by me he said:

In your treatment, about half the time you talk about Progressionism with a sneer, always illustrated with some detestable examples of racism or male chauvinism. I think much of your writing would be improved if you would admit that much of Progressionism was a rather noble philosophy. In fact a very good case could be made for the claim that the current mess is the result of the loss of this philosophy.[66]

What is going on here? When you think about it, the answer is really quite simple. Every one of these men had been trained as a professional scientist. Haldane was a biochemist, Fisher a statistician, Wright much the same, Dobzhansky a geneticist, Mayr an ornithologist and systematist, Simpson a paleontologist. They had all taken in the message that successful science, mature science, epistemic science, professional science, is culture-value free. As Karl Popper was wont to say, "Knowledge without a knower."[67] Science is objective, science is not dependent on people or society—a message very important in the 1930s, when the Nazis were persecuting people for practicing Jewish science (see Weindling, Chapter 8, this volume). It was not for nothing that this was the time when logical positivism was becoming the dominant philosophy of the West. So these men all knew that shoving progress—a token for cultural and social Progress—into their science would be the death of what they were doing. They wanted—they wanted desperately—to be able to do evolutionary studies in a professional way. Hence, even though a major (often *the* major) motivation was to prove biological progress and hence cultural Progress, they knew that being explicit was anathema to their aim. Evolutionary studies were in a bad enough state already.

So, positively, they had to do the sorts of things associated with professional science—at one level, being as epistemically strong as possible. Somewhat amusingly, this led to all sorts of contortions with mathematics, the mark of an epistemically successful science. Most of the new empiricists were hopeless at sums, so one finds them rather pathetically holding on to the theoreticians for help and respectability. Dobzhansky wrote several papers with Wright, not knowing a thing about the content between the first empirical paragraph and the last empirical paragraph.[68] At another level, the new evolutionists did the social things, like starting societies, and a journal (*Evolution*), getting jobs at universities (both Mayr and Simpson moved from the American Museum of Natural History to Harvard), going after grants (Dobzhansky persuaded the Atomic Energy Commission that studies of fruit flies tell us much about the effects of atomic bomb fallout;

Ford persuaded the Nuffield Foundation that studies of butterflies are perfect models for studies of human diseases), getting students, and the like.

Negatively, the evolutionists had to keep biological progress out of their studies. Mayr was the first editor of *Evolution*, and, in rejecting papers, never wrote a sentence when a paragraph would do. First, he emphasized the importance of being causal and epistemically spry.

> After all, even though there may be considerable agreement about the major outlines of the evolutionary field, there is still a great deal of disagreement as to many concrete problems. Furthermore, the field has reached a point where quantitative work is badly needed. Also, evolutionary research, as you realize, has shifted almost completely from the phylogenetic interest (proving evolution) to an ecological interest evaluating the factors of evolution.[69]

Then he emphasized the importance of keeping progress talk to a minimum, meaning totally absent. Orthogenesis, a vitalistic theory suggesting that organisms have a kind of innate momentum, driving them ever upward, was a particular bugbear.

> Your manuscript, "Orthogenesis in Evolution," has been studied by two members of the Editorial Board who have reported back to the Editor that they do not consider the manuscript suitable for EVOLUTION.[70]

Even the language was verboten.

> It might be well to abstain from the use of the word "orthogenesis" (harmless as it is, in my opinion), since so many of the geneticists seem to be of the opinion that the use of the term implies some supernatural force.[71]

Above all, one must stay away from discussions of values and the like. These are bad. They are worse. They are philosophy.

> It has so far been the editorial policy of EVOLUTION to present concrete facts in every paper followed by the conclusions to be drawn from these facts. This policy was adopted deliberately because the prestige of evolutionary research has suffered in the past because of too much philosophy.[72]

Very revealing in all of this is the story of Julian Huxley (1887–1975), the grandson of T. H. Huxley. A professional biologist, who wandered from academia into writing popular books, then running the London Zoo, being a well-known radio (later television) pundit, and for a time being secretary of UNESCO. He was always a fanatical Progressionist and for the life of him could not keep progress out of his biology writings. In 1942, he wrote a major overview of the field—*Evolution: The New Synthesis*—and there was biological progress, front, back, and throughout.[73] A no-nonsense, brutally strong progress, leading to the emergence of "one somewhat curious fact": not only are we humans the winners in the upward climb but also "it could apparently have pursued no other course than that which it has historically followed."[74] Not that we are the endpoint.

> The Ant herself cannot philosophize–
> While Man does that, and sees, and keeps a wife,
> And flies, and talks, and is extremely wise . . .
> Yet our Philosophy to later Life
> Will seem but crudeness of the planet's youth,
> Our Wisdom but a parasite of Truth.

This is from a sonnet, "Man the Philosophizer."[75] Huxley's mother was the niece of the great Victorian poet, Matthew Arnold. Not all change is Progress.

The reaction to Huxley's speculations about biological progress and the status of human beings was brutal. He was excluded from all of the professional perks. He was pipped at the post when he wanted to found and edit a new journal of evolution, he was refused all of the grants, and, when he took up the cause of Teilhard de Chardin, the review written by Nobel Prize winner Peter Medawar was as funny as it was cruel.[76] "Yet the greater part of it, I shall show, is nonsense, tricked out with a variety of metaphysical conceits, and its author can be excused of dishonesty only on the grounds that before deceiving others he has taken great pains to deceive himself."[77]

Of course, this was not—could not be—all of the story. The new evolutionists were professionals who kept biological progress out of their science, because it was antithetical to their aims as professionals. At the same time they were all ardent cultural Progressionists who to a person believed in evolutionary progress, and many of whom had got into evolution precisely to establish these links! What was to be done? The answer was simple. Two series of books running in parallel. The first, the professional, had not a mention of progress or Progress. The second was like the first but with the mathematics removed (usually a simple matter, since all that meant was taking out the Hardy–Weinberg law—a principle describing the equilibrium within a population), then with a couple of happy chapters on biological progress and cultural Progress added, and finally in the introduction the sanitizing claim that this book is "for the general reader." Simpson, the brightest of them all, was a master at this. *Tempo and Mode in Evolution* (1944), professional and no progress; *The Meaning of Evolution* (1949), popular and lots of progress; The *Major Features of Evolution* (1953; a revision of *Tempo and Mode*), professional and not a hint of biological progress.[78]

And so the story has persisted down to today. Professional evolutionists—intellectual grandchildren of the original professionals—would never dream of bringing biological progress into their discussions. Popular writers feel free to bring human-driven Progress and biological progress right into the mix. Look at any issue of the journal *Scientific American*. Take up any child's guide to the past. Go to any museum and look at the panoramas. My favorite is the wonderful museum in Paris in the Jardin des Plantes—La Grande Galerie du Muséum national d'Histoire naturelle. A magnificent nineteenth-century building, the displays are in galleries around the central open space. The evolution exhibit is on the third tier. You start with blobs at one end and, having gone round four walls, come out with human beings at the other—in Paris you come out with yourself on the television screen! And lest the message be missed or hard to fathom, there are always helpful hints about the significance of cultural (in this case, technological) Progress. In the gallery just above, on the fourth tier (which is where most people start), the story is that of human life, of human inventiveness, from stone tools, through agriculture, industrialism, and right through to the electronic age.

In short, I claim that it is indeed true that thoughts of progress (and hence of Progress) are absent from today's professional evolutionary biology. But the reason for the expulsion was not epistemic purity but its incompatibility with the desire to be professional scientists. Am I saying that all of today's professional evolutionists are ardent secret biological progressionists and cultural/social Progressionists? Probably not. They have the same motives as the rest of us, although I would note—as many have noted before me[79]—that, as a group, scientists tend to be bullish on human-driven Progress. So do not entertain too seriously an alternative explanatory hypothesis, that the reason that there is no talk of biological progress in today's professional evolutionary discussions is because the authors no longer believe in cultural Progress. In a way, scientists have to be supporters of some kind of Progress, otherwise they will not persist when the going is rough. They have to believe in truth and a better day tomorrow (in the sense of more understanding of reality), or they might as well give up doing science. They believe in science—they must believe in science—and with notable exceptions tend to believe that science is a good thing and that it helps humankind. How else can one persist in what one does?

Am I saying that everyone who thinks about evolution today—public or professional realm—supports biological progress, or at least does not deny it? Certainly not. This was not so in the past and there is no reason why it should be so today. The late Stephen Jay Gould was a notable opponent of biological progress, writing that it is a "noxious, culturally embedded, untestable, non-operational, intractable idea that must be replaced if we wish to understand the patterns of history."[80] He complained: "Why is it that people are always so hung up on the notion of progress? I don't think there's a great deal to it, other than that humans are latecomers."[81] Note, however, that the places where Gould opposed progress (and hence a simplistic notion of Progress) were in his popular writings, like *The Mismeasure of Man* and *Wonderful Life*.[82] When he was writing about snails, progress discussions were absent. Paradoxically, incidentally, Gould did believe in some notions of social Progress. It was just that he thought that thoughts of biological progress had often been used to stop social Progress—as when Jews were excluded from the USA because they were considered inferior. So he went after biological progress to aid his vision of social Progress.

Do I claim that all of today's evolutionists are careful at keeping the popular and the professional apart, especially with respect to biological progress? Absolutely not! E. O. Wilson has good claim to being today's most distinguished evolutionist. His greatest work is *Sociobiology: The New Synthesis*, which, as the title promises, looks in overview at the whole of the study of animal social behavior from an evolutionary perspective.[83] It was very controversial when it appeared, with the Marxists and others hating Wilson's application of his ideas to humankind. Marxists wanted social causes to be definitive in molding human nature, and others (like the feminists) had their axes to grind. Interesting, however, were the reactions of non-ideological professional evolutionists. Overall, they respected highly what Wilson had done, seeing his book rightfully as a major advance in understanding. However, Wilson—an ardent progressionist—had made such biological progress the backbone of his book and used it to pose what he saw as the fundamental problem of the science. Take the question of sociality. "Four groups occupy pinnacles high above the others: the colonial invertebrates, the social insects, the nonhuman mammals, and man. Each has basic qualities of social life unique to itself."[84] However, there is a paradox. Although "the sequence just given proceeds from unquestionably more primitive and older forms of life to more advanced and recent ones, the key properties of social existence, including cohesiveness, altruism, and cooperativeness, decline." Yet do not despair. "Man has intensified [the] vertebrate traits while adding unique qualities of his own. In so doing he has achieved an extraordinary degree of cooperation with little or no sacrifice of personal survival and reproduction. Exactly how he alone has been able to cross to this fourth pinnacle, reversing the downward trend of social evolution in general, is the culminating mystery of all biology."[85]

The professionals did not like this, and chided Wilson for it. Apparently their criticisms had effect, for when three years later Wilson came to apply his thinking exclusively to our species—in his Pulitzer Prize–winning *On Human Nature*—he was careful at the beginning of the book to explain that this was not an essay in professional science. It is about science and not in it. Wilson knew and respected the right gambits.[86]

Making sense of history

It is a wonderful story and it makes sense of evolutionary theory, from its beginning three centuries ago down to the present. Evolution and Progress do have a history together, but it is not straightforward and obvious. For the first century and a half, evolution was a pseudo-science, dependent for its existence and popularity—its unpopularity also—on the cultural notion of Progress. As went Progress, so went evolution. Charles Darwin's *Origin of Species* in 1859 changed things. No longer was the idea of evolution dependent on human-driven Progress. It could stand on its own. But contrary to Darwin's hopes, evolution did not become a mature professional science. It was stuck as a kind of popular science, a sort of secular religion, and Progress was an integral part of the picture. It was because evolution was taken to promote cultural and social Progress that it was cherished by so many. Around 1930, things changed again. With the coming of Mendelian genetics, with the building of population genetics and then the work of the empiricists fleshing out the theoretical skeleton, modern evolutionary biology was born. And this was a professional science with no place for biological progress, and hence supposedly no implication for Progress in the human world. But biological progress was expelled, not because people no longer believed in it or thought that their science supported it—in fact almost everyone did believe in Progress and thought that evolution was progressive—but because its presence was thought fatal to the aims of the new evolutionists in building a professional science. A new professional science, something that all agreed has no place for cultural values like Progress. In other words, the internal cultural value of being a professional trumped the external value of progress/Progress. A neat happening indeed, and something that leads one to reflect, if not cynically at least carefully, about evolutionary theorizing and to wonder whether similar stories can be told about other branches of science.

CHAPTER ELEVEN

Beauty and the beast? Conceptualizing sex in evolutionary narratives
Erika Lorraine Milam

Humans come in a variety of shapes, sizes, and colors (so do pigeons, fruit flies, peacocks, and many other animal and plant species). Perhaps the best example of stable biological variation within the human species is sex.[1] Each generation of humans contains males and females, and the perceived sex of individuals influences the way they are treated by their peers and the kinds of social networks in which they participate.[2] Although natural historians, philosophers, and scientists had long been interested in understanding differences between men and women in biological terms, the "woman question" became an object of especially intense scientific inquiry in the final decades of the nineteenth century.[3] Evolutionary theory was one of several explanations scientists used to discuss the origin of sex differences in biological terms: it was often suggested that femaleness was less evolved than, or a degeneration of, maleness—women were ascribed a more animalistic nature. Yet by the early twentieth century, eugenic rhetoric secured for women an evolutionary status equal to that of men, as it was recognized that they bore the racial future of Anglo-American society. The capacity of women to choose their husbands, thereby increasing the beauty and intelligence of the next generation, distinguished them from mere beasts. After World War II, scientists' attentions focused not on female love but on male aggression as the key to understanding humanity's evolutionary past. Equally loaded with cultural assumptions about the biological nature of sex differences, this frame served to create an evolutionary picture of man the hunter and woman the sexually available mother. In the 1970s, sociobiology absorbed this gendered division of society, and simultaneously returned to an evolutionary framework that emphasized biological continuities between animals and humans.[4] Not until the 1980s was the sociobiological story reevaluated through the addition of narratives about male–female friendships and female primates as manipulative and aggressive in their own right.[5]

Biologists and feminists have consistently redrawn and renegotiated the ideological relationship between sex and evolutionary theory as a function of a shifting boundary between human and animal, and also of changing conceptions of how best to use animals as models for understanding our own behavior. Charles Darwin (1809–82) derived his theories of natural and sexual selection with no knowledge of the genetic basis of sex or behavior that came to form the backbone of sociobiological and evolutionary psychological accounts of male and female difference over a century later. Yet he assumed that the differences he observed in male and female human and animal behavior were variations requiring a biological rather than cultural explanation.[6] Throughout the past century, social scientists have challenged the underlying assumption that sex differences must be biological by calling into question the amount and quality of the research on which these conclusions were based and constructing alternative accounts of women's nature based on their own data. In spite of such active resistance to politically conservative conclusions about women's nature, sociological and cultural explanations of human behavior have failed to replace biological explanations within the popular and scientific imagination.[7] Popular convictions about the genetic, instinctual basis of our sexual behavior continue to be reproduced in newspapers, films, and even Internet dating websites.[8] Simultaneously, among biologists and anthropologists interested in primate cultures, a new synthetic picture is emerging, one in which both nature and culture contribute to the social organization and experiences of women and other female primates.

In this chapter I reflect on the various resonances between cultural and social norms one the one hand, and biological theory on the other, rather than seeking to advance any single solution as the "correct" way of adjudicating the relationship between sex and evolution. I explore the ways in which biologists, by establishing equivalencies between types of human and animal behaviors (such as territoriality or courtship), have marshaled their behavioral data in animals to explore the bounds of natural behavior in humans, and have sought to understand the relative contributions of nature and culture in defining the boundary between woman and animal, beauty and beast.

Scientific explications of women's nature, or womanhood as a subject of science

Even Darwin was unable to use a single theory to explain both the Earth's remarkable biodiversity (the sheer number of species that have populated the planet) and the presence of semi-stable varieties of individuals within a species (like sex and, most importantly to Darwin, race). Although today we know that categories such as gender and race are not individual characteristics in a genetic sense, Darwin and many subsequent zoologists did consider them to be fundamentally biological traits. Darwin then accounted for the origin of new species with his theory of natural selection, or differential survival, and the origin of variations within species (like sex and race) with his theory of sexual selection, or differential reproduction. Darwin saw these two facets of the evolutionary process as complementary biological phenomena, and devoted a book to each.[9] The first book, *On the Origin of Species*, dealt with the process by which a single species could be transformed into multiple species over evolutionary time—his explication of the origins of interspecific variation. He needed another kind of explanation to account for differences in size, shape, coloration, and behavior of individuals within the same species. *The Descent of Man and Selection in Relation to Sex*, the second book, concerned the origins of such intraspecific variation and the persistence of those differences over time. Darwin suggested two mechanisms by which sexual selection could cause and maintain intraspecific variation in a population: male–male competition for access to reproductively available females; and female choice of the aesthetically most pleasing males with whom to mate. Whereas male–male competition would select for male weapons or armor, female choice would select for male beauty and decoration.[10]

Mechanisms for the transmutation and alteration of species over time acted as theoretical structures that scientists could use to explain what they saw as the facts of sexual and racial difference. Sexual and natural selection provided two of a wide array of scientific theories used for this purpose; yet simultaneously they reinforced and contributed to the general perception that differences between men and women were biological in nature.[11] For Victorian and Edwardian scientists, evolutionary change in humans was linear and hierarchical.¹² When biologists placed humans within an evolutionary hierarchy, humans as a whole resided at the top of an evolutionary "ladder," but not all varieties of humans were equal. As the predominantly white, predominately male biologists considered their own biological heritage, they sought to identify differences, not similarities, between themselves and women, other ethnic groups, and animals. They consequently placed themselves at the top of the evolutionary hierarchy and used women and people of color as a kind of buffer zone, insulating themselves from their own animal legacy.¹³

Humans presented a special problem for Darwin. In some ways, evolution seemed to act in humans in the same way as it did in animals. For example, Darwin described male animals and humans as more variable than their female counterparts. Because men lived at the cutting edge of the struggle for existence, he argued, they were likely to vary in traits important to their survival. In civilized societies, women were removed from that struggle, so consequently women were less likely to differ from each other. In animals also, sexual selection acted to increase male variability, as female choice and male–male competition endowed males with extravagant horns or antlers or outrageous plumage. In other ways, humans seemed exceptional. Darwin suggested that the process of sexual selection may have been reversed in some human societies; certainly Victorian women, not men, concocted colorful displays with which to garner the attention of the opposite sex, and men, not women, chose their mates.¹⁴

In addition to evolutionary theory, three other scientific principles describing the law-bound nature of life came together to define women's place in the order of Victorian nature: Ernst Haeckel's biogenetic law, that is the way an individual "grows up" mirrors that individual's evolutionary lineage (ontogeny recapitulates phylogeny); the first law of thermodynamics, that is energy can change form but can never be created or destroyed; and the physiological division of labor.¹⁵ Each field of inquiry contributed to the perception that women differed from men in their physical and mental capacities.

The idea that an individual's embryological development proceeded according to the evolutionary history of their lineage proved irresistible to some scientists. Psychologist G. Stanley Hall (1844–1924), for example, posited that the process by which a human child matures intellectually would follow the evolutionary history of humanity itself. Women and men of what he described as the "lower races" were not missing links between animals and man, but kinds of perpetual adolescents. Following the laws of embryological development elaborated in the nineteenth century, Hall argued that, as an individual grew from a baby to an adult, the body became more complex over time and the mind became more individuated and specialized. Individuals who remained closer to the embryological type, who retained childish characteristics as adults, were thus less developed than fully functioning, individuated adults. Women, Hall continued, lagged behind men developmentally. Female faces belied their incomplete development in wide-set eyes, just as the female tendency to aggregate in groups reflected an inability to venture forth as fully developed individual personalities.[16]

Most scientists, however, found Darwin's theories inadequate explanations of sex differences and turned instead to other physiological mechanisms. Even ardent defenders of natural selection, such as the theory's co-discoverer Alfred Russel Wallace (1823–1913), dismissed sexual selection in animals, claiming that the bright coloration and plumage of males during the mating season was due to their naturally high energetic state. Only the extraordinary dullness of females required explanation and here Wallace invoked the differential action of natural selection on the sexes. He suggested that, to avoid predation, female animals required greater levels of camouflage when nesting. Over evolutionary time, natural selection acted to decrease any brightly colored plumage or eye-catching accoutrements in females. Biologists Patrick Geddes (1854–1932) and J. Arthur Thompson (1861–1933) followed a different line of reasoning to account for sex differences without an evolutionary cause: they drew an analogy between the physiology of animals and the sex cells (gametes) of humans. Small animals exhibited high metabolic rates, moved around a lot, and were highly variable (as with sperm); large animals were passive, sluggish, and conservative, with much lower metabolic rates (as with eggs). The essential nature of masculinity and femininity in adult humans derived from these energetic qualities of the sex cells. Males were exuberant and brilliantly colored as adults because they metabolized energy quickly, not because of the different action of natural selection on the sexes and not because of female choice. Geddes and Thompson used the energetics of sex, coupled with the first law of thermodynamics, to author social prescriptions for the proper behavior of men and women. If energy could not be created or destroyed but merely changed form, and if each human possessed a finite amount of energy, then the creative output of any one individual involved certain trade-offs. Because a woman's total available energy was probably less than a man's to begin with, and, because women invested most of their energy in reproduction, little remained to devote to intellectual pursuits. When applied with reverse logic, Geddes and Thompson insisted women who went to college and became notable scholars did so at the expense of their reproductive duties to their race.[17]

Such theories contributed to the belief that men and women participated in the social body through a physiological division of labor.[18] A man's duty was intellectual and financial productivity, although men varied widely in their capacities; a woman's duty was to reproduce and maintain the moral health of her family. The physical, intellectual, and behavioral differences between the sexes were complementary within the family unit.[19]

These varied investigations into the biological nature of sexual differences not only were caused by the rising feminist challenge to Anglo-American social order in the latter half of the nineteenth century but also inspired further resistance among educated women and men who sought to undermine scientific support for separate spheres. Social activists with connections to the women's movement responded to the challenge of sex difference in an evolutionary context in a variety of ways, each designed to answer the "woman question" on their own terms.[20] Some early feminists, such as Antoinette Brown Blackwell (1825–1921), denied that sufficient amounts of data had been gathered to support the claims of women's inferiority to men. Charlotte Perkins Gilman (1860–1935) and Olive Schreiner (1855–1920), on the other hand, appropriated the rhetoric of evolutionary theory to support their claims that the sexes were complementary and equal, or even, in the cases of Elisa Burt Gamble (1841–1920) and Frances Swiney (1847–1922), to argue for women's moral superiority![21] Many of the early feminists who sought to elevate their own biological status reified racial and class distinctions through their appropriation of an evolutionary hierarchy.[22]

The increasing importance of biology in demarcating constitutional maleness from femaleness is further highlighted by the veritable explosion in the number of hermaphrodite cases at the end of the nineteenth century, as medical doctors increasingly sought to define a biological basis for human sex that could be universally applied to all individuals. When confronted with people of intermediate or ambiguous sex, medical doctors and scientists found it difficult to agree on what criteria should be used to assign to all individuals one of two dichotomous sexes. Yet whereas earlier efforts to classify intersex individuals relied on their external appearance and behavior—manner of dress, timbre of voice, and sexual preference—by the 1890s, doctors turned to the anatomical structure of internal gonads as the repository of true sex, and after World War I to chromosomes.[23]

By the end of the nineteenth century, male and female biological differences were inscribed within a wide variety of "new" biological theories, from physiology to development and the evolutionary history of humanity. Each theory contributed to an increasingly scientific picture of physical, intellectual, and emotional differences between the sexes.

Female choice and the animal/woman divide in the early twentieth century

In the early twentieth century, concerns over controlling the evolution of human culture gained far more precedence within the context of Anglo-American eugenics (see Larson, Chapter 7, this volume) than they had in earlier decades. Some biologists and social reformers feared that natural selection no longer acted in civilized cultures with the same speed or efficiency as it did when people lived in more primitive conditions, thanks to modern medicine and urban life. By the end of World War I, fears of national disorder and racial degeneration inspired efforts to control and direct the evolutionary development of British and American societies. Recent reevaluations of the positive eugenics movement in the USA, and the embedded connection of marriage selection to love, sexual desire, and eugenic value, have pushed its period of influence both backward to the 1880s and forward into the 1950s, and have demonstrated how social activists with a broad array of political agendas, from the political right wing to the left and members of the women's movement, appropriated the rhetoric of Darwin's sexual selection to advance their ideological positions on scientific grounds.[24]

When applied to people, Darwin's theory of female choice couched the social role of women in a slightly different context than that of contemporaneous scientific theories of sexual difference. Womanhood was not merely an impressionable clay upon which selection left its mark; women possessed the ability to indelibly proscribe the intellectual, moral, and physical attributes of future generations. Scientists posited that women, through proper selection of husbands, could change the evolutionary future of the human species. The power of female choice was unleashed through differential rates of reproduction. Those women who chose eugenically sound husbands raised larger families; the next generations contained a disproportionate number of their attractive, intelligent, and fertile offspring. This evolutionary coupling of women's aesthetic and moral ability to discern the "quality" of potential husbands and therefore the future of the race proved incredibly powerful within the growing movement to control society's development through the application of biological principles. It also distanced human evolution from animal evolution: women chose their mates; female animals merely submitted to the most attractive male.[25] Sources in popular literature testify to the resonance of female choice and male–male competition among audiences with a variety of political backgrounds. Short stories and novel-length fiction directed at the "new" woman or man touted the importance of rational reproduction, female choice based on emotional truth not economics, and strong men who could fight to survive for producing happy marriages with lots of offspring.[26]

Even biologists disinclined toward accepting sexual selection in animals nevertheless saw female choice as an important component of marriage selection in humans. They drew a sharp line between animalistic sexual instincts, where any mate would do, and the civilized reasoned choices of marriage partners. Toward the end of his long life, Alfred Russel Wallace, for example, suggested that female choice could act with greater effect in humans than in animals. Whereas animals lived in harsh environmental conditions and were still subject to natural selection, human urban existence had removed the ecological pressures on individual survival. For Wallace, the unique ecological environment of humans prevented the effects of natural selection from swamping the effects of mate choice, making humans the only species in which sexual selection had a significant effect.[27]

Two of the most influential scientists publishing on sexual selection in the early twentieth century were both interested in using biology to better understand the human predicament, but used rather different approaches: Ronald Aylmer Fisher (1890–1962), a statistician and mathematical population geneticist, used his mathematical skills to elaborate a universal of theory of evolutionary change in populations, whereas Julian Sorrell Huxley (1887–1975), grandson of "Darwin's bulldog" Thomas Henry Huxley and a zoologist in his own right, was infinitely fascinated by the natural behavior of animals. Their common vision of evolution as a directional process within a single population provided an easy means of generalizing between animal and human populations.

Although the action of natural selection could explain the physical structure of animals and people, in the 1910s Fisher suggested that only sexual selection might explain the mental, aesthetic, and moral evolution of humans. In particular, Fisher argued that two factors would determine the evolutionary future of a population: the relative quality of breeding individuals in a population, and the quantity of offspring they produced in each generation. People could most acutely perceive mate quality, "all the traits of human excellence," under the influence of burgeoning love. Fisher lamented, however, that the most attractive and intelligent couples were not producing as many babies as couples who lacked professional or artisanal skills. He insisted that, although the action of sexual selection in animals and man was very similar, the influence of sexual selection would be more keenly developed in humans because "the choice of mate is of more importance among mankind than among most other animals."[28] As Fisher reserved for sexual selection jurisdiction over the "power of beauty, form, colour, voice, expression and grace of movement," it is not surprising that he believed the effects of sexual selection were most pronounced in humans and not in animals.[29]

When it comes to sex, historians and biologists remember Fisher for his 1930 theory of "runaway" sexual selection. Runaway sexual selection occurred when the expression of a trait in males began to evolve in lock step with preference for that trait in females. If females preferred to mate with the male who possessed the longest tail or the most colorful display or the "most" of any arbitrary trait, then Fisher posited that their choice of mates would quickly drive the expression of that trait to an extreme condition within males. Runaway sexual selection provided Fisher with a mechanism for the evolution of traits that seemed to run counter to the effects of natural selection, like a long tail or bright color in birds, or in humans the tendency of young men to sacrifice their lives in battle for the good of the country or tribe to which they belonged. Fisher suggested that, as long as some of these altruistic young men returned safely home, their exploits would surely make them the focus of a great many women's attentions. He hoped that such female choice would correct for the dysgenic effects of war.[30]

In 1914, around the same time that Fisher sought to understand the role of marriage choice in humans, Julian Huxley published a long article on the "Courtship Habits of the Great Crested Grebe" (Figure 11.1). In this, Huxley insisted that, for most animals, mate choice existed in a primitive state of "unconscious mental activities" and "inherited sexual passions."[31] By 1938, he extended his critique of sexual selection even further. Only in highly polygamous species with a skewed sex ratio could the effects of

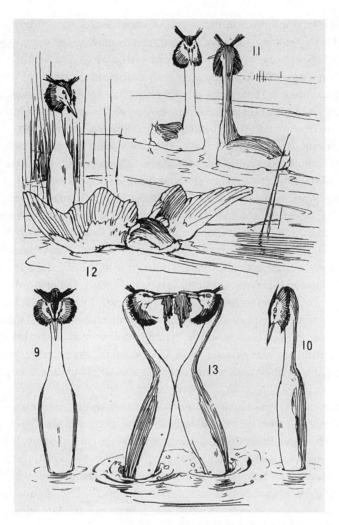

FIGURE 11.1 Great crested grebes exhibit many elements of their courtship displays after pairing up for the mating season. Huxley suggested that perhaps "love-habits" would be a better term for the grebes' mutual displays of affection. (From Julian S. Huxley, "The Courtship Habits of the Great Crested Grebe [*Podiceps cristatus*]; with an Addition to the Theory of Sexual Selection," *Proceedings of the Zoological Society* [1914] 35: 491–562 [491].)

sexual selection outweigh the effects of natural selection. Natural selection, Huxley contended, acted on courtship displays in the same way as it affected copulatory organs. The former acted on a psychological plane, the latter on a physiological plane, but both facilitated the union of sex cells and helped to keep the couple together after mating.[32] In other words, for Huxley, unlike for Fisher, the mating display itself served an evolutionary function (it wasn't arbitrary) and continued long after copulation.

In his model of mating behavior, Huxley argued that the most important function of courtship displays in animals was rather the promotion of pair-bonding between the male and female. Huxley's equation of mating display in animals with each individual sex act in humans (rather than marriage selection) resonated strongly with the sexology literature of the early twentieth century.[33] For example, Marie Stopes (1880–1958), the well-known paleobotanist turned suffragist and author of marriage manuals, suggested that courtship should not end with the marriage vows, but should instead be reenacted with every act of coitus. In her words, "wild animals are not so foolish as man; a wild animal does not unite with his female without the wooing characteristic of his race, whether by stirring her by a display of his strength in fighting another male, or by exhibiting his beautiful feathers or song."[34] In Stopes' hands, the proper equivalent of animal courtship was human foreplay and marital bonding, not spousal choice.

Both Fisher's and Huxley's models of female choice, from eugenically sound marriage partners to the decision of whether to engage in sex, paint a picture of women as sexual partners and physical embodiments of reproductive duty. Men and women might contribute equally to a productive society (however understood). As scientists like Fisher and Huxley explored the scientific bounds of natural sexuality in animals, they simultaneously redefined normal sexuality in humans in terms of their white middle- and upper-class sensibilities.[35]

Additionally, Fisher and Huxley emphasized the constitutive differences between woman and animal. Fisher's conception of female choice required that females possess the ability to discern minute differences among potential males and decide (whether consciously or not) which of these possible suitors was the best of the bunch, according to whatever criteria they employed. The process of comparison and aesthetic sensibility this implied made Fisher doubt whether the expression of female choice would ever be demonstrated in non-human animals. Huxley also insisted that animals were unlikely to be capable of true choice-based behavior, and suggested an alternative way of conceptualizing sexual selection in animals. Over the course of his scientific career, Huxley came to prefer thinking of female choice in animals as simply the ability to distinguish acceptable mates from inappropriate ones, on the basis of the aesthetic preferences of the species and not on individual tastes.[36] Fisher and Huxley's reactions to female choice as a potential mechanism of evolutionary change illuminate the historical divergence of two models of sexual behavior, one based on rational choice (reserved for humans) and one based on instinct, which could apply equally to humans and animals.

Animal and human models of "natural" sexual behavior after World War II

With the onset of World War II and concerns over the uses and abuses of biological theory in Germany by the Third Reich (see Weindling, Chapter 8, this volume), it is unsurprising that the rhetoric of social control through selective breeding was muted in Anglo-American communities of the 1940s and '50s. European scientists interested in human behavior instead concentrated their efforts on searching for those behavioral complexes shared by all humans. American scientists still hoped to use biological theory as a basis for social control, but through environmental conditioning and behaviorist psychology, rather than breeding.[37] Although their approaches to animal and human behavior differed considerably, biologists and psychologists on both sides of the Atlantic had a common goal. They sought, first, to understand the human capacity for mass murder, a trait that seemed to distinguish people from all other animals. Only in humans had the struggle between members of the same species become so deadly. Second, biologists and anthropologists strove to create an antiracist account of human evolution, unifying all human cultures into a single coherent family.[38] Ironically, these efforts combined to produce an evolutionary narrative emphasizing "man the hunter" (or man the killer) as a universal sign of humanity and reasserted sex differences as the biological basis of gender in all human societies.[39]

American and European approaches to the question of human behavior also emphasized different ways in which one could generalize from animal observations to theories about human behavior. In the USA, the predominantly behaviorist cast of psychological investigations into animal and human actions led biologists to use animals as models of simplified human behavior. Because behaviorists drew a strict line between the unconscious reactions of animals and the conscious, reasoned decisions of humans, animal behavior could be interpreted as roughly equivalent to human behavior without all that culture complicating things.[40] The popular writings of ethologist (animal behavior specialist) Konrad Lorenz (1903–89) in the 1950s and '60s inspired an entire genre of "pop ethology" that also generalized from animal to human. Many of his scientific peers critiqued his popular writings for their easy elisions between animals and people. Lorenz's friend and colleague Nikolaas Tinbergen (1907–88), for example, argued that, instead of extrapolating from the results of animal experiments, ethologists should apply their methods of behavioral analysis to humans. As organisms adapt their behavior to their environment, and the human environment differed radically from that of animals, observations of animal behavior provided only limited insight into the evolutionary origins of human behavior. Although this approach might seem as if it should have been appealing to anthropologists, in practice ethologists were interested in discovering the biological traits that united all humanity, while cultural anthropologists were interested in understanding the societies they studied as unique and valuable ways of life. Inevitably, these different research traditions, each invested in understanding the human predicament, came into conflict. At the core of their disagreements was the relative importance of innate human action and learned behaviors in determining the cultural and social structures of modern societies, and whether scientists should adjudicate their claims to naturalness or normality with reference to the non-human animal world or not.[41]

Before the creation of the National Science Foundation, the National Research Council–Committee for Research in Problems of Sex (NRC–CRPS) funded much of the sex research in the USA. Founded in 1921, the NRC–CRPS sponsored biological researchers who sought to uncover the genetic and hormonal basis of sex determination and differentiation. Although at its inception the NRC–CRPS had intended to conduct research on humans directly, during the 1930s the committee chose instead to fund scientists conducting basic research on animals as stand-ins for human subjects. These researchers argued that the short generation times and possible physiological manipulations of fish, mice, and birds allowed experimental investigations that were simply not experimentally or ethically viable for direct research on humans. Eventually, the committee did find a credible biologist interested in gathering direct information on human sexual habits, and today the NRC–CRPS is most often remembered as the agency that funded the infamous questionnaires of Alfred Kinsey (1894–1956).[42]

The Kinsey reports on male and female sexual habits, published in 1948 and 1954 respectively, entered the public consciousness in an age of heightened concern about American moral identity. The pages of *Sexual Behavior of the Human Male* and *Sexual Behavior of the Human Female* highlighted a notable gap between Americans' espoused moral ideals and their actual sexual practices. By wrestling with this discrepancy, sexual identity continued its status as a critical component of the American national character. Although the Kinsey reports may not have changed public perceptions of what constituted "normal" sexual behavior, they provided a platform for a national discussion about the sexual habits and implied moral fortitude of U.S. citizens. Americans' conflicting anxieties about sexual vigor and susceptibility mirrored their perception of democracy as simultaneously a powerful agent of international social reform and a delicate ideology in need of vigilant protection.[43]

Kinsey's books provided the English-speaking public with access to behaviorist and psychobiological ideas about the potential for conditioning of human actions, personal identity, and sexual habits. His emphasis on the importance of socioeconomic class and early childhood sexual encounters in producing his observed variation in adult sexual behavior arose from a more general contention that human behavior emerges from the environmental conditions in which an individual is raised. Through his discussion of men's sexual habits, Kinsey sought to dismiss sexual orientation and race as biologically valid categories. If humans participated in particular kinds of sexual behavior then, by definition, those behaviors must be natural and therefore normal. The publication of Kinsey's books corresponded with the high water mark for the "American experiment" as a way of predicting and controlling people's actions and behaviors through the science of human behavior.[44]

Yet even Kinsey subscribed to a view of biological differences between men and women. He suggested that men exhibited a greater range of sexual behaviors than did women, in part because early sexual experiences were more important in shaping adult male sexual habits than they were in determining the later expression of sexual behavior in women. Even if most readers were shocked by the range of sexual behaviors American mothers and daughters described to Kinsey in their sex histories, biology again seemed to support the claim that women were less variable than men.[45] The public response to his reports illustrated several sexual double standards in American sentiments about appropriate sexual behaviors in men and women. Healthy men should be sensitive to the needs of their wives inside the home and simultaneously aggressive in defense of their family and country. Inside the home, healthy women enjoyed sex and demanded foreplay from their sensitive husbands; yet not too much, lest they undermine the moral character of their husband and children outside the home. In an attempt to live up to these double standards, heterosexual couples emphasized the importance of marriage as the locus of sexual pleasure. Just as the family served as the cornerstone of postwar democratic capitalism, purchasing furniture and appliances for their home, married couples also acquired (or refamiliarized themselves with) new tricks in the bedroom.[46]

Meanwhile, on the eastern side of the Atlantic Ocean, ethologists explored the instinctive basis of animal and human behavior. Ethologists argued that environmental conditions shaped animal behavior in the same way that it shaped the physical structure of their limbs and organs. Individuals exhibiting behaviors that better enabled them to survive in their unique ecological surroundings were more likely to produce a greater number of offspring and pass along those behaviors to the next generation. Over time, natural selection shaped the behaviors characteristic of a species to the conditions of the species' existence. Although ethologists sought to understand the role of the environment in determining behavior, they typically thought in terms of generations changing in evolutionary time rather than individuals changing over the course of their life, as did their American counterparts.[47]

Methodologically, ethologists hoped to approach animal and human behavior with a similar set of analytical tools. Oxford-based Nikolaas Tinbergen synthesized these into four distinct questions that could be asked of any observed behavior. What prior experiences must an individual undergo in order to act like this now (ontogeny)? What immediate physiological or psychological machinery in the animal triggers the expression of the behavior (cause)? How does the behavior help the animal to survive in its natural environment (function)? What is the evolutionary history of the behavior and how is it related to similar behaviors in closely related species (evolution)? By asking these questions, ethologists were able to analyze behavior without recourse to hypothesizing about an inaccessible internal point of view of their subjects.[48]

One conceptual theory in particular, ritualization, enabled ethologists to search for similarities in behavioral patterns between large numbers of species.[49] For Tinbergen, ritualization denoted the process by which a movement or posture became a specialized signal over evolutionary time. He equated signals with "derived organs" that had evolutionarily been coopted from their original function to become adapted to a new purpose. (A classic morphological example of this process is feathers, which were originally used for insulation in dinosaurs, but were later coopted by evolution for bird flight.) Julian Huxley, years after his observations of mating in great crested grebes, extended the analogy of morphology and behavior by claiming that an animal's behavioral repertoire could be divided into functional units or "behaviour-organs." The same kinds of analytical techniques could then be used to discuss the exaggeration, reduction, or functional cooption of behavior-organs in some animals as compared with others.[50]

When ethologists applied the concept of ritualization to the evolution of courtship behavior, they emphasized both the aggressive tendencies of males and the coy sexual behavior of females. Ritualization was key to both reducing intraspecific fighting to a "tournament" in which no individuals were fatally injured, and redirecting male aggression into sexual stimulation.[51] Ritualized behaviors were "signals" to other individuals of the same species—and induced an appropriate behavioral response (often sex specific). The same ritualized behavior—a colorful display on the part of a male bird—induced aggressive displays in other males, and sexually acquiescent behavior in females. Thus species-specific sexual behavior was adaptive—in Tinbergen's terminology, the *result* of natural selection for both the individual and, sometimes, for the species as well. In the case of territorial disputes, males used ritualized displays rather than actually fighting, thereby reducing the number of fatal encounters and maintaining the species. In territorial animals, courtship behavior could also act to spread competing males out over a larger area, further reducing aggression.[52]

Driven by the desire to professionalize the study of animal behavior in the field, this new approach to synthesizing behavior and evolutionary theory had profound consequences for the ways in which ethologists investigated reproductive behavior. For example, most ethologists remained profoundly uninterested in the ways behavior could shape the evolutionary future of a species, and were therefore not interested in female choice as a mechanism of evolution. When they chose to investigate the causes of intraspecific variation in animals, they were far more interested in male–male competition as a biological root of aggression than in female choice as a mechanism for structuring social relationships. In his analysis of the elaborate male mating displays of bower birds (the decorating of nest sites with items of specific colors), Alan John Marshall (1911–67) argued that the mating rituals had not evolved as the result of female choice, but from the need of the species to coordinate female ovulation and male sperm production in response to unpredictable periods of drought and rain. He downplayed the importance of "choice-based" behaviors and instead described the mating rituals of these birds in terms of physiology and ecological necessity.[53]

When ethologists turned their attention to humans, in one sense they found it easy to include humans as simply another animal, uniquely adapted to its natural environment, exhibiting the same categories of behaviors as other species: foraging, courtship, territory acquisition, and defense. In another sense, humans represented a methodological problem in terms of identifying normal human behavior under natural conditions.[54] Yet books that purported to advance ethological interpretations of human behavior attracted a great deal of popular and scientific attention—not all of it positive.[55]

One of the first ethological treatises on human behavior took as its central theme the human tendency to make war and wreak violence on other humans, a problem of intra-specific aggression.[56] Konrad Lorenz argued in *On Aggression* (1963) that human behavioral evolution had been outstripped by technology, explaining why genocide seemed to be a uniquely human trait.[57] Modern weapons killed people at a distance and with such ruthless efficiency that they circumvented the victims' opportunities to offer signals of submission. Men were like doves; when crowded into a cage they lacked the common sense not to peck each other to death. Tinbergen and Lorenz were long-standing friends and scientific collaborators, but that didn't stop Tinbergen from suggesting in print that Lorenz had inappropriately generalized straight from animals to people. What we needed, suggested Tinbergen, were studies of man in his own right instead of "uncritically extrapolating the results of animal studies to man."[58] Humans, he suggested, were different from other animals because they exhibited an extraordinary ability to change their environmental conditions, and therefore the selective pressures, in which they lived. For Tinbergen, cultural evolution was outpacing the physical evolution of human society, as was evidenced by the ability of politicians to brainwash soldiers into thinking that fleeing was cowardly and despicable while simultaneously making killing easier through the development of long-range weapons that removed the possibility of their victims' avoiding death by signaling distress and appeasement. The parallels with the Christian doctrine of original sin were unmistakable: in order to live in an orderly and productive society, humans must overcome their innate nature.[59]

Rather than concentrating on fear and aggression, in *Love and Hate: The Natural History of Behavior Patterns* (1970), Irenäus Eibl-Eibesfeldt, founder of the field of human ethology, turned instead to courtship rituals as a structuring force of human social systems. Eibl-Eibesfeldt couched his analysis in terms of ritualized actions of humans that required no choice; the effect of courtship was to signal the emotional state of one individual to another. He suggested that "if a girl blushes we know she feels embarrassed, and if someone strikes a table with his clenched fist we know something has annoyed him." Eibl-Eibesfeldt used these examples to illustrate how human behaviors codified into rituals became both more simplified and more exaggerated "in the manner of a mime."[60] He argued that flirting in humans was a form of ritualized flight—when a woman caught the eye of a man and then quickly looked away, she was inviting him to pursue. Humans, as animals, exhibited ritualized behaviors that could be analyzed without presupposing a particular cognitive state in the individual expressing those behaviors.[61] The professionalized mechanomorphic language of the ethologists, when applied to humans as well as animals, removed agency and personality from the individuals they studied.[62]

In the postwar decades, both American psychobiologists and European ethologists emphasized sex differences as the root cause of gender distinctions in modern society. Resonating with the social mores of the period, they painted a picture of universal man as an aggressive hunter, and woman as naturally coy. Part of this description depended on biologists' removal of "agency" as a motivating cause of actions in animals, and, when they extended their analysis to humans, in human behavior as well. When combined with the sexual stereotypes implicit in assumptions that sexual behavior was more highly variable in men than in women, behavioral research of the post–World War II era provided the reading public with powerful reasons, etched in science and evolutionary theory, to believe in the primacy of biological sex differences in structuring human social organization and cultural interactions.

Sexual selection and biological determinism in the 1970s

The successes of animal film stars in the 1960s and '70s facilitated the easy slippage between human and animal social behavior in the popular imagination, from television series featuring animals as boyhood pals (like Lassie the dog and Flipper the dolphin in the USA, or Skippy the kangaroo in Australia) to the heroic story of Elsa the lion's return to the wild in *Born Free* on the silver screen. By the mid-1970s scientific debates over the biological basis of human behavior also began to grace the pages of the *New York Review of Books*, *Time* magazine, and books aimed at non–scientifically trained audiences. The intensity of this media attention to the animalistic basis of human behavior made it seem that the construction of human behavior within an evolutionary framework was an innovative idea. Most of the outcry centered around two books. In *Sociobiology: The New Synthesis*, Edward O. Wilson interpreted patterns of social and sexual behavior in humans through the lens of evolutionary theory, which assumed that in order for selection to act on a trait, the trait in question must comprise heritable genetic differences. In *The Selfish Gene*, Richard Dawkins further suggested that organismal bodies were simply "survival machines" designed to perpetuate the genetic code they contained (see McGrath, Chapter 13, this volume). Together these books painted much the same version of men's and women's biology as before, albeit with a return to behavior as a mechanism of evolutionary change in the form of sexual selection and active female choice.[63]

Although his application of biological theory to human nature built on the work of psychologists, sexologists, and biologists before him, Wilson extended their "biologization" of human behavior and culture. He argued that "scientists and humanists should consider together the possibility that the time has come for ethics to be removed temporarily from the hands of the philosophers and biologicized."[64] Wilson's claim that even the morality of human culture should be analyzed as a biological trait engendered considerable lamentation and consternation among biologists and social scientists alike. As a result of the highly visible scholarly debates over the validity of the (socio)biological approach to human behavior, many of the earlier Anglo-American attempts to use evolutionary theory to understand human behavior were forgotten.[65]

In his review of *Sociobiology* published in the *New York Review of Books* in August 1975, developmental geneticist Conrad Hal Waddington (1905–75) argued that the core issue Wilson hoped to explain was how the altruistic basis of social interactions had arisen in a biological world defined by individual fitness and survival.[66] Waddington suggested that the answer to this question had already been solved conceptually in the 1930s, and then elaborated mathematically in the 1960s—kin selection.[67] Because our relatives are genetically related to us, helping them survive and reproduce actually helps to spread our own genes in the population. A far more interesting question for Waddington was how to determine the relative significance of the following factors in determining someone's behavior: communication among members of a social group, individual experience, and genetics. Waddington suggested that Wilson ought to have paid more attention to resolving this nexus of influences. In fact, the relative importance of these factors became a central issue in subsequent debates over Wilson's book.

Three months later, a second review of *Sociobiology* appeared in the same journal, signed by sixteen people, some of whom worked down the corridor from Wilson at Harvard. Entitled "Against 'Sociobiology'" this review took Wilson to task for supporting biological determinism, a tradition they extended back to Spencer's cruel social policies, American eugenic beliefs in racial differences in IQ, and Hitler's policies of extermination. They claimed that Wilson's suppositions failed for five reasons that could be applied with equal efficiency to these earlier biological determinists. First, the behavior and social structure of organisms were not "organs" that could be analyzed as simple extensions of their genetic makeup. Second, biological determinism ignored the cultural transmission of ideas through human social interactions. Third, simple generalizations from non-human animals to humans on the basis of carefully crafted metaphors and analogies were scientifically flawed. Fourth, there was no direct evidence that genes for behavior exist at all. Fifth, Wilson's analyses were based on speculative reconstructions of human prehistory. Wilson responded immediately, claiming that the attack was "self-righteous vigilantism," and that he certainly did not endorse the "naturalistic fallacy" that what is, should be! Wilson's efforts to assuage his critics, however, did not stop others from joining the antisociobiology bandwagon.[68]

Additionally, in *Sociobiology* Wilson had described male and female sex roles as complementary opposites. Where males were aggressive, females were coy. Sex itself he touted as "an antisocial force in evolution. Bonds are formed between individuals in spite of sex and not because of it."[69] Wilson accounted for such fundamental differences between men and women with Darwinian sexual selection theory. He contended that these differences were the result of selection acting differently on men and women, encoding sex differences in the human genome. Males, he posited, were salesman, and females had evolved to resist their efforts. In other words, by playing hard to get, females would land themselves a more vigorous mate.

Appearing only a few months later, Dawkins' *Selfish Gene* added fuel to an already blazing fire. Dawkins provocatively asserted that animal and human bodies were simply genetic-replication devices. Although Dawkins was careful to assert in the first few pages of the book that he was "not advocating a morality based on evolution," in the *Selfish Gene* he framed sexual stereotypes as biological, not cultural, entities.[70] Dawkins argued that the nature of maleness and femaleness arose because the gametes of males (sperm) were significantly smaller and more numerous than the gametes of females (eggs). All sex-based behavioral differences, Dawkins continued, stemmed from this one significant fact, lending men the ability to reproduce at a virtually unlimited rate, while females were limited by their capacity to bear only one child at a time.[71] Although women might choose males by different strategies (by looking for a good father or the male with the highest-quality genes), Dawkins maintained they were the choosy sex, whereas males were more profligate in their sexual attentions. His analysis of sex differences based on gamete physiology could almost have been lifted from Geddes and Thompson's treatise on sex differences eighty-six years earlier. By 1989, when a second annotated version was released, Dawkins had softened this claim, and instead suggested the relative size of gametes did not determine behavioral differences in adults, but that both were a consequence of the two-sex solution to genetic recombination.[72] The sex differences themselves remained. Much like Darwin, Dawkins added a final note to his chapter on sex: if a biologist were to analyze human sexual systems based on appearance alone, he (or she) would be likely to determine that in humans the process of sexual selection was reversed as it is the females of the species who spend so much time and attention on their appearance and the men who hold the power of mate selection![73]

Following the publication of *Sociobiology* and *The Selfish Gene*, many second-wave feminists objected strenuously to Wilson and Dawkins' characterizations of men as sexually vigorous and women as passive or coy.[74] The rise of second-wave feminism had begun much earlier. In her 1963 bestseller *The Feminine Mystique*, Betty Friedan urged suburban housewives to recognize their frustrations with the social proscriptions defining their lives.[75] Women needed to break down the stereotypes that defined femininity, she claimed. Also published in 1963, the psychological studies of William Masters and Virginia Johnson simultaneously illuminated the psychology and physiology of the female sexual response, including women's capacity for multiple orgasms.[76] The sexual revolution drew even more popular attention to the variety of women's sexual choices in society.

By the 1970s, then, the debates over the stereotypical depictions of sex in Wilson and Dawkins' books helped to crystallize an anti-evolutionary-theory contingent among feminists.

A young feminist primatologist, Sarah Blaffer Hrdy, was one of the few people to read Wilson's chapter 15, "Sex and Society," before its publication; at the time, Hrdy was a Ph.D. student in anthropology at Harvard, where Wilson worked.[77] Whether or not Hrdy's objections to Wilson's book coalesced when she first read his manuscript, several years later she published a scathing critique of the enduring role of the "coy" female and "active" male in evolutionary theory.[78] In this book, Hrdy identified sexual selection as "one of the crown jewels of the Darwinian approach basic to sociobiology," and argued that the theory of sexual selection was based on "partially true assumptions" that portrayed females as sexually passive, yet highly discriminating, and males as sexually aggressive and undiscriminating.[79] Wilson had based much of his analysis of sexual selection on two publications by a young biologist, Robert Trivers.[80] Hrdy contended that Trivers had uncritically imported sexual stereotypes about passive women and active men into his work. She argued that when Trivers' essay on sexual selection became "the second most widely cited paper in all of sociobiology," other biologists, in turn, integrated the myth of the "coy female" into their evolutionary theorizing about the sexual relations between males and females.[81] Hrdy further suggested that the coy-female stereotype gained acceptance within the biological community because of male scientific bias, and that only the recent advent of women into the biological sciences had begun to alter the application of sexist stereotypical human behavior to animals.

Hrdy intended her analysis of the "coy female" as a critique of experimental practice and theorizing in 1980s sexual selection research, yet it also served to highlight how central the ethological model of mate choice had become to the animal behavior community. Sociobiologists had combined a pre–World War II conception that behavior (especially female choice) could describe the evolutionary past and help to predict the evolutionary future of human sociality, with an ethological mechanomorphic model of female mating behavior devoid of true "choice." Although this model allowed ethologists to professionally distance themselves from their animal subjects, when applied to human mating behavior it depicted women as socially and sexually passive while promoting men as instinctually aggressive in their social and sexual behavior.

Bridging the gap between feminism and evolutionary theory

In recent decades primatological research has highlighted the cultural complexities of primate social interactions, thereby causing a reconsideration of boundaries demarcating nature from culture, animal from human. As the controversies over sociobiology continued to develop, a younger generation of biologists and primatologists began to create new scientific narratives about sexual identity and evolutionary theory, seeking to provide depth and reason to female social and sexual behavior. Primatologists such as Jeanne Altman, Barbara Smuts, Sarah Blaffer Hrdy, and Patricia Gowaty offered critiques of female passivity and sought to provide alternative evolutionary narratives with their research.[82] They hoped their research would not only redress what they identified as male bias in the primatological literature on sex but also convince feminists that biological theory was not antithetical to their goals. Simultaneous efforts to understand the capacity for language in primates, especially in bonobo chimpanzees, have strengthened biologists' convictions that a simple dichotomy between animal nature and human culture is untenable.

By the 1990s, primatological research helped to destabilize the "man the hunter" paradigm by adding narratives about sexually assertive, polyandrous, strategizing females, producing a new "female perspective" within evolutionary biology.[83] Jeanne Altman's research on baboons in Amboseli National Park, Kenya, helped to break down cultural assumptions about the universal nature of mother love.[84] She showed how baboon mothers changed their behavior toward their offspring as a result of changes in habitat, environmental variation, or their social relationships within the troop (Figure 11.2). Also working on baboons, Barbara Smuts elucidated the close associations that form between male and female baboons that can last lifetimes, not solely as sexual partners, but as "allies."[85] Smuts' research provided insights into why the top male doesn't always get to mate with the reproductively available female, and described the ways in which female baboons strategized and manipulated their social relationships. The assertive female baboons central to her research varied in their predilections for lovers and allies. In Altman's and Smuts' descriptions, female baboons were far from mere baby incubators, the passive recipients of male attention and behavioral structure. Published several years later, Hrdy's *Mother Nature* offered a broad synthesis of maternal behavior in a variety of primates, and carried much the same message—primate societies were not structured solely by competition between males.[86] Fe-

FIGURE 11.2 Baboons have served as models of human social behavior among anthropologists for almost fifty years. During this period, biologists' interpretations of baboon behavior have shifted from mostly emphasizing aggressive interactions between large males as the basis of social structure to incorporating both male and female behavior as important to understanding group dynamics. (Photo by Dorothy Cheney.)

males define and redefine their reproductive strategies in order to gain the assistance of other individuals in raising their offspring. Whether help in offspring care and provisioning comes from grandmothers, sisters, unrelated females, or multiple males in the group, a single female almost never raises her infants alone. Simultaneously, biologists have been paying more attention to situations in which male and female desires come into conflict. Patricia Gowaty's research has demonstrated the crucial importance of male coercion in limiting and defining the options available to females; females, she argues, behave differently when they can freely chose their mates and reproductive strategies than when those choices are constrained by their social situation.[87]

These female scientists travel a narrow and arduous path between the objective stance from which they analyze their animal subjects and a simultaneous recognition that their research, like the research of their male counterparts, is affected by the political and social climate in which they work. Many of them have self-consciously tried to produce valuable and insightful contributions on the place of women in nature and to appreciate the complexity and nuances of human social structures more generally.

The rise of these new narratives about female primates, including women, couched in the language of sociobiology corresponded to the rise of a new subdiscipline in the history of science devoted to women's encounters with science as producers and subjects of biological research. Female scientists, like their male counterparts, draw on their own experiences and political climate in the decisions they make to follow the leads they find interesting, and in framing the lives of their female subjects as active creators of their biological and social destinies. Their scientific narratives are accessible to both professional and popular audiences, and are often coupled with calls for the preservation of primate habitats and ecologies that are in the process of being destroyed.[88]

Bonobos, also known as pygmy chimpanzees, began to replace baboons as the primate species thought to most closely resemble our early hominid ancestors in the 1990s. Their apparent facility with language, ease in walking upright, and tendency to use sexual relations as a tool for conflict resolution brought the bonobo to popular attention. From his interactions with bonobos, Frans de Waal, for example, has argued that both learned experiences and innate capacities for social interactions define chimpanzee and human culture. Sue Savage-Rumbaugh has similarly suggested that bonobos are capable of abstract thought, empathetic understanding of another's experiences, recognizing themselves in a mirror, and communicating their emotional experiences to other individuals. A simple dichotomy between animal nature and human culture is becoming increasingly less tenable as we learn more about non-human animal cultures, making scientific theories about sex differences and social relations in ape communities seem even more relevant for understanding our own culture.[89]

Perhaps as a result, a true rapprochement between feminism and evolutionary biology has been elusive. Evolutionary psychologists and popular writers continue to produce biological narratives that reproduce old sexual stereotypes: men seek large numbers of young, vibrant, fertile partners in order to maximize their total number of offspring, and women seek older, economically secure husbands to provide their offspring with financial support and stability. In one extreme case, Randy Thornhill and Craig Palmer argued in *A Natural History of Rape* that social scientists misunderstood biological imperatives when they described rape as fundamentally about power, not sex. Rape, Thornhill and Palmer suggested, may have been selected for as an alternative reproductive strategy for men who cannot gain offspring any other way. Their account of male and female sexuality flew in the face of vast quantities of social science research on humans, but accorded well with male reproductive strategies in scorpion flies (Thornhill's model organism of choice). Perhaps not surprisingly, many reviewers criticized the book, noting that scientists should not generalize conclusions from one animal model directly to humans, especially when that model is only distantly related to humans.[90]

In a nutshell

The strategy of using "natural" animal models in the search for explanations of sex differences has a long-standing tradition in twentieth-century biological science, from Darwin's theory of sexual selection, to Fisher's eugenic politics in the 1920s, and even Kinsey's research on the variability of human sexuality. Yet the hardening of evolutionary biology into a rigid adaptationist framework after World War II lent new credibility to biologists who sought to understand the biological basis of human behavior through an evolutionary lens.[91] Despite recent efforts by primatologists to describe behavioral development without recourse to a nature–nurture dichotomy, their message is just beginning to find its way into popular discourse. The power of "nature" as the arbiter of difference between woman and man, human and animal, continues to endure independently of whether scientific explanations of difference are evolutionary, physiological, or genetic in nature.

CHAPTER TWELVE

Creationism, intelligent design, and modern biology
Ronald L. Numbers

Charles Darwin's revolutionary attempt "to overthrow the dogma of separate creations," announced in 1859 in *The Origin of Species*, provoked a broad spectrum of responses among the religious, ranging from mild enthusiasm to anger.[1] Christians sympathetic to his effort tried to mold Darwinism into a shape that made it appear compatible with their religious beliefs. Typical in this respect were two of Darwin's most prominent defenders in the USA, the Calvinists Asa Gray (1810–88), a Harvard botanist, and George Frederick Wright (1838–1921), a cleric-geologist. Gray, who assumed the lead in ensuring that Darwin received a fair hearing in America, welcomed natural selection as the primary mechanism in the production of most species, but for years he favored a "special origination" in connection with the appearance of humans and expressed skepticism about the ability of natural selection "to account for the formation of organs, the making of eyes, &c." In addition, he proposed attributing the inexplicable variations on which natural selection acted to divine providence, not randomness. Wright, a sometime collaborator, went even further, limiting the scope of Darwinism to the descent of plants and animals from several originally created organisms. In doing so, he claimed to be following Darwin himself. In a widely quoted passage near the end of the *Origin of Species*, Darwin had summarized his argument: "I believe that animals have descended from at most only four or five progenitors, and

FIGURE 12.1 Patrick Morris as Asa Gray (left) and Terry Molloy as Charles Darwin (right) in *Re:Design*, produced by the Menagerie Theatre Company, Cambridge. Darwin's correspondent and confidante, Asa Gray, a committed Christian, was Professor of Natural History at Harvard University, and helped popularize Darwinian evolution in North America. (Photo by Sam Pablo Kuper; courtesy of the Menagerie Theatre Company.)

plants from an equal or lesser number," adding that analogy would lead him to believe "that probably all the organic beings which have ever lived on this earth have descended from some one primordial form, into which life was first breathed." In later editions the British naturalist explicitly attributed the first breath to "the Creator."[2]

In 1868, when Darwin published *Variation of Animals and Plants under Domestication*, he politely rejected Gray's suggestion of attributing the variations in Darwin's scheme to divine providence (see Figure 12.1). "However much we may wish it," he wrote, "we can hardly follow Professor Asa Gray in his belief that 'variation has been led [by God] along certain beneficial lines.'"[3] This public refusal to allow God a role in the process of evolution understandably prompted some critics to equate Darwinism with atheism. After reading Darwin's comment, Charles Hodge (1797–1878), the leading Calvinist theologian in America, devoted a small book to answering the question *What Is Darwinism?* (1874). His verdict:

"It is Atheism." Hodge, who had long ago embraced the evolution of the solar system and the antiquity of life on earth, freely granted that orthodox Christians such as Gray could embrace a theistic version of organic evolution—but not Darwin's godless scheme.[4]

Benjamin B. Warfield (1851–1921), one of Hodge's best students, who later held the Charles Hodge chair at Princeton Theological Seminary, tried to reconcile the doctrine of biblical inerrancy with a "modified theory of evolution." Although he once described himself as a youthful "Darwinian of the purest water," he rejected "thoroughgoing evolutionism" and limited it to modification within originally created "types." Evolution, in his view, allowed "not only the directing hand but also the *interfering* hand of God." Like the Southern Presbyterian James Woodrow (1828–1907), who also tried to harmonize evolution and Calvinism, Warfield could identify only one biblical passage at odds with the theory of creation by evolution: "the very detailed account of the creation of Eve," in which God took one of Adam's ribs to help to make Eve. "The upshot of the whole matter is that there is no necessary antagonism of Christianity to evolution, *provided that* we do not hold to too extreme a form of evolution," concluded Warfield. "To adopt any form that does not permit God freely to work apart from law and that does not allow *miraculous* intervention (in the giving of the soul, in creating Eve, etc.) will entail a great reconstruction of Christian doctrine, and a very great lowering of the detailed authority of the Bible."[5]

Liberal Anglican divines in England employed similar tactics. Among the most prominent was Frederick Temple (1821–1902), who spent the last years of his life as archbishop of Canterbury. In writings such as *The Relations between Religion and Science* (1884), he attached a sufficient number of "limitations and modifications" to Darwin's theory to make it theologically acceptable as God's method of creation. The creator, he explained, "impressed on certain particles of matter which, either at the beginning or at some point in the history of His creation He endowed with life, such inherent powers that in the ordinary course of time living creatures such as the present were evolved." Perhaps even more influential among Anglicans in reconciling evolution and Christian theology was Aubrey L. Moore (1848–90), an Oxford-based lecturer, priest, chaplain, and botanist. Rejecting special creation as "an exploded scientific theory . . . which has neither Biblical nor patristic, nor mediaeval authority," he argued that Darwinism was "infinitely more Christian" because it implied "the immanence of God in nature, and the omnipresence of his creative power."[6]

Just as the advocates of evolution massaged Darwinism to fit their ideological and theological needs, so too did its adversaries highlight its real and imagined liabilities. Although it is impossible to document quantitatively, most Christians, at least in the USA, remained skeptical about, if not downright hostile to, Darwinism, however defined. They damned it not only for its scientific shortcomings but for its materialistic implications, its incompatibility with traditional readings of the Bible, and its apparent advocacy of ethics at odds with the teachings of Jesus.[7] Representative of such critics was the Calvinist cleric-geologist Edward Hitchcock (1793–1864). Although he saw no theological barriers to admitting the antiquity of life on earth, a local flood, or even an Edenic creation limited to particular species, he listed a number of religious objections to developmental hypotheses, including the allegations that they made God unnecessary and promoted materialism. "But after all," he added, "the real question is, not whether these hypotheses accord with our religious views, but whether they are true." It seemed preposterous to him—and to many other men of science—that anyone would claim that humans were "merely the product of transformation of the radiate monad through the mollusk, the lobster, the bird, the quadruped, and the monkey."[8]

The issue of human evolution understandably provoked the greatest outrage. One American premillennialist, H. L. Hastings, wrote a tract called *Was Moses Mistaken? or, Creation and Evolution* (1896), in which he addressed this "delicate" topic:

> I do not wish to meddle with any man's family matters, or quarrel with any one about his relatives. If a man prefers to look for his kindred in the zoological gardens, it is no concern of mine; if he wants to believe that the founder of his family was an ape, a gorilla, a mud-turtle, or a monar [*moner*], he may do so; but when he insists that *I* shall trace *my* lineage in that direction, I say No *sir!* . . . I prefer that my genealogical table shall end as it now does, with "Cainan, which was the son of Seth, which was the son of Adam, which was *the son* of God," rather than invent one which reads, "Which was the son of skeptic, which was the son of monkey, which was the son of oyster, which was the son of monar, which was the son of *mud!*"—a genealogical table which begins in the mud and ends in the gravel, which has a monar at the head, a monkey in the middle, and an infidel at the tail.[9]

The aggressive declarations of a few biologists, who announced their determination to drive God from science, aroused fear and anger among the orthodox. "Every preacher of the present hour is compelled to deal with the theory of evolution," complained an irritated Baptist cleric in 1909. "Its advocates have invaded his realm." As evidence of the invasion he cited a text written by Maynard M. Metcalf (1868–1940), professor of biology in the Woman's College of Baltimore, who, though a Christian himself, applauded efforts to cleanse religion from the life sciences, "the last stronghold of the supernaturalist."[10]

Although the overwhelming majority of biologists after about 1880 accepted some form of organic evolution, by the late nineteenth century many were expressing skepticism about the ability of Darwin's theory of natural selection to account for the origin of species. By the early twentieth century, criticism of natural selection had swelled to such proportions that a Christian botanist in Germany, Eberhard Dennert, brought out a book titled (in English) *At the Deathbed of Darwinism* (1903). An American edition appeared a year later. In 1905 the Methodist minister and author Luther T. Townsend (1838–1922), a longtime professor at the Boston Theological Seminary, published *Collapse of Evolution*, in which he told of the mounting anti-Darwinian sentiment in Europe and challenged American evolutionists to be "as honest and manly" as their German colleagues in admitting the failure of Darwinism.[11]

By 1907 rumors of the demise of Darwinism had spread so widely that the Stanford biologist Vernon L. Kellogg opened his book *Darwinism To-Day* with a chapter devoted to its alleged death. Kellogg contrasted the situation in Germany, where even reputable biologists had joined the chorus of criticism, to that in America, where Kellogg saw little evidence of intellectual unrest: "Our bookshop windows offer no display, as in Germany, of volumes and pamphlets on the newer evolutionary study; our serious minded quarterlies, if we have any, and our critical monthlies and weeklies contain no debates of discussions over '*das Sterbelager des Darwinismus* [the deathbed of Darwinism].'" But, he predicted, "just as certainly as the many material things 'made in Germany' have found their way to us so will come soon the echoes and phrases of the present intellectual activity in evolutionary affairs." As word spread of the decline of Darwinism, some American antievolutionists came to suspect that perhaps liberal Christians had capitulated to evolution too quickly. In view of the recent turn of events it seemed likely to one observer that those who had "abandoned the stronghold of faith out of sheer fright will soon be found scurrying back to the old and impregnable citadel, when they learn that 'the enemy is in full retreat.'"[12]

The antievolution crusade in America

The Great War in Europe and the breakdown of social norms focused increasing attention on the evils of evolution. No one contributed more to this backlash than the Presbyterian layman and thrice-defeated Democratic candidate for the presidency of the USA William Jennings Bryan (1860–1925). Early in the 1920s Bryan helped to launch a crusade aimed at driving evolution out of the churches and schools of America. Since early in the century he had occasionally alluded to the silliness of believing in monkeys as human ancestors and to the ethical dangers of thinking that might makes right, but until the outbreak of World War I he saw little reason to quarrel with those who disagreed with him. The war, however, exposed the darkest side of human nature and shattered his illusions about the future of Christian society. Obviously something had gone awry, and Bryan soon traced the source of the trouble to the paralyzing influence of Darwinism on the human conscience. As he explained to one young correspondent, "The same science that manufactured poisonous gases to suffocate soldiers is preaching that man has a brute ancestry and eliminating the miraculous and the supernatural from the Bible." By substituting the law of the jungle for the teaching of Christ, evolution threatened the principles he valued most: democracy and Christianity. Two books in particular confirmed his suspicion. The first, Kellogg's *Headquarters Nights* (1917), recounted firsthand conversations with German officers that revealed the role Darwin's biology had supposedly played in persuading the Germans to declare war. The second, Benjamin Kidd's *Science of Power* (1918), purported to demonstrate the historical and philosophical links between Darwinism and German militarism. Both fundamentalists and scientists took these revelations seriously. But whereas Bryan responded by attempting to quash the offending doctrine, a number of American biologists sought to restore the tarnished reputation of science by stressing the importance of cooperation rather than conflict in the evolutionary process.[13]

About the time that Bryan discovered the alleged Darwinian origins of the war, he also became aware, to his great distress, of unsettling effects the theory of evolution was having on America's own young people. From frequent visits to college campuses and from talks with parents, pastors, and Sunday-school teachers, he heard about an epidemic of unbelief that was sweeping the country. Upon investigating the cause, his wife reported, "he became convinced that the teaching of Evolution as a fact instead of a theory caused the students to lose faith in the Bible, first, in the story of creation, and later in other doctrines, which underlie the Christian religion." Again Bryan found confirming evidence in a recently published book, *Belief in God and Immortality* (1916) by the Bryn Mawr psychologist James H. Leuba (1868–1946), who demonstrated statistically that college attendance endangered traditional religious beliefs. After the much-publicized 1924 trial of Nathan Leopold and Richard Loeb for the kidnapping and murder of Robert Franks, Bryan and other antievolutionists commonly added this heinous crime to their list of charges against the teaching of Darwinism, prompting one frustrated critic to point out that the notorious outlaw Jesse James was a fundamentalist and that "Memphis, Tennessee, where Evolution is outlawed, ... has more murders in proportion to population than any other city in the United States."[14]

Armed with information about the cause of the world's and the nation's moral decay, in 1921 Bryan launched a nationwide crusade against the offending doctrine. In one of his most popular and influential lectures, "The Menace of Darwinism," he summed up his case against evolution, arguing that it was both un-Christian and unscientific. Darwinism, he declared, was nothing but "guesses strung together," and poor guesses at that. Borrowing an illustration almost certainly taken from Alexander Patterson's *Other Side of Evolution* (1903), Bryan explained how the evolutionist allegedly accounted for the origin of the eye:

> The evolutionist guesses that there was a time when eyes were unknown—that is a necessary part of the hypothesis. . . . A piece of pigment, or, as some say, a freckle appeared upon the skin of an animal that had no eyes. This piece of pigment or freckle converged the rays of the sun upon that spot and when the little animal felt the heat on that spot it turned the spot to the sun to get more heat. The increased heat irritated the skin—so the evolutionists guess, and a nerve came there and out of the nerve came the eye!

"Can you beat it?" he asked incredulously—and that it happened not once but twice? As for himself, he would take one verse in Genesis over all that Darwin wrote.[15]

Bryan was far from alone in balking at the evolutionary origin of the eye. Christian apologists had long regarded the intricate design of the eye as "a cure for atheism," and Darwin himself had readily conceded his vulnerability on this point. "To suppose that the eye, with all its inimitable contrivances for adjusting the focus to different distances . . . could have been formed by natural selection, seems, I freely confess, absurd in the highest possible degree," he wrote in the *Origin of Species*. But logical consistency impelled him to extend "the principle of natural selection to such startling lengths." When an American correspondent, the Harvard botanist Asa Gray, wrote a letter candidly describing the section dealing with the making of the eye as "the weakest point in the book," Darwin confided in reply that "the eye to this day gives me a cold shudder." That Bryan, too, shuddered says nothing about his scientific sophistication.[16]

Like Bryan, most of the antievolutionists of the 1920s paid little attention to developments in biology. Thomas Theodore (T. T.) Martin (1862–1939), a Baptist preacher from Mississippi, claimed to be a long-time student of the sciences; however, his knowledge of organic evolution was such that he identified the inheritance of acquired characteristics and natural selection as the "two great pillars" of the evolutionary edifice, apparently unaware that biologists had come to view these two theories as describing competing rather than complementary mechanisms. Convinced that evolution was poisoning the minds of the nation's youth, he campaigned tirelessly to eradicate "the scourge" from the classrooms of America. In a diatribe titled *Hell and the High Schools* (1923), written while alleged wartime atrocities were still fresh in people's minds, he argued that "the Germans who poisoned the wells and springs of northern France and Belgium" and fed little children poisoned candy were angels "compared to the text-book writers and publishers who are poisoning the books used in our schools." Martin could hardly find words strong enough to express the enormity of evolution. "Next to the fall of Adam and Eve," he insisted, "Evolution and the teaching of Evolution in tax-supported schools is the greatest curse that ever fell upon this earth."[17]

Antievolutionists typically went out of their way not to appear antiscience. The Northern Baptist minister William Bell Riley, founder of the World's Christian Fundamentals Association, outlined the reasons why fundamentalists opposed the teaching of evolution. "The first and most important reason for its elimination," he explained, "is the unquestioned fact that evolution is not a science; it is a hypothesis only, a speculation." Bryan often made the same point, defining true science as "classified knowledge . . . the explanation of facts." This view came straight from the dictionary. "What is science?" asked one Holiness preacher who opposed evolution. "To the dictionaries!" he answered. There one learned that science was "certified and classified knowledge." By narrowly drawing the boundaries of science and emphasizing its empirical nature, creationists could at the same time label evolution as false science, claim equality with scientific authorities in comprehending facts, and deny the charge of being antiscience. "It is not 'science' that orthodox Christians oppose," a fundamentalist editor insisted defensively. "No! no! a thousand times, No! They are opposed only to the theory of evolution, which has not yet been proved, and therefore is not to be called by the sacred name of *science*."[18]

To support their claim that evolution was unscientific, creationists compiled impressive rosters of scientists who allegedly shared their point of view. Shortly after the turn of the century, Townsend assembled one of the earliest—and most frequently cribbed—lists in order to prove that "the most thorough scholars, the world's ablest philosophers and scientists, with few exceptions, are not supporters, but assailants of evolution." In addition to such late worthies as Louis Agassiz, Arnold Guyot, and John William Dawson, Townsend cited the English physiologist Lionel S. Beale (1828–1906) and the German pathologist Rudolf Virchow (1821–1902), both of whom had expressed doubts about the evidence for human evolution, as well as a dozen or so lesser lights, including a "Dr. Etheridge, of the British Museum, one of England's most famous experts in fossilology," and "Professor Fleischmann, of Erlangen, one of the several recent converts to anti-Darwinism." In 1920 Martin, responding to claims that all responsible scientists were evolutionists, borrowed generously from Townsend's booklet to draw up his own list of "twenty-one really great scientists in the world" who rejected evolution. The target of Martin's scholarship, the Southern Baptist evolutionist William Louis Poteat (1856–1938), replied with a withering analysis of the credentials of the twenty-one:

> Two do not appear in the biographical dictionaries, five are misrepresented: seven won reputation in other than biological fields, and six have been in their graves more than forty years, two of these having died long before Darwin's great book was published. One lone biologist is left to support the thesis that the doctrine of evolution is discarded by the science of today. And this man's position is so peculiar that he is usually mentioned as the single exception to the universal acceptance of evolution by biologists of responsible position.[19]

The lone biologist identified by Poteat was Albert Fleischmann (1862–1942), a reputable but relatively obscure German zoologist who taught for decades at the University of Erlangen in Bavaria. In 1901 he published a scientific critique of organic evolution, *Die Descendenztheorie*, in which he rejected not only Darwinism but all theories of common organic descent.

This gave him a unique position among biologists. As Kellogg noted in 1907, Fleischmann seemed to be "the only biologist of recognised position ... who publicly declared a disbelief in the theory of descent." The German creationist apparently remained of the same mind for the rest of his life. In 1933, the year of his retirement from Erlangen, he presented a paper to the Victoria Institute in London in which he dismissed the notion of a "genealogical tree" as a "fascinating dream." "No one can demonstrate that the limits of a species have ever been passed," he asserted. "These are the Rubicons which evolutionists cannot cross." In his declining years, Fleischmann informed English acquaintances that he was writing a book "that will wipe evolution off the slate," but the work never appeared.[20]

The widely touted "Dr. Etheridge, of the British Museum," who always appeared in creationist literature without a given name, was quoted by Townsend as saying: "In all this great museum there is not a particle of evidence of transmutation of species. Nine-tenths of the talk of evolutionists is sheer nonsense, not founded on observation and wholly unsupported by fact. This museum is full of proofs of the utter falsity of their views." The content of Etheridge's statement varied from work to work, and its source remained unidentified, except for Alexander Patterson's comment that Etheridge was answering a question put to him by a Dr. George E. Post. When curious parties in the 1920s inquired about the identity of Etheridge, the director of the British Museum surmised that the man in question was "Robert Etheridge, Junr., who was Assistant Keeper of Geology in this Museum from 1881 to 1891," at which time he left for Australia, where he died in 1920. The director hastened to add that "Mr. Etheridge's opinion on this subject should not be considered as in any way representing scientific opinion in this Museum."[21]

The antievolutionists received an important psychological boost at the beginning of their campaign from an address in 1921 to the American Association for the Advancement of Science (AAAS). At that meeting the distinguished British biologist William Bateson (1861–1926) declared that scientists had *not* discovered "the actual mode and process of evolution." As knowledge of living things had multiplied, he said, biologists had grown increasingly agnostic about questions of origin. Although he warned against misinterpreting his comments as a rejection of evolution, creationists applauded his speech while evolutionists rushed to control the damage. The council of the AAAS issued a statement affirming that "the evidences in favor of the evolution of man are sufficient to convince every scientist of note in the world," and Henry Fairfield Osborn (1857–1935), head of the American Museum of Natural History, contemptuously dismissed Bateson's remarks as having come from someone "living the life of a scientific specialist, out of the main current of biological discovery." To say that "we have failed to discover the origin of species," declared Osborn, "is valueless and directly contrary to the truth." Creationists, however, paid no more attention to the warnings of the AAAS and Osborn than they did to Bateson's own caveat.[22]

Creationists celebrated again in the late 1920s when the distinguished Smithsonian zoologist Austin H. Clark (1880–1954), struck by the absence of intermediate forms between the major groups of animals, challenged the conventional view of evolution that represented species as branches of a single tree. The *Christian Fundamentalist* quoted him as saying that "so far as concerns the major group of animals, the creationists seem to have the better of the argument. There is not the slightest evidence that any one of the major groups arose from any other. Each is a special animal-complex, related more or less closely to all the rest, and appearing, therefore, as a special and distinct creation." But both in correspondence with creationists and in his controversial book, *The New Evolution: Zoogenesis* (1930), Clark pointedly refused to support a supernatural view of origins. As one confused and disappointed creationist explained, "Dr. Clark . . . denies the most pointed evidence of evolution, but he sticks to evolution just the same." Thus during the heyday of the antievolution movement, the German zoologist Fleischmann remained the only well-credentialed biologist to lend his authority to the creationist attack on evolution.[23]

Unfortunately for the antievolutionists, no one active in their movement possessed so much as a master's degree in biology. S. James Bole (1875–1956), professor of biology at fundamentalist Wheaton College, Illinois, stood virtually alone as a creationist with advanced training in biology. After earning a master's degree in education from the University of Illinois, he had gone on to study pomology (fruit culture) at the Illinois school of agriculture—and abandoned evolution for creation and biblical inerrancy. Unlike most of his fellow-believers, Bole granted that "from the point of view of specialized knowledge it would seem that the greater authority rested with the evolutionists." But because organic evolution dealt with theory, not facts, choosing between creation and evolution became a matter of faith: "The evolutionist has faith in his hypothesis; the Christian has faith in his God."[24]

Two other prominent antievolutionists had exposed themselves to science while studying medicine. Arthur I. Brown (1875–1947) was an

American-born surgeon who had settled in Vancouver, British Columbia, where he became, in his own estimation, "one of the leading surgeons of the Pacific Coast." His handbills touted him as "one of the best informed scientists on the American continent," and hosts routinely introduced him as a famous scientist, perhaps the "greatest scientist in all the world." By the early 1920s he was publishing antievolution pamphlets with titles such as *Evolution and the Bible* (1922) and *Men, Monkeys and Missing Links* (1923). Despite his superior knowledge of science, at least for a creationist, he contributed few novel ideas to the cause; instead, he tended to popularize the works of others and to use illustrations of apparent design in the natural world to argue for the necessity of special creation. Brown's most original contribution to the arsenal of antievolution polemics was a critique of arguments for evolution based on laboratory analyses of blood serum from animals and humans, which provided scientists with a means of identifying lines of descent that turned out to be identical to those suggested by morphological studies. In *Evolution and the Blood-Precipitation Test* (1925), Brown argued that biologists had no justification for assuming that animals having chemically similar blood shared a common ancestor, because similarities in blood types could also be explained by appealing to "an Omniscient and Omnipotent Creator and Designer."[25]

The other erstwhile medic was Harry Rimmer (1890–1952), a young Presbyterian minister and self-styled "research scientist" who had attended San Francisco's Hahnemann Medical College, a small homeopathic institution that required only a high-school diploma or its "equivalent" for admission. Although he did not remain long enough to earn a degree, he did pick up a vocabulary of "double-jointed, twelve cylinder, knee-action words" that later served to impress the uninitiated. During the early 1920s, Rimmer served as an itinerant speaker for the YMCA while maintaining a home in Los Angeles. Between trips he built a little workshop in the backyard next to the garage, which he liked to call his laboratory. As described by his wife, this "little laboratory . . . contained a darkroom, a sink and running water, a microscope, centrifuge and test tubes." Although he rarely conducted actual experiments, he spent considerable time in the laboratory taking pictures of microscopic animals and other objects, which he used to illustrate his books and lectures, including one on "The Collapse of Evolution." In his public talks he liked to pose as a research scientist, but he often merely displayed his ignorance of contemporary biology. Following a lecture at Kansas State Teachers College (now Pittsburg State University), the student paper give the following description:

Dr. Rimmer offered as the greatest argument against evolution the fact that acquired characteristics are not inherited. For instance, he related an experiment performed on the rabbit family.

[A] rubber vest was strapped to the breast of the female rabbit to prevent the use of fur for the lining of the nest. In its place carded wool was used. After ten generations it was found that the rabbit did not inherit the acquired habit that was thrust upon its ancestors.

Dr. Rimmer cited as another instance that acquired characteristics are not inherited, the age-old Chinese custom of binding of the feet. After three thousand years of this custom we find the Chinese children are born with perfectly shaped feet.

Regardless of his arguments, local antievolutionists hailed him as "a scientific Moses or Lincoln," and the president urged him to join the faculty.[26]

According to the editor of *Science*, "the principal scientific authority of the Fundamentalists" was the Canadian George McCready Price.[27] As an adolescent he and his widowed mother had joined the Seventh-day Adventist Church, founded by the prophetess Ellen G. White. During one of her trance-like visions God had allowed her personally to witness the creation, which she saw had occurred in six literal days of twenty-four hours. On other occasions God revealed the role played by Noah's Flood in laying down the fossil-bearing rocks—in a single year, not millions. Collapsing the geological column into the work of one year deprived evolutionists of the time they needed and focused attention on the deluge rather than on the creation. Price repeatedly insisted that the idea of successive geological ages represented the crux of the evolution problem. He thus devoted comparatively little attention to biological questions. "What is the use of talking about the origin of species," he asked as early as 1902, "if geology can not prove that there has actually been a succession and general progress in the life upon the globe?" Besides, the development of Mendelian genetics, which seemingly allowed for only "definite and predictable" variations, had so eroded confidence in the efficacy of natural selection, Price felt he needed only to write the "funeral oration" for Darwinism. As he observed in *The Phantom of Organic Evolution* (1924), his most extensive critique of the *biological* arguments for evolution, "A dead lion needs no bullets." Price sat in the audience in 1921 when Bateson confessed to the AAAS that the origin of species remained a scientific mystery. Price left the lecture hall convinced that he had just heard "the swan song of Darwinism."[28]

In truth, Price had little quarrel with evolutionists regarding the origin of *species*, as the term was commonly used by biologists. Although he preferred to equate "species" with the originally created Genesis "kinds," and in later life regretted that he had conceded so much to evolutionists, he at times freely admitted that new species, narrowly defined, had evolved from "the great stocks, or families," initially created by God—and at a rate far more rapid than most evolutionists demanded. The more variation he allowed, the easier it was for him to sidestep "a real difficulty," namely, explaining "how the great diversity of our modern world may have come about after the world disaster of the Deluge, from a comparatively few kinds which were salvaged from that great cataclysm." The point at issue between creationists such as Price and evolutionists was not variation but its extent and direction: "whether the general run of these changes have . . . all been in the direction of degeneration, not development."[29]

In private, while addressing fellow Adventists, Price followed Ellen G. White in substituting demonic manipulation for natural selection in order to explain the origin of species. Publicly, however, he tended to attribute the appearance of new species to environmental influences and *divine* intervention. In the chapter "Species and Their Origin" in *The Phantom of Organic Evolution,* he outlined a theory of common descent with modifications characterized by a process of "degeneration downward" rather than "development upward" and by quick rather than slow change. The most rapid change had probably occurred in the immediate post-Flood period, when both humans and animals had encountered a novel environment. In a textbook for high-school students, he explained the process:

> Very radical changes in the environment of plants or of animals tend to make the species vary; and if they survive in such new environments, their size or color or habits, or other physical "characters," will be different. This is the reason why the living forms are so different, in some instances, from their ancestral forms found as fossils in the rocks; for there has been a very radical change in their environment in passing from the antediluvian world to the modern one. Some species have changed so considerably that scientists do not recognize them as the same, but give them different specific or even different generic names, calling the older form "extinct," and the modern form a "new" species.

New postdiluvian species arose not so much by means of Darwinian natural selection or by a Lamarckian inheritance of acquired characteristics but because "the great superintending Power which is over nature, adapted these men and these animals and plants to their strange world." In accounting for the geographical distribution of plants and animals, which he admitted was a "very difficult" problem for creationists because nearly identical environments did not always possess the same flora and fauna, he similarly appealed to divine intervention, speculating that after the deluge animals spread out from Mount Ararat "under the direct guidance of the Creator."[30]

Price sincerely believed that evolution was scientifically unsound, but he admitted from the beginning that his primary objections to the theory were "philosophical and moral." Like other religious fundamentalists, he repeatedly warned that evolution not only made a shambles of Christian theology and ethics but threatened political freedom as well. The scope of the threat is conveyed by a query with which he began one of his antievolution tracts: "Do you know that the theory of evolution absolutely does away with God and with His Son Jesus Christ, and with His revealed Word, the Bible, and is largely responsible for the class struggle now endangering the world?" In two of his books, *Poisoning Democracy* (1921) and *Socialism in the Test-Tube* (1921), he explicitly linked evolution and "Marxian" socialism. With Bryan, he thought that World War I, during which Germany put "the ruthless ethics of Darwinism . . . into actual practice," provided ample evidence of the threat evolution posed to human freedom.[31]

Evolution within created kinds

During the period between the 1920s and the 1960s the most pressing biological question centered on the fixity of species since Eden. Most Flood geologists, as Price's followers came to be known, eagerly welcomed "every bit of *modification during descent* that can reasonably be asserted." The underlying problem was the shortage of space on Noah's Ark. "If we insist upon fixity of species we make the Ark more crowded than a sardine can," argued one Flood geologist. "If we agree to all [variation] that can be demanded, we simplify the Ark problem greatly," because all present-day species could have descended from relatively few passengers on the ark. In *Genes and Genesis* (1940) Harold W. Clark, a student of Price's who had gone on to earn a master's degree in biology from the University of California, Berkeley, defended limited Darwinian natural selection—within genera, families, and even orders—against the "extreme creationism" of those who insisted that God had created every species.[32]

Clark was not alone in pushing for greater acceptance of microevolution within the originally created kinds. In the early 1940s another university-trained Price protégé, Frank Lewis Marsh (1899–1992), joined Clark in advocating post-Edenic speciation. While teaching at an Adventist school in the Chicago area, Marsh took advanced work in biology at the University of Chicago and obtained an M.S. in zoology from Northwestern University in 1935, specializing in animal ecology. Later, in 1940, after joining the faculty of Union College in Lincoln, he completed a Ph.D. in botany at the University of Nebraska, where he wrote his dissertation on plant ecology and became the first Adventist to earn a doctoral degree in biology. Like Clark, Marsh never deviated from a literal, recent creation and universal Flood, but the more he learned, the more he questioned the notion that all species had originated by separate creative acts. Zoologists, he noted, had identified thousands of species of dry-land animals alone, yet Adam had been able to name all of them in a single day. Thus it seemed unreasonable to equate the Genesis kinds with the multitudinous species of the twentieth century.[33]

In his first book, *Fundamental Biology* (1941), written from the point of view of a "fundamentalist scientist," Marsh portrayed the living world as the scene of a cosmic struggle "between the Creator and Satan." Confusion about the nature of the originally created "kinds" often arose, he warned, because Satan, "a master geneticist," had "built up within the kinds, different races, strains, and types which look quite unrelated to other members of the kind." Taking his cue from Ellen G. White, whom he regarded "as essential to the biologist today as an aid in understanding the written Word and the book of nature," Marsh speculated that amalgamation or hybridization had been "the principal tool used by Satan in destroying the original perfection and harmony among living things." It explained "the origin of multitudinous species, varieties, and races of plants and animals on the earth today." The black skin of Negroes was only one of many "abnormalities" engineered in this diabolical way. Despite generally agreeing with Clark on the natural origin of species, he thought his California friend had gone too far in allowing for the crossbreeding of "kinds." In contrast to Clark, who believed that the biblical reference to plants and animals producing "after their kind" was merely a moral principle, Marsh regarded the Genesis statement as a biological law that forever separated the different "kinds," which he called "baramins" (from the Hebrew words *bara*, "created," and *min*, "kind") to denote the original units of creation.[34]

Hoping that a defense of creationism by a credentialed biologist might cause the scientific world to take note, he had his publisher send out complimentary copies of *Evolution, Creation, and Science* (1944) to prominent evolutionists, including the Harvard zoologist Ernst Mayr (1904–2005) and the Russian-born Columbia geneticist Theodosius Dobzhansky (1900–1975). The former declined to comment, preferring to take the advice of "the rabbit in Walt Disney's film *Bambi*: 'Don't say anything if you don't have anything nice to say.'" Dobzhansky, however, believing that "the majority should at least consider the minority view and subject it to criticism," engaged in an extended correspondence with Marsh that vividly reveals the issues dividing creationists and evolutionists. Shortly before the appearance of Marsh's book, the Orthodox Russian émigré had remarked in his *Genetics and the Origin of Species* (1937) that "among the present generation no informed person entertains any doubt of the validity of the evolution theory in the sense that evolution has occurred." But reading Marsh's book convinced him otherwise. In reviewing it for the *American Naturalist,* Dobzhansky announced that Marsh had written what he had previously thought to be impossible: a sensibly argued defense of special creation. Dobzhansky expressed particular surprise at discovering how much evolution (within "kinds") a creationist such as Marsh was willing to grant: "He outbids evolutionists on the score of the speed of the changes, for he assumes that all dogs, foxes, and hyenas are members of a single 'kind,' and, therefore, must have descended from a common ancestor in any event less than 6000 years ago." However, in rejecting macroevolution, Marsh's book taught the valuable lesson that "no evidence is powerful enough to force acceptance of a conclusion that is emotionally distasteful."[35]

Although Dobzhansky found Marsh's ideas scientifically invalid and religiously subversive, he respected the church-college biologist for being "the only living scientific anti-evolutionist." The obscure creationist, for his part, could scarcely conceal his delight at having the unexpected chance to argue his case before one of the leading evolutionists in the world. In response to Dobzhansky's comment about outbidding the evolutionist, Marsh amended and clarified his position. He explained that God in originally stocking the world with plants and animals had created not only "kinds" but "varieties" within those kinds, capable of crossbreeding. Thus he professed not to claim, for example, that all dogs and foxes had descended from a single ancestor—ignoring for the moment the problem of overcrowding on Noah's Ark. And though he had said virtually nothing about the mechanism of change in his book, he now admitted that the concepts of natural selection and survival of the fittest were "really extremely important" in explaining the present distribution of species.[36]

The central issue separating the two biologists hinged on the nature of scientific proof. Marsh, who assimilated all the evidence of microevolution into his creationist paradigm of changes within "kinds," demanded nothing less than laboratory-based demonstrations of macroevolution. But, as Dobzhansky pointed out, the evidence for such large-scale evolution rested on inference, not direct observation. Because macroevolution took place in geological time, he patiently explained, it could "be proven or disproven only by inference from the available evidence." Marsh, predictably, found this argument unconvincing. "Alas! Inferential evidence again!" he exclaimed. "Is there no *real* proof for this theory of evolution which we may grasp in our hands?" Eventually, explanation gave way to frustration, with Dobzhansky finally brushing Marsh's concerns aside with the quip "If you demand that biologists would demonstrate the origin of a horse from a mouse in the laboratory then you just can not be convinced."[37]

Marsh rejected the inferential evidence for macroevolution primarily because of his prior commitment to the scientific and historical veracity of the Bible. In justifying such an allegiance, he maintained that "in not one single instance" had the Bible been proven false. "That very real fact should mean something to us as scientists," he argued. "In the light of that fact, the Genesis statement regarding the origin of living things must likewise be tested if we are to make wise use of the sources of truth at our disposal." Dobzhansky found this line of reasoning unpersuasive, but he credited Marsh for at least stating candidly in his book that "the account given by the Bible is settled for you before you begin to consider the biological evidence."[38]

After more than two months of nearly weekly exchanges, Marsh trusted Dobzhansky sufficiently to expose some of his innermost fears and feelings. In what he suspected would be his final letter, he assured the geneticist that he was not "a chronic grouch who goes about looking for something to argue about." Nor was he "looking for ease, comfort, and a good name." Though he disliked being at odds with his scientific brethren, he was "willing to take it on the chin" if principle required it and if he could get mainstream scientists to accept special creation as a legitimate alternative to evolution. In closing, he expressed the hope that Dobzhansky would find "some diversion in these letters, some pleasant mental gymnastics, and possibly experience a broader acquaintance with unusual ideas so that the benefits will not all be going one way." Six years later, in the third edition of his *Genetics and the Origin of Species*, Dobzhansky cited Marsh as the exception to the rule that "an informed and reasonable person can hardly doubt the validity of the evolution theory, in the sense that evolution has occurred." The creationist biologist proved "only that some people have emotional biases and preconception[s] strong enough to make them reject even completely established scientific findings." It wasn't much of an acknowledgment, but Marsh appreciated the recognition.[39]

The leading British creationist at this time was Douglas Dewar (1875–1957), a barrister and amateur ornithologist who had taken a first-class honors degree in natural science at Jesus College, Cambridge, before entering the Indian Civil Service. After returning to England in 1924, he turned his attention to evolution. He had left school "a rather half-hearted" Darwinist, but, he later explained, "subsequent study and field work led me first to reject Darwinism and finally evolution." He also grew increasingly committed to the infallibility of the Bible and concerned about the great harm evolution was doing "to the morality of the white races." By 1931 he had completed a critique called *Difficulties of the Evolution Theory*, a catalog of alleged problems ranging from geographical distribution to blood reactions. Swayed in particular by the scarcity of transitional forms in the fossil record and by his inability to identify natural forces capable of originating new organic types, he recommended adopting "a provisional hypothesis of special creation . . . supplemented by a theory of evolution." The biologist, he argued, had only two choices: "Every kind of living organism must have originated by one or the other of two processes—evolution or special creation." Unable to find a publisher for his book, Dewar underwrote the printing himself.[40]

About the same time, Dewar and a collaborator submitted a short paper on the distribution of mammalian fossils for publication in the proceedings of the Zoological Society of London, of which both authors were fellows. Their statistics, showing a greater number of mammalian genera in the lower Oligocene than at present, appeared to undermine evolutionists' claims about the incompleteness of the fossil record and thus "to shatter the whole case for evolution." The society, understandably, declined to publish it. "We got the opinion of a first-rate paleontologist and geologist about it," explained the secretary, "and he told us that although it must have taken a very long time to compile it, he thought this kind of evidence led to no valuable conclusion." The distinguished geneticist J. B. S.

Haldane (1892–1964), who questioned the suitability of the proceedings for such a paper, offered a simple explanation of Dewar's findings: The fossil record represented many successive faunas over millions of years, while the comparatively few existing mammalian fauna reflected "a single moment of evolution." Dewar, nevertheless, interpreted the rejection as an attempt by those who had made evolution "a scientific creed" to muzzle independent-thinking creationists. Articulating what would soon become a *creationist* dogma, he argued that "those who do not accept this creed, are deemed unfit to hold scientific offices; their articles are rejected by newspapers or journals; their contributions are refused by scientific societies; and publishers decline to publish their books except at the author's expense." To overcome such prejudice, Dewar and a handful of like-minded British creationists founded the Evolution Protest Movement, publicly launched in 1935.[41]

Scientific creationism

In 1961 John C. Whitcomb, Jr., an Old Testament scholar, and Henry M. Morris, a hydraulic engineer, published *The Genesis Flood*, which carried Price's hitherto marginal Flood geology to masses of Bible-believing Christians. In attacking evolution, the authors followed Price in focusing on its geological and paleontological foundations rather than on biology. The success of *The Genesis Flood* led to the formation in 1963 of the Creation Research Society (CRS), dedicated to promoting a "young-earth" version of creationism, which limited the history of life on earth to no more than 10,000 years. Of the ten founding members, five possessed earned doctorates in biology from major universities; a sixth had earned a Ph.D. degree in biochemistry from the University of California, Berkeley; and a seventh held a master's degree in biology. The leader of the group was Walter E. Lammerts (1904–96), a Berkeley-trained plant geneticist and Missouri Synod Lutheran who as an undergraduate in the 1920s had embraced Flood geology. Although he went on to earn a doctorate in cytogenetics from the University of California—making him one of the first strict creationists to hold a Ph.D. in biology—and won national acclaim as a plant breeder, for decades he maintained a low profile as an antievolutionist. As he explained in the early 1940s, while teaching ornamental horticulture at the University of California, Los Angeles, "I keep my anti-evolution views to myself there . . . it will do no good for me to chirp until I've become a big enough 'bird' to at least make a loud squawk!" By the early 1960s he felt financially and professionally secure enough to create a stir.[42]

The presence of a credentialed geneticist in the creationist camp led to a dispute over the repopulation of the earth after the Flood: Could it have occurred naturally or were miracles required? Like most other young-earth creationists, Morris permitted considerable organic development after the deluge, while Lammerts attributed post-Flood diversity to divine genetic engineering. Morris viewed such miraculous intervention—amounting, in his opinion, to a second creation—as theologically suspect because it violated the "economy of the miraculous in God's orderly world"; he openly condemned Lammerts's "imaginative reconstruction of the variegated activity of the one creation and cataclysm clearly described in the Bible." Lammerts denied that God's genetic tinkering after the Flood could be considered "creations," though it had produced new forms of plant, animal, and human life. "I will say that if Morris can explain how Noah could have been heterozygous for all the distinctive characteristics of human beings, and then how the races of mankind could have originated by *natural means* I will be most happy to accept such an explanation," he replied. "Evolutionists are hard put to account for this complexity in hundreds of thousands of years, yet Morris would have this occur in the 5300 or so years since the Flood, by natural selection of the variation potential which existed in the survivors of the Flood."[43]

The phrase "creation research" in the society's name bordered on the oxymoronic. Even Lammerts, who coined the term, admitted the impossibility of researching creation because "we were not there to watch God do it!" As he explained to one perplexed questioner, the CRS investigated extant evidence of God's creative activity, not the act of creation itself. For years creationists tended to expend their limited resources on library research (to find fallacies or inconsistencies in the writings of evolutionists) or on low-cost field studies (to find evidence that might support Flood geology). The early volumes of the *Creation Research Society Quarterly* featured a few experimental reports from Lammerts himself intending to cast doubt on the likelihood of beneficial mutations required by evolutionary theory. Using the facilities of the Lawrence Radiation Laboratory in Livermore, California, he induced mutations in roses by irradiating buds with high-energy neutrons. He found that the resulting changes stayed within fixed limits and rarely, if ever, produced commercially desirable features. "The fact that they change only within the limits of their variability potential and that most mutations are harmful should make evolution minded scientists reconsider their basic assumptions," he concluded.[44]

In 1967 the CRS established a research committee to encourage original investigations, but the minuscule grants produced little of value. The chair of the committee in the late 1960s and early 1970s, Larry G. Butler, a biochemist at Purdue University, sought to "present an image of scientific respectability as much as possible without Biblical compromise." Above all, he hoped "to exclude authentic psychopaths, cranks, & kooks" looking for a forum to present their far-fetched ideas. As Butler quickly discovered, too many creationists suffered from a weakness for the sensational: "We make astonishing observations (human footprints contemporary with dinosaurs); we postulate dramatic upheavals (sudden deposits of masses of ice from a planetary visitor); we propose sweeping scientific generalizations (negation of the entire system of 14C dating)." He grew increasingly horrified by the continuing influence of what he called "the lunatic fringe" of creationism. In 1972 he circulated a stunning "Critique of Creationist Research" to CRS board members. In it he damned creationists for their "negative, critical perspective" toward evolution, for their delusional search "for some dramatic discovery which will completely overthrow the theory of evolution once and for all at a single blow," and for their emphasis on historical and descriptive evidence rather than experimental and manipulative studies. "Investigations relying on evidence involving affidavits, sworn statements, etc., are not scientific," he observed. Given the difficulties associated with investigating processes no longer in operation, he found it remarkable that "creationist research" existed at all. Although a few CRS leaders congratulated Butler on his jeremiad, it produced little tangible change, and the crusading biochemist increasingly felt like "a thorn in flesh" of his brethren. Discouraged by his failure to raise the scientific standards of creationist research, he resigned from the board of directors in 1975 and later abandoned creationism altogether.[45]

Creationists relished exposing instances of evolution being turned to questionable social and political ends. Nell Segraves, founder of the Creation Science Research Center in San Diego, reported in 1977 that research conducted by the center had "demonstrated that the results of evolutionary interpretations of science data result in a widespread breakdown in law and order," leading to "divorce, abortion, and rampant venereal disease."[46] The link between evolution and abortion, regarded by conservative Christians as two of the greatest evils, seemed especially striking. "Why worry about disposing of 'unwanted' babies in the womb," asked one creationist, "since they are simply baby primates?" Other creationists attempted to connect evolution with feminism. "Feminism is anticreation and antiChristian," declared a contributor to the *Creation Research Society Quarterly*:

> There can be no such thing as Christian feminism. They are conceptual contradictions like capitalistic communism and evangelical lesbianism. The divine order of the sexes is a corollary doctrine of the order of creation. It establishes the family as the natural unit within society and the church.... The husband/father is given the headship in the family and the church with the wife/mother as an assistant and chief educator of the children.

In a non-Christian society governed by "survival of the fittest," men and women would depart from this ideal to the extent of competing equally for jobs.[47]

In the early 1970s the partisans of Flood geology, hoping to gain entrance into public-school classrooms, began referring to their Bible-based views as "scientific creationism" or "creation science." In his popular textbook *Scientific Creationism* (1974) Morris argued that creationism

FIGURE 12.2 An exhibit in the $27 million Creation Museum in Kentucky. (Courtesy of Answers in Genesis.)

could "be taught without reference to the book of Genesis or to other religious literature or to religious doctrines." Although Morris and his creationist colleagues occasionally appealed to scientific and philosophical authorities, for the most part they worked outside the context of established science and philosophy. References to the names of scientists who questioned evolutionary orthodoxy served largely as literary ornaments (Figure 12.2).[48]

Intelligent design

Scientific creationism continued to flourish at the turn of the millennium, but by the mid-1990s the focus of attention among creationists and evolutionists alike was shifting to a new form of antievolutionism called "intelligent design" (ID). From a small but well-funded base at the Discovery Institute in Seattle, Washington, these antievolutionists sought to foment a veritable "intellectual revolution." Unlike creation scientists, the advocates of ID largely ignored geological arguments associated with the age of the earth and the extent of Noah's Flood and focused instead on finding biological evidence of a god-like designer. Most controversial of all, they tried "to reclaim science in the name of God" by rewriting the rules of science to allow the inclusion of supernatural explanations of phenomena, insisting that *"The ground rules of science have to be changed."*[49]

Leading this effort was a Presbyterian lawyer on the faculty of the University of California at Berkeley, Phillip E. Johnson. Although he targeted evolution, he identified the real enemy as scientific naturalism, which ruled all god-talk out of science. Most troubling of all was the widespread acceptance of so-called methodological naturalism among even evangelical Christians, who, despite their beliefs, left God at the laboratory door and refrained from invoking the supernatural when trying to explain the workings of nature. In contrast to "metaphysical naturalism," which denied the existence of a transcendent God, methodological naturalism implied nothing about God's existence and activities. But to Johnson, it smacked of thinly veiled atheism, and he set out to discredit it using a strategy called "the wedge." As he explained it:

> A log is a seeming solid object, but a wedge can eventually split it by penetrating a crack and gradually widening the split. In this case the ideology of scientific materialism is the apparently solid log. The widening crack is the important but

seldom-recognized difference between the facts revealed by scientific investigation and the materialist philosophy that dominates the scientific culture.

Johnson fancied himself making "the initial penetration into the intellectual monopoly of scientific naturalism."[50]

In one of the canonical ID texts, *Darwin's Black Box: The Biochemical Challenge to Evolution* (1996), the biochemist Michael J. Behe argued that biochemistry had "pushed Darwin's theory to the limit ... by opening the ultimate black box, the cell, thereby making possible our understanding of how life works." The "astonishing complexity of subcellular organic structure" led him to conclude—on the basis of scientific data, he asserted, "not from sacred books or sectarian beliefs"—that intelligent design had been at work. "The result is so unambiguous and so significant that it must be ranked as one of the greatest achievements in the history of science," he declared. "The discovery [of intelligent design] rivals those of Newton and Einstein, Lavoisier and Schroedinger, Pasteur and Darwin." As newspapers and magazines spread the news of Behe's discovery of what he called "irreducibly complex" organic structures—such as the bacterial flagellum, which propels microscopic organisms—he won recognition as a modern-day William Paley (1743–1805), the famous natural theologian of the early nineteenth century (see Topham, Chapter 4, this volume).[51]

Unlike Behe, who flirted with theistic evolution and the notion of common descent from a single organism, William Dembski, another leading ID theorist, unambiguously rejected both ideas. "*Design theorists are no friends of theistic evolution*," he declared.

> As far as design theorists are concerned, theistic evolution is American evangelicalism's ill-conceived accommodation to Darwinism. What theistic evolution does is take the Darwinian picture of the biological world and baptize it, identifying this picture with the way God created life. When boiled down to its scientific content, theistic evolution is no different from atheistic evolution.

Although he acknowledged that organisms had "undergone some change in the course of natural history," he maintained "that this change has occurred within strict limits and that human beings were specially created."[52]

Both theistic and atheistic evolutionists expressed horror at the ID movement's radical agenda. Having long since come to terms with doing science naturalistically, reported the Gordon College chemist J. W. Haas, Jr., editor of the evangelical journal *Perspectives on Science and*

Christian Faith, "most evangelical observers—especially working scientists—[remained] deeply skeptical." Though supportive of theistic world views, they balked at being "asked to add 'divine agency' to their list of scientific working tools." The Christian Haas agreed with the British atheist Richard Dawkins (see McGrath, Chapter 13, this volume) that relying on intelligent design to explain complex biological organisms was "a pathetic cop-out of [one's] responsibilities as a scientist." The phenomena ID advocates deemed "irreducibly complex," their critics regarded as a scientific problems to be explained by future research. Besides, Haas noted, ID theorists rarely appealed to disciplines outside of biology, leaving "the rest of us physicists, chemists, mathematicians, or geologists ... to go our 'godless' ways in spite of the complexities we face at the quantum level or with the weather."[53] In 2004 the ID camp published its first work in a peer-reviewed scientific journal, the less-than-prestigious *Proceedings of the Biological Society of Washington.* In it the Cambridge-trained philosopher of science Stephen C. Meyer, who headed up the Discovery Institute's ID initiative, argued that "purposive or intelligent design" best explains "the origin of the complex specified information" involved in the so-called Cambrian explosion, when, an estimated 530 million years ago, a host of new animal phyla appeared on the scene. The merits of Meyer's achievement evaporated when officers of the society discovered editorial irregularities associated with the acceptance of the article.[54]

The first major legal test of intelligent design began with the Dover (Pennsylvania) Area School District Board's decision to make students "aware of gaps/problems in Darwin's theory and of other theories of evolution including, but not limited to, intelligent design." To implement this action, the board instructed ninth-grade biology teachers to read a statement to their classes describing Darwinism as "a theory ... not a fact." At a subsequent trial to determine the constitutionality of this directive, in 2005, Behe, the star witness for the defense, acknowledged that intelligent design lacked a mechanism for explaining how designed structures arose. Shortly before Christmas Judge John E. Jones III handed down his verdict, excoriating the Dover school board for its actions, which he described as a "breathtaking inanity." Although a conservative Republican and a practicing Lutheran, Jones ruled that ID was "not science" because it invoked "supernatural causation" and failed "to meet the essential ground rules that limit science to testable, natural explanations." The board's promotion of it thus violated the establishment clause of the First Amendment to the U.S. Constitution, requiring the separation of church and state. He rejected as "utterly false" the assumption "that evolutionary theory is antithetical to a belief in the existence of a supreme being and to religion in general." His conclusion: "it is unconstitutional to teach ID as an alternative to evolution in a public school science classroom."[55]

After more than a decade of effort the Discovery Institute proudly announced in 2007 that it had got some 700 doctoral-level scientists and engineers to sign "A Scientific Dissent from Darwinism." Though the number may have struck some observers as rather large, it represented less than 0.023 percent of the world's scientists. On the scientific front of the much ballyhooed "Evolution Wars," the Darwinists were winning handily.[56] The ideological struggle between (methodological) naturalism and supernaturalism continued largely in the fantasies of the faithful and the hyperbole of the press.

CHAPTER THIRTEEN

The ideological uses of evolutionary biology in recent atheist apologetics
Alister E. McGrath

One of the most remarkable developments during the opening years of the twenty-first century has been the appearance of a number of high-profile populist books offering an aggressively atheist critique of religion.[1] This "clustering" of prominent works of atheist apologetics in the period 2004–7 is of no small historical interest in its own right, and is widely taken to reflect a cultural reaction against "9/11"—the suicide attacks in New York in September 2001, widely regarded as being motivated by Islamic extremism.[2]

Yet the appearance of these works is of interest for another reason. A central theme of two of them is that developments in biology, especially evolutionary biology, have significantly negative implications for belief in God. Daniel Dennett's *Breaking the Spell* and Richard Dawkins' *The God Delusion*, both published in 2006, express the fundamental belief that the Darwinian theory of evolution has such explanatory power that it erodes many traditional metaphysical notions—such as belief in God—through its "universal acid." This represents an extension of the basic lines of argument found in earlier works, in which an appeal to biological understandings of human origins,[3] subsequently amplified to include accounts of the origins of human understandings of purpose and value based on evolutionary psychology,[4] which was made in order to erode the plausibility of belief in God.

From its first appearance, some saw Darwinism as a potential challenge to at least some aspects of the traditional Christian view of creation. Yet it is important to appreciate that most early evolutionists, including Charles Darwin himself, did not consider that they were thereby promulgating or promoting atheism.[5] Since the beginning of the nineteenth century, serious Christian thinkers had come to realize that at least some metaphorical interpretation was demanded in considering the early chapters of Genesis, so that their possible incompatibility with evolution was not the major stumbling block for the intelligentsia that might be expected (see also Harrison, Chapter 1, this volume).[6] Nor is there any shortage of later significant evolutionary biologists who held that their science was consistent with their faith, such as Ronald A. Fisher, author of *The Genetical Theory of Natural Selection* (1930), and Theodosius Dobzhansky, author of *Genetics and the Origin of Species* (1937).[7] The emphasis upon Darwinism as an acid that totally erodes religious belief, though anticipated in earlier periods, appears to have reached a new intensity in the first decade of the twenty-first century.

This chapter sets out to explore the emergence of this focused appeal to evolutionary biology in Dennett's and Dawkins' recent works of atheist apologetics, both considering it in its historical context and offering an assessment of its impact on the popular understanding of Darwinism in the early twenty-first century.[8] This appeal to biology in the defense of atheism is complex and nuanced, and there are significant differences of substance and emphasis between atheist writers who adopt such an approach. Nevertheless, some common factors emerge, which suggest that this is an appropriate line of inquiry to pursue, of no small intrinsic intellectual interest to both historians and evolutionary biologists.

As my concern in this chapter is specifically with biological issues, I shall not engage with the more general argument, also embedded within some recent atheist writings, that the natural sciences as a whole make faith in God intellectually irresponsible or risible.[9] This argument occasionally reflects an implicit presumption, generally not defended by an appeal to historical scholarship, of the permanent validity of a "warfare" or "conflict" model of the historical interaction of science and religion.[10] It is clear that this model has continuing cultural authority, especially at the popular level. It may have been radically revised, even discredited, by academic historians; it is, however, clear that this development has yet to filter down to popular culture. While this atheist argument merits close attention, as it has relevance for the calibration of traditional Christian approaches to evidence-based apologetics, it is not a topic that I propose to address further here. My main theme is the manner in which Darwinism has been transposed in recent atheist apologetics from a provisional scientific theory to an antitheistic ideology. My focus is on the ideological use of the biological sciences, especially evolutionary biology, in recent atheist apologetics, a topic which I believe is best considered under three broad categories: (1) the elevation of the status of Darwinism from a provisional scientific theory to a worldview; (2) the personal case of Charles Darwin as a role model for scientific atheism; and (3) the use of the concept of the "meme"—a notion that reflects an attempt to extend the Darwinian paradigm from nature to culture—as a means of reductively explaining (and hence criticizing) belief in God.

It is clear that this list of themes could easily be expanded; however, limits on space have caused me to focus on what I consider to be the most important elements in these works, illuminating the manner in which Darwinism is increasingly being presented as an ideology, rather than as a provisional scientific theory.

Darwinism as an ideology

One of the most interesting developments of the twentieth century has been the growing trend to regard Darwinian theory as transcending the category of provisional scientific theories, and constituting a "worldview."[11] Darwinism is here regarded as establishing a coherent worldview through its evolutionary narrative, which embraces such issues as the fundamental nature of reality, the physical universe, human origins, human nature, society, psychology, values, and destinies. While being welcomed by some, others have expressed alarm at this apparent failure to distinguish between good, sober, and restrained science on the one hand, and non-empirical metaphysics, fantasy, myth and ideology on the other.[12] In the view of some, this transition has led to Darwinism becoming a religion or atheist faith tradition in its own right.[13]

Richard Dawkins

One of the most important advocates of Darwinism as a universal atheist theory or worldview is Richard Dawkins (born 1941), who until 2008 was professor of the public understanding of science at Oxford University (Figure 13.1).[14] Dawkins went up to Oxford to read zoology in 1959. After

FIGURE. 13.1 Richard Dawkins (born 1941). (Photo by Matthias Asgeirsson.)

graduating in 1962, he went on to undertake research in Oxford University's department of zoology under the supervision of Professor Nikolaas Tinbergen (1907–88), joint winner of the Nobel Prize in Medicine and Physiology for 1973. Tinbergen and his Austrian colleague Konrad Lorenz (1903–89) are widely credited with pioneering ethology—the study of the behavior of animals in their normal environment. After completing his doctorate, Dawkins spent a year doing postdoctoral research, along with some lecturing in the department of zoology. Tinbergen was on sabbatical leave during the academic year 1966–7 and asked Dawkins to cover some of his lectures while he was at the same time writing up his Ph.D. thesis for publication. Dawkins' lectures allowed him to explore some aspects of W. D. Hamilton's theory of kin selection, including the question of how certain apparently cooperative forms of behavior arise.[15] An individual behaves in such a way that the reproductive capacity of another individual is enhanced, even to its own detriment. The phenomenon may be observed in aspects of the social, parental, and mating behavior of animals. So how could this have evolved?

Dawkins came to the conclusion that the "most imaginative way of looking at evolution, and the most inspiring way of teaching it" was to see the entire process from the perspective of the gene. The genes, for their own good, are manipulating and directing the bodies that contain them, and carry them about. Throughout his writings, Dawkins has developed the rhetoric of a gene's eye view of things—not simply of the individual, but of the entire living world. Animals were thus to be seen as "machines carrying their instructions around" with them, using their every aspect as "levers of power to propel the genes into the next generation." Dawkins can be regarded as the first, and still the most systematic, ethologist of the gene.

From Oxford, Dawkins went on to become assistant professor of zoology at the University of California at Berkeley in 1967, returning to Oxford as lecturer in zoology and fellow of New College in 1970. It was during this time that his most influential and creative works were published, including *The Selfish Gene* and *The Blind Watchmaker*.[16] In 1995, he was appointed to a new academic position at Oxford University, made possible by the generosity of Charles Simonyi, then one of the Microsoft Corporation's foremost software architects, who went on to co-found the Intentional Software Corporation in August 2002. Dawkins was appointed as the first Charles Simonyi Reader in the Public Understanding of Science. The following year, Oxford University conferred on Dawkins the additional title of "professor of the public understanding of science," by which he is now generally known. He was elected as a Fellow of the Royal Society—the supreme accolade for a British scientist—in 2001.

Because of his influence as a popularizer of biological themes,[17] especially those relating to evolutionary biology, and his linking of these themes with the pursuit of atheist proselytizing,[18] Dawkins is a particularly important figure in any discussion of the recent use and abuse of biology. For Dawkins, Darwinism is more than a provisional scientific theory; it is a universal theory, capable of accounting for the world of empirical observation—including the observation that many people believe in God.

In an important essay of 1996, Dawkins argued that there are at present only three possible ways of seeing the world: Darwinism, Lamarckism, or God. The last two fail to explain the world adequately; the only option is therefore Darwinism. "I'm a Darwinist because I believe the only alternatives are Lamarckism or God, neither of which does the job as an explanatory principle. Life in the universe is either Darwinian or something else not yet thought of."[19]

Now the rhetoric of this argument would seem to suppose that Darwinism, Lamarckism, and belief in God are three mutually exclusive views, so that commitment to one necessarily entails rejection of the others. This is, of course, highly questionable, even if it has become part of the settled assumptions within Dawkins' circle. While a good case can be made for the mutual incompatibility of Darwinism and Lamarckism, neither of these is actually incompatible with belief in God.

Yet Dawkins' comments are interesting for another reason, in that they illustrate his firm conviction that Darwin's theory of evolution—as developed in the light of Mendelian genetics and our understanding of the place of DNA (deoxyribonucleic acid) in the transmission of inherited information—is more than just a scientific theory, on the same epistemological level as other such theories. It is a worldview, a total account of reality. Darwinism, he suggests, is a "universal and timeless" principle, capable of being applied throughout the universe. In comparison, rival worldviews such as Marxism are to be seen as "parochial and ephemeral."[20] These rival worldviews lack the grounding in a scientific understanding of reality that is so characteristic of Darwin's theory of natural selection, and the understandings of the world and human nature that emerge from it.

Now, at one level, this might seem to be little more than a reassertion of the role of the natural sciences in understanding our place within the world. Its obvious exaggerations might be explained away as reflecting the need to restate the legitimate role of the sciences in the face of the rival claims of the humanities or religion, especially in the light of postmodern attempts to deny any epistemic advantages to the natural sciences.[21] Yet a closer reading of Dawkins indicates that this is not the case. Darwinism is not being presented as a representative element of the scientific enterprise, with a legitimate place at the round table of ethical and social debate. It is clearly understood as *the* defining account of reality.

Where most evolutionary biologists would argue that Darwinism offers a *description* of reality, Dawkins goes further, insisting that it is to be seen as an *explanation* of things.[22] Darwinism is a worldview, a *grand récit*, a metanarrative—a totalizing framework, by which the great questions of life are to be evaluated and answered. For this reason, we should not be surprised to learn that Dawkins' account of things has provoked a response from postmodern writers, for whom any metanarrative—whether Marxist, Freudian, or Darwinian—is to be resisted as a matter of principle.[23]

Daniel Dennett

A similar position is taken in the recent writings of Daniel Dennett (born 1942), currently the director of the Center for Cognitive Studies and the

FIGURE 13.2 Daniel Dennett (born 1942). (Photo by David Orban.)

Austin B. Fletcher Professor of Philosophy at Tufts University (Figure 13.2). Dennett is primarily concerned with providing a philosophy of mind that is grounded in empirical research informed by the capacity of a Darwinian theory of evolution to explain some of the content-producing features of consciousness. Dennett defends a theory of "neural Darwinism," similar to that developed in the late 1970s by Gerald Edelman.[24] During the 1990s, Dennett set out to explore a series of issues from an evolutionary standpoint, including the question of what distinguishes human minds from animal minds,[25] how human free will may be reconciled with a naturalist view of the world,[26] and how the origins and successes of religion may be accounted for.[27] Like Dawkins, Dennett has sought to make these ideas accessible to a wider public, arguing that evolutionary theory allows naturalist accounts to be given of human belief in the transcendent in general, and the concept of God in particular.

In *Darwin's Dangerous Idea* (1995), Dennett aims to show "why Darwin's idea is so powerful, and why it promises—not threatens—to put our most cherished visions of life on a new foundation."[28] Darwinism, for Dennett, is the "universal acid" that erodes outdated, superfluous metaphysical notions, from the idea of God downward. Darwinism, he asserts, achieves a correlation of "the realm of life, meaning, and purpose with the realm of space and time, cause and effect, mechanism and physical law."[29] The Darwinian world is devoid of purpose and transcendence, in that all can and should be explained by the "standard scientific epistemology and metaphysics." The Darwinian worldview demystifies and unifies our experience of the world, and places it on more secure foundations.[30]

The Darwinian debate

Dawkins and Dennett are important contributors to, and exhibitors of, an important cultural and historical debate over the status of Darwinism. Is it a scientific theory, similar to Einstein's theory of general relativity, which, though clearly important in a specific domain, has limited relevance to a broader cultural agenda? Or is it to be compared with Marxism, a way of understanding the world that, if correct, has major implications for a much broader cultural and social agenda?[31] Although this debate was initiated by the publication of Darwin's *Origin of Species* itself, it has emerged as a significant point of contention following the reemergence of creationism within conservative Protestant groups in the USA. Part of the debate occasioned by the emergence of the "intelligent design" movement concerns whether Darwinism, as taught within the science curriculum of public schools, is considered as a scientific theory or a worldview.[32]

This transition from Darwinism, considered as a provisional scientific theory applying to one aspect of reality, to Darwinism, considered as a universal worldview—"reductionism incarnate" (Dennett)[33]—encounters some serious difficulties. One is the obvious point that all scientific theories are to be regarded as provisional, open to correction—and possibly even requiring to be abandoned—in the light of accumulating observational evidence and advancing theoretical understanding. Dawkins is aware of this problem, and is quite explicit about its consequences. "Darwin may be triumphant at the end of the twentieth century, but we must acknowledge the possibility that new facts may come to light which will force our successors of the twenty-first century to abandon Darwinism or modify it beyond recognition." Dawkins suggests that it may be possible to isolate a "core Darwinism" that is relatively resistant to this kind of historical erosion.[34] The religious and metaphysical implications of this will be obvious. Yet Dawkins generally seems curiously reluctant to incorporate this note of caution in his bold statements about the triumph of Darwinism as a worldview.

This point needs further comment. It is perfectly correct to say that "evolutionary biologists currently believe that Darwinism is the best theoretical explanation of earth's present life-forms." But this does not mean that future evolutionary biologists will share this judgment. We may *believe* that Darwinism is right, but we do not *know* that this is so. To do this, we need to stand in a hypothetical position, which allows us to see the observational evidence and have access to the theoretical judgments of the future. On the basis of what we currently know, we may hold this theoretical position; however, the history of science makes it clear that new evidence has a habit of arising, causing radical revision—perhaps even abandonment—of many long-held theories. Will Darwinism be one of them? And what then will happen to any worldviews, atheist or theist, that are founded on Darwinism? The only honest answer is that we don't know. We may indeed—as I would—suggest that Darwinism is sufficiently robust to allow us to expect its future endorsement. Yet we must at least allow for the possibility of radical theory change here, even if we may regard this as improbable, on the basis of the evidence currently at our disposal.

The key point is that scientific theorizing is *provisional*. It offers what is believed to be the best account of the experimental observations currently available. Radical theory change takes place either when it is believed that there is a better explanation of what is currently known, or when new information comes to light that forces us to see what is presently known in a new light. Unless we know the future, it is impossible to take an absolute position on the question of whether any given theory is "right." What can—and, indeed, must—be said is that this is believed to be the best explanation currently available. History simply makes fools of those who argue that every aspect of the current theoretical situation is true for all time. The problem is that we don't know which of today's theories will be discarded as interesting failures by future generations.

So how do Dawkins and Dennett defend this deeply problematic transition from "scientific theory" to "worldview"? I am not persuaded that they perceive the need for any such defense, in that they tend to assume that this is self-evident. However, others have noted the problem, and sought to account for the transition. One particularly helpful explanation is offered by Timothy Shanahan, who pointed out that Dawkins' method proceeded "by elaborating the logic of 'adaptationist philosophy' for Darwinian reasoning."[35] Dawkins' broader conclusions are determined by a set of logical premises, which are ultimately—yet indirectly—grounded in the empirical data. In effect, Dawkins uses an essentially inductive approach to defend a Darwinian account of evolution, and then infers from this a Darwinian worldview, in the form of a set of premises from which secure conclusions may be deduced. Having inferred that Darwinism is the best explanation of what may be observed, Dawkins appears to transmute a provisional theory into a certain worldview. Atheism is thus presented as the logical conclusion of a series of axiomatic premises, having the certainty of a deduced belief, even though its ultimate basis is actually inferential.

So what are the leading features of such a Darwinian worldview? Clearly, it is assumed to be atheist by those who deploy it in their apologetic writings. While Darwinism qua scientific theory clearly has no decisive implications for belief in God, Darwinism qua worldview holds that belief in God arises through an essentially Darwinian mechanism, in that this somehow conveys survival value on individuals or communities. There is clearly a debate within evolutionary circles over both the role of religion in cultural development, and the level at which selection takes place. While some believe that religious beliefs confer advantages on groups, Dawkins holds that "religion" is an epiphenomenon (a random by-product of something else) that has selective advantage.[36] In this chapter, I note two other features of a Darwinian worldview: the critique of any notion of purpose, and the rejection of any transcendent involvement in the world.

The notion of purpose

What impressed Thomas Henry Huxley (1825–95) most forcibly on his first reading of Darwin's *Origin of Species* was the "conviction that teleology, as commonly understood, had received its deathblow at Mr Darwin's hands."[37] Yet, for some writers, Darwinism was perfectly capable of preserving the notion of "purpose." Two influential writers who took such a view may be noted: Henri Bergson (1859–1941) and Pierre Teilhard de Chardin (1881–1955), both of whom developed philosophies of life that were founded on the acceptance of biological evolution, yet interpreted this as having some kind of purpose.[38] This was subjected to a withering criticism by their fellow Frenchman Jacques Monod (1910–76),[39] who argued that *teleonomy* had displaced *teleology*. The term "teleonomy" was introduced into biological use in 1958 by the Princeton biologist C. S. Pittendrigh (1918–96) "in order to emphasize that recognition and description of end-directedness does not carry a commitment to Aristotelian teleology as an efficient causal principle."[40] In using this term, Monod wished to highlight that evolutionary biology was concerned with identifying and clarifying the mechanisms underlying the evolutionary process. While the mechanisms that governed evolution were of interest, they had no goal. One thus could not speak meaningfully of "purpose" within evolution.

This point was developed further by Dawkins, who conceded the "appearance of design" or purpose within the biological realm, while insisting that these "appearances" were simply semblances, understandable illusions that must now be abandoned. This "strong illusion of purposeful design" can easily be explained on the basis of the outcome of chance mutations over huge periods of time. A Darwinian world has no purpose, Dawkins argues, and we therefore delude ourselves if we think otherwise. "The universe we observe had precisely the properties we should expect if there is, at bottom, no design, no purpose, no evil and no good, nothing but blind pitiless indifference."[41]

There are, however, some difficulties with this position. In his discussion of the notion of teleology in biology, Ernst Mayr (1904–2005) emphasized that many biological phenomena could legitimately be termed teleological, in that they are characterized by two significant features: (1) they are seemingly purposeful, directed toward a goal (such as finding food); and (2) they consist of active behavior.[42] For Mayr, some criticisms of the notion of "teleology" in a biological context actually rest on philosophical precommitments, rather than on biological observations. It is clear that the discussion of the origins and significance of goal-directed behavior and phenomena within the biological realm has some considerable way to go.[43]

Furthermore, Dawkins moves with remarkable alacrity from the debatable (yet entirely possible) statement "Darwinism discloses no purpose to human life" to the rather more problematic assertion that "there is therefore no purpose to life." Many would argue that the natural sciences (including Darwinism, if considered as a scientific theory), although entirely competent at clarifying and correlating the complex relationships that exist within the material world, are nevertheless out of their depth when it comes to questions of meaning or value. When it comes to "ultimate questions" (qua the philosopher Karl Popper), science simply does not possess the intellectual resources necessary for answers. This does not mean that there are no answers; it is simply to note that the natural sciences cannot determine them.

It is clear that there are many questions that, by their very nature, have to be recognized as lying beyond the legitimate scope of the scientific method. In their philosophically sophisticated account of the foundations of the neurosciences, Max Bennett and Peter Hacker pointed out that the natural sciences are not in a position to comment upon such questions, if their methods are applied legitimately.[44] These questions cannot be lightly dismissed as illegitimate or nonsensical; they are simply to be recognized as lying beyond the scope of the scientific method. If they can be answered, they must be answered on other grounds.[45] But the fact that they need to be answered on non-scientific grounds does not mean that they are unanswerable or inappropriate.

This point was made repeatedly by Peter Medawar (1915–87), the noted Oxford immunologist who won the Nobel Prize for Medicine for his discovery of acquired immunological tolerance. In a significant publication entitled *The Limits of Science*, Medawar considered the question of how science was limited by the nature of reality. Emphasizing that "science is incomparably the most successful enterprise human beings have ever engaged upon," he distinguished between what he held to be "transcendent" questions, which are better left to religion and metaphysics, and questions about the organization and structure of the material universe. With regard to the latter, he argues, there are (and can be) no limits to the possibilities of scientific achievement.

So what of other questions? What about the question of God? Or of whether there is purpose within the universe? Though a self-confessed rationalist, Medawar was clear that certain limits exist. "That there is indeed a limit upon science is made very likely by the existence of questions that science cannot answer, and that no conceivable advance of science would empower it to answer." Such questions, he suggested, included the fundamentally teleological issues of "What are we all here for?" or "What is the point of living?" Whereas doctrinaire positivism tended to dismiss such questions as non-questions or pseudo-questions "such as only simpletons ask and only charlatans profess to be able to answer," Medawar affirmed both their validity and importance.[46]

Elimination of transcendent causes

The other issue, especially associated with Daniel Dennett, has to do with the elimination of transcendent causes and presences within the natural realm in general, and from scientific explanation in particular. This is the essential point of his ingenious distinction between "cranes" and "skyhooks." Cranes are natural, yet complex, intermediary mechanisms that arise in the course of the process of evolution itself, and contribute to that process by enabling the emergence of still more complex structures. "Skyhooks" are arbitrary, imaginary inventions, devised to evade purely naturalist explanations and accounts of the natural world and its processes. God is perhaps the most obvious example of such a "skyhook." For Dennett, "good reductionism," which is fundamentally a "commitment to non-question-begging science," holds that "everything in nature can all be explained without skyhooks."[47]

In developing this point, Dennett adopts a strategy that is common in recent atheist apologetics. Darwin has solved the riddles that led previous generations to conclude that divine causality was the only explanation of the orderedness of the world. Even such luminaries as John Locke or David Hume held that it was simply inconceivable that design could arise without the influence of a creative mind. After Darwin, however, it became possible to explain how design can emerge by way of a purposeless, mindless, mechanical process.[48]

This naturally leads us to consider how Darwin himself has featured in recent atheist apologetics.

The religious views of Charles Darwin

Richard Dawkins attributes a pivotal role to Darwin in the elimination of any explanatory necessity for God. For Dawkins, the advent of Darwin has changed everything. Before Darwin, scientists (such as Isaac Newton) believed in God; afterwards, they do not—or at least they ought not to. Not if they are "real" scientists, that is. Darwin, Dawkins insists, has shown belief in God to be little more than "cosmic sentimentality," endowing life with a "saccharine false purpose," a naïve and dangerous belief that natural science has a moral mission to purge and debunk. Such naïve beliefs, he argues, might have been understandable before Darwin came along. But not now. Darwin has changed everything. Newton, Dawkins argues, would be an atheist if he had been born after Darwin. Before Darwin, atheism was just one among many religious possibilities; now, it is the only serious option for a thinking, honest, and scientifically informed person. To believe in God nowadays is to be "hoodwink'd with faery fancy."[49]

Once upon a time such religious beliefs would have been understandable, perhaps even forgivable. But not now. Humanity was once an infant. Now we have grown up and discarded infantile explanations. And Darwin is the one who marks that decisive point of transition. Intellectual history is thus divided into two epochs—before Darwin, and after Darwin. James Watson, the Nobel Prize winner and co-discoverer of the structure of DNA, once stated his belief that Charles Darwin will eventually be seen as a far more influential figure in the history of human thought than either Jesus Christ or Mohammed. According to Dawkins, Darwin impels us to atheism. It is not merely that evolution erodes the explanatory potency of God; it eliminates God altogether.

It will be clear that if Darwin could be shown to have abandoned faith in God as a consequence of his theory of evolution, the case for suggesting that Darwinism was intrinsically atheist would be considerably enhanced. If everything altered after Darwin, it is clearly important to determine what Darwin himself believed to have changed as a result of his new ideas. Yet Dawkins' discussion of the complex and fascinating interaction of Darwin's scientific and religious views is most disappointing, and in my opinion fails to deal satisfactorily with the issues involved.[50]

The view that Darwin was indeed an atheist on account of his evolutionary doctrine was vigorously advocated in Edward Aveling's pamphlet *The Religious Views of Charles Darwin* (1883).[51] The evidence brought forward in this short work is far from persuasive, and it is unclear what weight should be attached to it. Darwin had earlier refused Aveling permission to dedicate his *Student's Darwin* to him. Aveling was one of Karl Marx's most dedicated English followers, and regarded Darwin's evolutionary views as reinforcing the basic ideas of Marxist materialism. Darwin did not wish to endorse such an association.

There are several important passages in Darwin's writings that can be interpreted to mean that Darwin ceased to believe in an orthodox Christian conception of God on account of his views on evolution. The problem is that there are other passages that variously point to Darwin maintaining a religious belief, or to his losing an orthodox Christian faith—which is not the same as ceasing to believe in God!—for reasons quite other than evolutionary concerns. The tragic death of Darwin's daughter Annie in

1851 at the age of ten, for example, seems to have eroded any lingering belief he may have had in a benevolent divine providence.[52] However, a note of caution must be injected: on the basis of the published evidence at our disposal, it is clear that Darwin himself was far from consistent in the matter of his religious views. It would therefore be extremely unwise to draw any confident conclusions on these issues.[53]

There can be no doubt that Darwin abandoned what we might call "conventional Christian beliefs" at some point in the 1840s, although the dating of this must probably remain elusive. Yet there is a substantial theoretical gap between "abandoning an orthodox Christian faith" and "becoming an atheist." Christianity involves a highly specific conception of God; it is perfectly possible to believe in a god other than that of Christianity, or to believe in God and reject certain other aspects of the Christian faith. On this reading, Darwin can be placed within a transition, particularly within cultured circles, away from the specifics of Christianity toward a more generic concept of God, largely determined by the ethical values of the day.[54] While it is certainly fair to suggest that Darwin, like many others, appears to have moved toward what might reasonably be described as a "deist" conception of God, rather than the more specific divinity of the Christian faith, this needs to be set in its historical context. Many would argue that, as a matter of historical fact, many scientists have held religious views that were heterodox, at least to some degree.[55]

Darwin is often situated within, both as spectator and participant, the narrative of the "Victorian crisis of faith." While this has been dominant in recent accounts of this formative period in British cultural history, it is open to challenge at a number of points.[56] Furthermore, the alleged "crisis of faith" resulting from Darwin appears to have been limited in England to working-class radicals and biblical literalists. Within the Victorian intelligentsia of the period, the revolt against Christian orthodoxy appears to have been fundamentally *ethical* rather than *intellectual*.

There is no doubt that Darwin reflects the ambiguous attitude toward religious faith of his age. In 1879, while working on his autobiography, Darwin commented on his personal religious confusion: "My judgement *often fluctuates* . . . In my most extreme fluctuations I have never been an Atheist in the sense of denying God. I think that generally (and more and more as I grow older), *but not always*, that an Agnostic would be the more correct description of my state of mind." That seems entirely consistent with his stated views on the matter, and is generally considered to be the safest interpretation of his position.

Many have pointed out that some high-profile attempts to establish an annual "Darwin Day" were initially driven by an antireligious agenda that has threatened to overshadow Darwin's importance as a scientist. It is perhaps important to recall that, like his contemporary T. H. Huxley—often referred to as "Darwin's bulldog"—Darwin actually had little time for either religious or antireligious dogmatism. The Christian church quickly learned that it could live with, even value, Darwin's ideas, and I see no overwhelming reasons—historical, scientific, or theological—for dissenting from that judgment.[57] Darwin, I believe, like all good natural scientists, helps us to appreciate the complexity, the beauty, and the wonder of nature. And by doing so, he prompts us to ask deeper questions—questions of meaning. What is the point of life? Why are we here? If Peter Medawar is right, science may not be able to answer those questions. It may clarify processes; yet illumination of mechanisms is not the same as establishing meaning.

The meme as a biological explanation of belief in God

We have already noted the view that Darwinism is too big a theory to be restricted to the biological domain. Richard Dawkins has been one of the most outspoken advocates of the application of Darwinism to human cultural development as a whole. "Universal Darwinism" possesses an explanatory capacity that is capable of being extended far beyond these limits.[58] An example of such an application of Darwinian theory to a broader range of cultural issues is found in evolutionary psychology.[59]

Dawkins here reflects a consensus within a significant section of the evolutionary biological community in the opening decade of the twenty-first century. The Darwinian paradigm, it is argued, offers a magisterial explanatory model, capable of accounting for developments far beyond the realm of the purely biological. This markedly upbeat assessment contrasts sharply with the situation a century ago. At the beginning of the twentieth century, the notion of "cultural evolution" temporarily began to lose its appeal, possibly on account of growing anxiety within professional biological circles concerning the plausibility of the Darwinian evolutionary schema.[60] However, in the period following World War II, the neo-Darwinian synthesis began to emerge, and proved intellectually resilient. This created new interest in the possibility of cultural evolutionary theory, which was given a new injection of energy by the work of American anthropologists Julian H. Steward (1902–72) and Leslie A. White (1900–75).[61]

The case for proposing that cultural evolution is, at least in some sense of the word, Darwinian is eminently plausible.[62] Darwin himself often used analogies drawn from the world of culture to illuminate his theory of natural selection. In particular, he was clearly intrigued by the parallels between biological and linguistic evolution[63]—a theme that would be explored with enthusiasm during the later nineteenth century[64] before being abandoned as something of a dead end.

More recently, however, the possibility of a neo-Darwinian mechanism for cultural evolution has been the subject of fresh discussion.[65] Some have argued that fundamentals of a Darwinian mechanism can be discerned within cultural development, thus reopening the possibility that Darwinian evolutionary theory may have a much wider applicability than once thought.[66] The development of human culture, it is argued, demonstrates variation, competition, inheritance, and the accumulation of successive cultural modifications over time. A reasonable case for Darwinian cultural evolution thus seems to have been established.

The evolutionary psychologist Donald T. Campbell (1916–96) developed such ideas as early as 1960,[67] and introduced the term "mnemone" to refer to the cultural replicators that his theory required.[68] Yet the serious exploration of the question of the mechanism of cultural evolution is generally thought to have begun in 1981, when the Stanford biologists Luigi Luca Cavalli-Sforza and Marcus Feldman proposed a quantitative approach that permitted the analysis of the dynamics of change within a fixed population of specified forms of cultural traits.[69] This study suggested that some form of non-Darwinian process was implicated, at least in part, suggesting that it might be necessary to propose a mechanism of "cultural evolution" that was distinct from natural selection.

A quite distinct approach was developed around the same time by Charles Lumsden and Edward O. Wilson, who proposed that human cultural transmission was fundamentally determined by a genetic unit they termed the "culturgen."[70] This concept was derived from the notion of the "artefact," already widely used in archaeology to refer to operational units of culture. According to Lumsden and Wilson, genetic and cultural evolution are interconnected. While culture is ultimately shaped by biological processes, those biological processes are simultaneously modified in response to ensuing cultural change.

There would therefore seem to be a case for the use of evolutionary theory—again, remember that this is a theory derived from biology—to a broader range of intellectual, cultural, and religious issues. If Darwinism is indeed the grand theory that Dawkins and others propose it to be, it should have the capacity to offer reductive explanations of why people believe in God. One such explanation—that of the "meme"—has featured prominently in recent atheist apologetics. In 1976, Dawkins introduced the notion of the "meme" as "a unit of cultural transmission, or a unit of imitation." The "god-meme" survives particularly well because it has "high survival value, or infective power, in the environment provided by human culture."[71] The meme is thus a hypothesized cultural analog of the biological gene.

Because Darwinism cannot be limited to biology, Dawkins argued for a memetic account of the development and transmission of ideas, including belief in God:

> Examples of memes are tunes, ideas, catch-phrases, clothes, fashions, ways of making pots or of building arches. Just as gene types propagate themselves in the gene pool by leaping from body to body via sperm or eggs, so memes propagate themselves in the meme pool by leaping from brain to brain by a process which, in the broad sense of the term, can be called imitation.[72]

Memetic evolution is not just analogous to genetic evolution; it is essentially a manifestation of the same phenomenon. Cultural evolution simply depends upon a different unit of transmission evolving in a different medium at a faster rate. Evolution by natural selection occurs wherever conditions of variation, replication and differential fitness arise.

The memetic model is, however, deeply problematic, for a variety of reasons. As Dawkins himself appreciated, his canonical statement of the concept of the meme in *The Selfish Gene* was deficient. Dawkins proposed the meme as a hypothetical cultural replicator to explain cultural development within a Darwinian framework—"a unit of cultural transmission, or a unit of *imitation.*"

Yet on Dawkins's "gene's eye" view of evolution, it is the *gene* that is the unit of selection, even though it is the *phenotype* that is actually subject to the process of selection. The gene is the replicator, or the set of instructions; the phenotype is the physical manifestation of the organism, the visible characteristics or behavior resulting from that set of instructions. Yet curiously, the examples of "memes" that Dawkins offers in *The Selfish Gene* are the *result* of such instructions, not the instructions themselves. They are all examples of what Cloak terms "m-culture"—in other words, things which arise through the impact of ideas on the environment—where one would expect them to be "i-culture" (again, in Cloak's terms).[73] While Dawkins proposed an analogy between *meme* and *gene*, he actually illustrated this by appealing to the cultural equivalent of *phenotypes*, not genes.[74] On this approach, the meme appears to be a Lamarckian, not a Darwinian, concept.

Furthermore, genes can be "seen," and their transmission patterns studied under rigorous empirical conditions. But what about memes? The simple fact is that they are, in the first place, *hypothetical constructs*, inferred from observation rather than observed in themselves, and in the second place, *unobservable*. This makes their rigorous investigation intensely problematic, and fails to enable a *meme* and an *idea* to be satisfactorily distinguished. How, many have asked, can the meme be considered a viable scientific hypothesis when there is no clear operational definition of a meme, no testable model for how memes influence culture, no explanation for why standard selection models are not adequate, a general tendency to ignore the sophisticated social science models of information transfer already in place, and a high degree of circularity evident in the explanation of the power of memes?[75]

Yet in *The God Delusion*, published thirty years after his introduction of the notion of the "meme," Dawkins seems to present memes as if they were established scientific orthodoxy,[76] making no mention of the inconvenient fact that the mainstream scientific community views it as of marginal interest, of questionable explanatory value. Dawkins presents the "meme" as if it were an actually existing entity, with huge potential to explain the origins of religion. Regrettably, Dawkins does not confront the arguments and concerns of leading critics of memetics within the scientific community and assess their significance. The 2005 decision to close down the *Journal of Memetics*, founded in 1997, was a telling sign of the failure of this discipline to gain scholarly acceptance. To its critics, memetics has been a short-lived fad whose effect has been to obscure more than it has been to enlighten mechanisms of information transfer. Yet Dawkins seems reluctant to concede the implausibility of memetics.

To illustrate the difficulties of his approach, we may consider his characteristically bold statement "memes can sometimes display very high fidelity."[77] This is a creedal statement posing as a statement of scientific fact. Dawkins is here translating an observation into his own theoretical language, which is not spoken elsewhere within the scientific community.

The *observation* is that ideas can be passed from one individual, group, or generation to another; Dawkins' *theoretical interpretation* of this observation—which in his writings is presented simply as fact—involves attributing fidelity to what most regard as a non-existent entity.

The idea also plays an important role in the thinking of Daniel Dennett. For Dennett, human brains provide shelter for memes, which proceed to create human minds. Far from being "godlike creators of ideas" who can manipulate, judge, and control them from an independent "Olympian standpoint," human beings are who they are, and think what they think, on account of "infestations of memes." Dennett also suggests that the human mind is "a sort of dungheap in which the larvae of other people's ideas renew themselves, before sending out copies of themselves in an informational diaspora," indicating that he expects his readers to find this an objectionable idea.[78] The idea of a human mind that somehow transcends both its genetic and memetic creators is nothing more than an outmoded myth.[79]

In *Breaking the Spell*, Dennett sets out a naturalist account of religion, based largely on an appeal to the meme. His highly developed account of the meme makes up for its lack of empirical foundations by a flamboyant account of its metaphysical importance, above all in spreading beliefs—such as belief in God. So are *all* beliefs spread by memes? Or just the ones that antireligious critics don't like? Is there a meme for atheism? Dennett's "Simple Taxonomy" certainly suggests so.[80] Dawkins, in contrast, seems to hold that there is no need to posit a meme for ideas that are self-evidently correct—such as atheism. Ideas that he clearly regards as indefensible, on the other hand, he holds to require memetic explanation.

As a matter of fact, there are no compelling reasons to believe that a cultural entity such as an idea or cultural artifact either contains or consists of self-replication instructions.[81] Ideas rely on the mechanisms of the human brain to remember and, when appropriate, express or embody them. Biological replication similarly relies on the presence of certain environmental conditions in order to function properly. Furthermore, cultural entities undergo transformation during the process of transmission.[82] Yet perhaps the most fundamental problem of all is that neither ideas nor cultural artifacts can conceivably be said to be, or to contain, a self-assembly code. They are not "replicators," as is required by the accounts of cultural transmission and development offered by Dawkins and Dennett. It is interesting to speculate about how Dawkins and Dennett might settle their differences over the role of the meme *empirically*, given the significant evidential underdetermination of the notion. Indeed, since there is no compelling scientific evidence for these entities, some writers have playfully—though not without good reason—concluded that there might even be a meme for believing in memes.[83]

This appeal to the meme is a telling sign of the importance of biology in a broader, often more popular, debate about how we are to understand the development of human culture, and the transmission and modification of its determinative ideas and values. Yet it is important to appreciate that, whereas Darwin's theory of natural selection was able to offer a coherent account of what had hitherto been regarded as largely disparate phenomena—such as vestigial organs, the uneven distribution of species, and evidence of extinct species—the appeal to biology as a source of models for interpreting culture in evolutionary terms has, as yet, failed to match Darwin's theory in its explanatory elegance and parsimony.[84] While it is clear—indeed, beyond dispute—that cultural entities (including ideas, habits, mannerisms, attitudes, and languages, in addition to artifacts such as tools and art) evolve in the general sense of incremental change reflecting the constraints and affordances of an environment, this does not necessarily entail that this occurs by or through a Darwinian mechanism.[85]

The intellectual battle

This brief engagement with atheist apologetics in the first decade of the twenty-first century confirms the general impression that there are many who remain firmly wedded to the "warfare" model of the relationship between science and religion, and see evolutionary biology as a strategic partner in the intensifying struggle against the incomprehensible (from a particular Enlightenment perspective, that is) persistence of religion. Yet this new approach goes beyond the traditional argument that evolution disproves Genesis, which therefore disproves Christianity. This argument ultimately amounts to a challenge to traditional biblical interpretation, a challenge that can be met by a revisionary hermeneutic.[86] Darwinism is here understood as a "universal acid," eroding faith in God and the transcendent, and offering a reductive, atheistic explanation of why some continue to believe in God, when all rational people know there is no God to believe in.

When stripped of their speculative elements and constrained by observational evidence, the evidence assembled by Dawkins and Dennett ultimately leads to little more than the conclusion that God need not be invoked as an explanatory agent within the evolutionary process. This is consistent with various atheist, agnostic, and Christian understandings of the world, but necessitates none of them. While it is clear that Christians have disagreed, and continue to disagree, over the implications of their faith for their attitudes to evolutionary biology, it is evident that, as a matter of historical fact, many Christians have been able to accommodate evolution within their worldview without undue difficulty.[87] Yet the suggestion that evolution erodes Christian belief clearly resonates with many secular thinkers and activists in North America, who are concerned about the growing public profile and political power of evangelical groups during the presidency of George W. Bush (2001–9). The hostile attitude toward any notion of biological evolution that is characteristic of many (but certainly not all) of these groups may well have contributed to this growing appeal to Darwinism as a weapon against their intellectual and cultural credibility. The amalgam of antievolutionary theology and neoconservative political values that characterized many conservative evangelicals during this period may thus account, at least to some extent, for the cultural origins, emotional animus, and intellectual direction of this renewed antireligious movement.

The primary concern of this chapter is the documentation of an emerging trend in recent atheist writing—namely, the identification of Darwinism as the ultimate eliminative explanation of any form of religion. Darwinism is here understood more in terms of an ideology, a worldview, extending beyond the limited domain of the biological to embrace the cultural and intellectual realms. Nature, many now argue, is religiously ambivalent,[88] making the manner of its interpretation of critical importance. This, as many have noted, means that the detached or neutral study of the natural order would seem to be capable of leading as much to a natural *a*theology as to a natural theology.[89] Writers such as Dawkins and Dennett have produced what some have termed "the Bridgewater treatises of the twentieth century," advocating a natural atheism (see also Topham, Chapter 4, this volume).[90] Where once a scientific account of the world was used to advocate a Christian perspective, the late twentieth century witnessed a growing trend to use it as a weapon against religious belief.[91] The recent surge of works of atheist apologetics that make a fundamental appeal to the natural sciences clearly point to the religious ambiguity of nature, a fact that needs to be taken into account in any attempt to reconstruct a viable contemporary natural theology.[92]

Yet perhaps the most significant of the developments noted in this chapter is the uneasy relationship that we have noted between two visions of Darwinism: a modest, provisional, and revisable scientific theory, an ally in our understanding of the world; and a triumphant worldview, which will sweep its rivals from the field of intellectual battle. As history makes depressingly clear, worldviews have an unhappy tendency to create in-groups and out-groups, orthodoxies and heresies, leading to dogmatic assertion where tentative questioning may be more appropriate.[93] Darwinism appears to have an innate propensity to attract those who wish to attach it to decidedly non-empirical ethical, philosophical, and religious systems. We have, as this volume makes clear, seen this in the past. Recent trends suggest it will continue in the future.

Religious belief remains deeply embedded in today's world, having obstinately refused to die the death that secularist readings of evolutionary biology predicted. In this chapter, I have explored recent metaphysical expansions of Darwinism, advocated by those wishing to adopt it as an ally in their attempt to achieve the final elimination, whether metaphysical or physical, of this irritatingly persistent phenomenon of religion. Yet to its critics, such metaphysically inflated versions of Darwinism seem to come close to becoming religions themselves. I hope I have raised important questions about how this expanded variant of Darwinism will develop biologically, theologically, and philosophically. Is this expanded vision of the Darwinian paradigm a useful and legitimate extension of approaches and methods from biology to human culture in general? Or does it represent an abuse of biology, an improper claim of biological hegemony over other disciplines, such as cultural anthropology? The debate continues, at both the scholarly and popular levels. We shall have to wait and discover its outcome.

Acknowledgments

The editors would like to thank the John Templeton Foundation for their financial support for the Authors' Workshop held in Cambridge September 17–19, 2007, organized by the Faraday Institute, St. Edmund's College, Cambridge, which led to the present volume. The editors would also like to thank Mrs. Sandi Irvine for her diligent assistance in preparing this volume for press, Mrs. Polly Stanton for her help in obtaining permissions for figures, and Karen Darling and her team at the University of Chicago Press for their hard work and support during the publishing process.

 Faraday Institute for Science and Religion

Notes

Introduction

1. "Science as the Norm of Truth," in Susan Faye Cannon, *Science in Culture: The Early Victorian Period* (New York: Science History Publications, 1978), 1–28; George Sarton, *The History of Science and the New Humanism* (New York: George Braziller, 1956), 10 (cumulative), 14 (discovery of truth), 116 (disinterested), 166 (struggle).

2. Albert E. Moyer, *A Scientist's Voice in American Culture: Simon Newcomb and the Rhetoric of Scientific Method* (Berkeley and Los Angeles: University of California Press, 1992), 85. See also Richard R. Yeo, "Scientific Method and the Rhetoric of Science in Britain, 1830–1917," in *The Politics and Rhetoric of Scientific Method: Historical Studies*, ed. J. A. Schuster and Richard R. Yeo (Dordrecht: D. Reidel, 1986), 259–97; Daniel Patrick Thurs, *Science Talk: Changing Notions of Science in American Culture* (New Brunswick, NJ: Rutgers University Press, 2007), 74–9.

3. David A. Hollinger, "Inquiry and Uplift: Late Nineteenth-Century American Academics and the Moral Efficacy of Scientific Practice," in *The Authority of Experts: Studies in History and Theory*, ed. Thomas L. Haskell (Bloomington: Indiana University Press, 1984), 142–56 (Eliot quotation on 142); David A. Hollinger, "Justification by Verification: The Scientific Challenge to the Moral Authority of Christianity in Modern America," in *Religion and Twentieth-Century American Intellectual Life*, ed. Michael J. Lacey (Cambridge: Cambridge University Press, 1989), 116–35.

4. Boris Hessen, "The Social and Economic Roots of Newton's *Principia*," in *Science at the Cross Roads: Papers Presented to the International Congress of the History of Science and Technology Held in London from June 20th to July 3rd, 1931, by the Delegates of the U.S.S.R.*, ed. Nicolai I. Bukharin (London: Kniga, 1931), 151–212. On the ideological context of Hessen's paper, see Loren R. Graham,

"The Socio-political Roots of Boris Hessen: Soviet Marxism and the History of Science," *Social Studies of Science* (1985) **15**: 705–22. For a useful overview of science and ideology, see Dorinda Outram, "Science and Political Ideology, 1790–1848," in *Companion to the History of Modern Science*, ed. R. C. Olby, G. N. Cantor, J. R. R. Christie, and M. J. S. Hodge (London: Routledge, 1990), 1008–23; John F. M. Clark, "Ideology," in *Reader's Guide to the History of Science*, ed. Arne Hessenbruch (London: Fitzroy Dearborn, 2000), 367–9.

5. Robert M. Young, "Evolutionary Biology and Ideology: Then and Now," *Science Studies* (1971) **1**: 177–206 (180–1); Roy Porter, "The History of Science and the History of Society," in Olby et al., ed., *Companion to the History* 32–46 (41); Robert M. Young, "Science, Ideology and Donna Haraway," www.human-nature.com/rmyoung/papers/paper 24h.html (science *as* ideology); Daniel Bell, *The End of Ideology: On the Exhaustion of Political Ideas in the Fifties* (Glencoe, IL: Free Press, 1960). For an influential collection of essays, see Robert M. Young, *Darwin's Metaphor: Nature's Place in Victorian Culture* (Cambridge: Cambridge University Press, 1985). See also Robert M. Young, "Marxism and the History of Science," in Olby et al., ed., *Companion to the History*, 77–86.

6. Norwood Russell Hanson, *Patterns of Discovery: An Inquiry into the Conceptual Foundations of Science* (Cambridge: Cambridge University Press, 1958), 19; Thomas Kuhn, *The Structure of Scientific Revolutions* (Chicago: University of Chicago Press, 1962), 93 (relevant community), 112 (Hanson), 137 (ideology); Thomas S. Kuhn, *The Road Since Structure: Philosophical Essays, 1970–1993, with an Autobiographical Interview*, ed. James Conant and John Haugeland (Chicago: University of Chicago Press, 2000), 307–8 (relativism, conservative).

7. Robert K. Merton, "Science, Technology and Society in Seventeenth Century England," *Osiris* (1938) IV, pt. 2: 360–632; Barry Barnes, *Scientific Knowledge and Sociological Theory* (London: Routledge & Kegan Paul, 1974), 2; David Bloor, *Knowledge and Social Imagery*, 2nd ed. (Chicago: University of Chicago Press, 1991), 166. See also Barry Barnes, "Sociological Theories of Scientific Knowledge," in Olby et al., ed., *Companion to the History*, 60–73; Barry Barnes, David Bloor, and John Henry, *Scientific Knowledge: A Sociological Analysis* (Chicago: University of Chicago Press, 1996). For another influential sociological study, see Thomas F. Gieryn, "Boundary-Work and the Demarcation of Science from Non-Science: Strains and Interests in Professional Ideologies of Scientists," *American Sociological Review* (1983) **48**: 781–95. For an excellent introduction to constructivism, see Jan Golinski, *Making Natural Knowledge: Constructivism and the History of Science* (Cambridge: Cambridge University Press, 1998).

8. Bloor, *Knowledge and Social Imagery*, 183–5.

9. Steven Shapin and Simon Schaffer, *Leviathan and the Air-Pump: Hobbes, Boyle, and the Experimental Life* (Princeton, NJ: Princeton University Press, 1985), 15 (sociology), 339 (dependent). For one flattering review, see Christopher Hill,

"'A New Kind of Clergy': Ideology and the Experimental Method," *Social Studies of Science* (1986)**16**: 726–35. Four years later Adrian Desmond brought out another iconic study that stressed the role of ideology: *The Politics of Evolution: Morphology, Medicine, and Reform in Radical London* (Chicago: University of Chicago Press, 1989). See also Adrian Desmond and James Moore, *Darwin* (London: Michael Joseph, 1991), and Adrian Desmond, *Huxley: From Devil's Disciple to Evolution's High Priest* (Reading, MA: Addison-Wesley, 1997).

10. For example, see such critiques of sociobiology as Steven Rose, Leon J. Kamin, and R. C. Lewontin, *Not in Our Genes: Biology, Ideology, and Human Nature* (New York: Pantheon Books, 1984); and Richard Lewontin, *Biology as Ideology: The Doctrine of DNA* (New York: Harper Perennial, 1993). For spirited rebuttals, see Richard Dawkins, "Sociobiology: The Debate Continues," *New Scientist* (1985), 24 January: 59–60; and Edward O. Wilson, "Science and Ideology," *Academic Questions* (1995) **8**: 73–81. For the context, see Ullica Segerstråle, *Defenders of the Truth: The Battle for Science in the Sociobiology Debate and Beyond* (New York: Oxford University Press, 2000).

Chapter One

1. Francesco Petrarch, *On His Own Ignorance and That of Many Others*, transl. Hans Nachod, in *The Renaissance Philosophy of Man*, ed. Ernst Cassirer, Paul Oskar Kristeller, and John Herman Randall, Jr. (Chicago: University of Chicago Press, 1948), 58–59.

2. John Johnston, *An History of the Wonderfull Things of Nature* (London, 1657), sig. A2r. See also Edward Topsell, *The Historie of Foure-Footed Beastes* (London, 1607), epistle dedicatory.

3. William A. Wallace, "Traditional Natural Philosophy," in *The Cambridge History of Renaissance Philosophy*, ed. Charles Schmitt and Quentin Skinner (Cambridge: Cambridge University Press, 1991), 201–35 (213).

4. Quoted in Ann Blair, *The Theatre of Nature: Jean Bodin and Renaissance Science* (Princeton, NJ: Princeton University Press, 1997), 35.

5. Johnston, *An History*, epistle dedicatory.

6. John Ray, *The Wisdom of God Manifested in the Works of Creation* (London, 1691), 123f.

7. John Ray and Francis Willoughby, *The Ornithology of Francis Willughby* (London, 1678), preface.

8. Alessandro Piccolomini, *Della institution morale libri XII*, (Venetia, 1543), quoted in Eugenio Garin, *Italian Humanism, Philosophy and Civic Life in the Renaissance*, transl. P. Munz (Oxford: Blackwell, 1965), 174. See also W. H. Woodward, ed. and transl., *Vittorino da Feltre and Other Humanist Educators* (Cambridge:

Cambridge University Press, 1987); E. F. Rice, *The Renaissance Idea of Wisdom* (Cambridge MA: Harvard University Press, 1958).

9. Philip Sidney, *Defence of Poesy* (London, 1595), sig. Dr. For a similar discussion in the French context, see Charles de Saint-Evremond, "Judgement sur les sciences," in *Oeuvres en prose*, ed. René Ternois (Paris: Marcel Didier, 1965), vol. 2, 6–12.

10. Sidney, *Defence of Poesy*, sigs. Dr, D4r; sig. D3r. Cf. Aristotle, *Poetics* 1451b.

11. Sidney, *Defence of Poesy*, sig. D3r.

12. For example, see Thomas Hobbes, *Leviathan* I, chap. 9, para. 2/4, ed. C. B. Macpherson (Harmondsworth: Penguin, 1982), 71.

13. Blair, *The Theatre of Nature*, 35.

14. J. B. Birks, *Rutherford at Manchester* (London, Heywood, 1962).

15. Pierre Gassendi, *The Mirrour Of True Nobility & Gentility Being The Life Of Peiresc*, transl. William Rand for John Evelyn (1657), chap. 7, unpaginated [*Vita Peireskii* (1641)].

16. The classic example was Diogenes Laertius, *The Lives of Eminent Philosophers*. See also Mark Edwards, *Neoplatonic Saints: The Lives of Plotinus and Proclus by the Students* (Liverpool: Liverpool University Press, 2000).

17. John Sellars, *The Art of Living: The Stoics on the Nature and Function of Philosophy* (Aldershot: Ashgate, 2003), esp. chap. 3.

18. John Cottingham, *Philosophy and the Good Life* (Cambridge: Cambridge University Press, 1998), 52–75, and Matthew Jones, *The Good Life in the Scientific Revolution: Descartes, Pascal, Leibniz, and the Cultivation of Virtue* (Chicago: University of Chicago Press, 2006), 4–7.

19. Another biblical passage frequently cited in this context was Job 12:7f.: "But ask now the beasts and they will teach thee; and the fowls of the air, and they shall tell thee. Or speak to the earth, and it shall teach thee, and the fishes of the sea shall declare unto thee."

20. Ambrose, *Hexaemeron*, V.x.29, VI.iv.21, V.xv.50, in *Fathers of the Church*, transl. J. J. Savage (Washington: Catholic University of America Press, 1947–), vol. 62, 184, 240, 200).

21. Ambrose, *Hexaemeron*, V.xxiii.77–9 (218f.). In the seventeenth century, Nicolas Malebranche argued similarly that "God willed to represent Jesus Christ by the changes in insects." Nicolas Malebranche, *Dialogues on Metaphysics and Religion*, XI.xiv, ed. Nicholas Jolley and David Scott (Cambridge: Cambridge University Press, 1997), 213.

22. Malebranche, *Dialogues*, II.xvi.24 (50). See also Francesco Zambon, "*Figura bestialis*: Les fondements théoretiques du bestaire médiéval Bestial figures [The theoretical foundations of medieval bestiary]," in *Epopeé animale, fable, fabliau*, ed. G. Bianciotto and M. Salvat (Paris: Publications de l'Université de Rouen, 1984), 709–19.

23. Bartholomaeus Anglicus, *De proprietatibus rerum* [On the properties of things] (Heidelberg, 1488), Prologue. See also Peter Harrison, *The Bible, Protestantism, and the Rise of Natural Science* (Cambridge: Cambridge University Press, 1998); William B. Ashworth, Jr., "Natural History and the Emblematic World View," in *Reappraisals of the Scientific Revolution*, ed. David C. Lindberg and Robert S. Westman, (Cambridge: Cambridge University Press, 1990), 303–32.

24. Harold J. Cook, "Physicians and Natural History," in *Cultures of Natural History*, ed. N. Jardine, J. Secord, and E. Spary (Cambridge: Cambridge University Press, 1996), 91–105.

25. John Riddle, *Dioscorides on Pharmacy and Medicine* (Austin: University of Texas Press, 1985).

26. See Topsell's translation in *Historie of Foure-Footed Beastes*, sigs. ¶2r–¶3r.

27. Levinus Lemnius, *An Herbal for the Bible*, transl. Thomas Newton (London, 1587), title page. See also William Westmacott, *Historia vegetabilium sacra* (London, 1695).

28. Francis Bacon, "Preparative towards a Natural and Experimental Philosophy," in *The Works of Francis Bacon*, ed. James Spedding, Robert Ellis, and Douglas Heath, 14 vols. (London, 1857–74), vol. 4, 261. On "book of nature" metaphors during this period, see *The Book of Nature in Early Modern and Modern History*, ed. K. van Berkel and Arjo Vanderjagt (Leuven: Peeters, 2006).

29. Michel de Montaigne, *The Essayes of Michael Lord of Montaigne*, transl. John Florio, 3 vols. (Oxford: Oxford University Press, 1951), vol. 3, 316, 305. For additional examples, see Peter Harrison, "The Virtues of Animals in Seventeenth-Century Thought," *Journal of the History of Ideas* (1998) **59**: 463–85.

30. Pierre Charron, *Of Wisdom* (London, 1697), 263; tr. of *De la sagesse* (Bourdeaux, 1601).

31. Quoted in Kenneth J. Howell, *God's Two Books: Copernican Cosmology and Biblical Interpretation in Early Modern Science* (Notre Dame: University of Notre Dame Press, 2002), 113.

32. Ralph Austen, *Spiritual Use of an Orchard* (Oxford, 1653), preface; Jacob Boehme, *The Second Booke. Concerning the Three Principles of the Divine Essence* (London, 1648), sig. A3r.

33. Noël Antoine Pluche, *Spectacle de la Nature: or Nature Display'd*, 5th ed., 7 vols., (London, 1770), vol. 1, vii–viii.

34. Topsell, *Historie of Foure-Footed Beastes* (1607), epistle dedicatory; René Antoine Ferchault de Réaumur, *The Natural History of Ants*, transl. W. M. Wheeler (London: Knapp, 1926), preface; Robert Boyle, "A Discourse touching Occasional Meditations," in *The Works of the Honorable Robert Bouyle*, ed. Thomas Birch, 6 vols. (Hildesheim: Georg Olms, 1966), vol. 2, 340.

35. Thomas Moffett, *Insectorum sive minimorum animalium theatrum* (1634), chap. 16.

36. Thomas Pope Blount, *A Natural History* (London, 1693), preface.

37. John Edwards, *A Demonstration of the Existence and Providence of God* (London, 1696), pt. I, 206–15, pt. II, 150.

38. René Antoine Ferchault de Réaumur, *The Natural History of Bees* (London, 1744), 297.

39. John Milton, *Complete Prose Works* (New Haven, CT: Yale University Press, 1971), vol. 7, 427. See also Karin Edwards, *Milton and the Natural World* (Cambridge: Cambridge University Press, 2000), 128–38.

40. Edwards, *A Demonstration*, pt. I, 206f.

41. Nicolas Malebranche, *Oeuvres Complètes*, ed. Genevieve Rodis-Lewis (Paris: J. Vrin, 1958–70), vol. 2, 394. For Descartes' views see "More to Descartes," 11 December 1648, in *Oeuvres de Descartes*, ed. C. Adam and P. Tannery (Paris: L. Cerf, 1897–1913), vol. 5, 243: Letter to More, 5 February 1649, in *The Philosophical Writings of Descartes*, ed. John Cottingham, Robert Stoothoff, Dugald Murdoch, and Anthony Kenny (Cambridge: Cambridge University Press, 1991), vol. 3 (cited hereafter as CSMK), 245. See also J. Cottingham, "'A Brute to the Brutes?': Descartes' Treatment of Animals," *Philosophy* (1978) **53b**: 551–61; Peter Harrison, "Descartes on Animals," *Philosophical Quarterly* (1992) **42**: 239–48.

42. Anita Guerrini, *Experimenting with Humans and Animals: From Galen to Animal Rights* (Baltimore: Johns Hopkins University Press, 2003), 33–45.

43. René Descartes, "Principles of Philosophy," 2.37, in *The Philosophical Writings of Descartes*, ed. J. Cottingham, Robert Stoothoff, and Dugald Murdoch, 2 vols. (Cambridge: Cambridge University Press, 1985) (cited hereafter as CSM), vol. 1, 241; Letter to Reneri, April 1638, CSMK, 99. See also Harrison, "Virtues of Animals," esp. 479–80.

44. Descartes, "The Passions of the Soul," I.50, in CSM, vol. 1, 348.

45. Descartes, "Principles of Philosophy," preface to the French ed. of 1647, CSM, vol. 1, 186.

46. Kenelm Digby, *Two Treatises* (London, 1645), 399–400

47. Plato, *Laws* X; Epictetus, *Discourses*, 1.16.8; Cicero, *On the Nature of the Gods*, 2.54.133–58.146.

48. Thomas's "fifth way" relies upon the *directing* of natural things—the operations of nature—according to a design. It is the orderliness of the motions of things, rather than their structures, that provides the premise of the argument. Thomas Aquinas, *Summa theologiae* 1a. 2, 3.

49. Henry More, *An Antidote against Atheisme* (London: 1653), 5.

50. Ray and Willoughby, *Ornithology*, preface.

51. On physico-theology generally, see Charles Webster, *The Great Instauration* (London: Duckworth, 1975), 150–1, 507–10; C. E. Raven, *Natural Religion and Christian Theology* (Cambridge, Cambridge University Press, 1953); Neal Gillespie, "Natural History, Natural Theology, and Social Order: John Ray and the 'Newtonian Ideology,'" *Journal of the History of Biology* (1987) **20**: 1–49.

52. Boyle, *Works*, vol. 1, clxviii.

53. Bernard Nieuwentijt, *The Religious Philosopher: or the Right Use of Contemplating the Works of the Creator,* transl. John Chamberlayne (London, 1718; originally published in Dutch, Amsterdam, 1715).

54. See Joseph Glanvill, "Modern Improvements of Useful Knowledge," in *Essays on Several Important Subjects in Philosophy and Religion* (London, 1676).

55. Robert Hooke, *Micrographia: or Some Physiological Descriptions of Minute Bodies made by Magnifying Glasses* (London, 1665), 8. On Hooke and the argument from design, see John Harwood, "Rhetoric and Graphics in *Micrographia*," in *Robert Hooke: New Studies,* ed. Michael Hunter and Simon Schaffer (Woodbridge: Boydell, 1989), 119–48.

56. See, for example, Joseph Glanvill, *Scepsis Scientifica: or Confest Ignorance the way to Science* (London, 1665), 3; Edwards, *A Demonstration,* pt. I, 220; Ray, *Wisdom of God,* 41; Henry Power, *Experimental Philosophy* (London, 1664), preface; Patrick Cockburn, *An Enquiry into the Truth and Certainty of the Mosaic Deluge* (London, 1750), 40.

57. Pluche, *Spectacle,* vol. 1, 1, 34. This seems to have become a commonplace. See, for example, Edwards, *A Demonstration,* pt. I, 205.

58. Nicolas Malebranche, "Éloge du P. Malebranche," in *Oeuvres Complètes,* ed. Genevieve Rodis-Lewis (Paris: J. Vrin, 1958–70), vol. 5, 461.

59. Johnston, *An History,* sig. a3v. Cf. R. Franck, *A Philosophical Treatise of the Original and Production of Things* (London, 1687), epistle dedicatory.

60. Edwards, *A Demonstration,* preface, 262.

61. Power, *Experimental Philosophy,* 183.

62. Robert Boyle, "Some Considerations touching the Usefulness of Experimental Natural Philosophy," in *Works,* vol. 2, 62f. Cf. Johannes Kepler, *Mysterium cosmographicum,* transl. A. M. Duncan (Norwalk, CT: Abaris, 1999), 53.

63. Thomas Sprat, *History of the Royal Society of London* (London, 1667), 394f.

64. Ray, *Wisdom of God,* 127.

65. Ray, *Wisdom of God,* 124.

66. Johannes Kepler, *Johannes Kepler Gesammelte Werke* [Johannes Kepler: Collected works], ed. Walter von Dyck and Max Caspar, 20 vols. (Munich, 1938–), vol. 7, 25; Letter to Herwath von Hohenburg, 26 March 1598, in Kepler, *Werke,* vol. 8, 193.

67. Robert Boyle, *A Disquisition about the Final Causes of Natural Things* (London, 1688), 34; See also H. Fisch, "The Scientist as Priest: A Note on Robert Boyle's Natural Theology," *Isis* (1953) **44**: 252–65

68. On this issue, see Mordechai Feingold, "Science as a Calling: The Early Modern Dilemma," *Science in Context* (2002) **15**: 79–119; Peter Harrison, "'Priests of the Most High God, with Respect to the Book of Nature': The Vocational Identity of the Early Modern Naturalist," in *Reading God's World,* ed. Angus Menuge (St. Louis: Concordia, 2004), 55–80.

69. Meric Casaubon, *A Letter of Meric Casaubon, D.D. &c. to Peter du Moulin D.D., concerning Natural Experimental Philosophie* (Cambridge, 1669), 32, and *passim*. For similar criticisms, see Henry Stubbe, *Campanella Revived* (London, 1670), and *Legends no Histories: or a Specimen of some Animadversions upon the History of the Royal Society . . . together with the Plus Ultra reduced to a Non-Plus* (London, 1670).

70. Robert South, quoted in Margery Purver, *The Royal Society: Concept and Creation* (London: Routledge and Kegan Paul, 1967), 71.

71. James Harrington, *The Prerogative of Popular Government* (London, 1658), epistle dedicatory.

72. Peter Harrison, "'The Fashioned Image of Poetry or the Regular Instruction of Philosophy?': Truth, Utility, and the Natural Sciences in Early Modern England," in *Science, Literature, and Rhetoric in Early Modern England*, ed. D. Burchill and J. Cummins (Aldershot: Ashgate, 2007), 15–36.

73. Jonathan Swift, *Gulliver's Travels*, ed. Paul Turner (Oxford: Oxford University Press, 1974), 171–5. For similar examples, see Steven Shapin, "The Image of the Man of Science," in *The Cambridge History of Science*, Roy Porter (Cambridge: Cambridge University Press, 2003), vol. 4, 159–83 (esp. 172f.).

74. See, for example, Sprat, *History*, 341, 355f.; Joseph Glanvill, "The Usefulness of Real Philosophy to Religion," 25, in *Essays;* Joseph Glanvill, *Philosophia Pia: or, A Discourse of the Religious Temper, and Tendencies of the Experimental Philosophy, which is profest by the Royal Society* (London, 1671), 16, 86. Cf. John Wilkins, *An Essay toward a Real Character and a Philosophical Language* (London, 1684), sig. B1r.

75. Francis Bacon, *Redargutio philosophiarum* (1608), in B. Farrington, *The Philosophy of Francis Bacon* (Liverpool: Liverpool University Press, 1964), 92f. See also Stephen Gaukroger, *Francis Bacon and the Transformation of Early Modern Natural Philosophy* (Cambridge: Cambridge University Press, 2001), 5; Antonio Pérez-Ramos, "Bacon's Legacy," in *The Cambridge Companion to Bacon*, ed. Markku Peltonen (Cambridge: Cambridge University Press, 1996), 311–34.

76. Bacon, "Valerius terminus," in *The Works of Francis Bacon*, vol. 3, 222.

77. Bacon, "Novum organum," II, §52, in *The Works of Francis Bacon*, vol. 4, 247. On the importance of this motif in seventeenth-century thought, see Peter Harrison, *The Fall of Man and the Foundations of Science* (Cambridge: Cambridge University Press, 2007), esp. 139–85.

78. Bacon, "Advancement of Learning," in *The Works of Francis Bacon*, vol. 3, 294. These ideas are anticipated in the earlier unpublished work "Valerius terminus" (*ca.*1603), in *The Works of Francis Bacon*, vol. 3, esp. 221–3. On charity in Bacon's work, see John Canning Briggs, "Bacon's Science and Religion," in *The Cambridge Companion to Bacon*, ed. Markku Peltonen (Cambridge: Cambridge University Press), 181–5; Masao Watanabe, "Francis Bacon: Philanthropy and the Instauration of Learning," *Annals of Science* (1992) **49**: 163–73; Benjamin Milner,

"Francis Bacon: The Theological Foundations of *Valerius Terminus*," *Journal of the History of Ideas* (1997) **58**: 245–64; Brian Vickers, "Bacon's So-Called 'Utilitarianism: Sources and Influences,'" in *Francis Bacon: Terminologia e fortuna nel XVII secolo*, ed. Marta Fattori (Rome: Edizione dell'Ateneo, 1984), 281–314.

79. Bacon, "De augmentis," in *The Works of Francis Bacon*, vol. 4, 419. See also Topsell, *Historie of Four-footed Beastes*, sig. ¶3v; Boyle, "Some Considerations," in *The Works of Francis Bacon*, vol. 2, 309; Joseph Glanvill, "The Usefulness of Real Philosophy to Religion," 41, in *Essays*.

80. Bacon, "Advancement of Learning," in *The Works of Francis Bacon*, vol. 3, 286; "Great Instauration," in *The Works of Francis Bacon* vol. 4, 27. Cf. Joseph Glanvill, *The Vanity of Dogmatizing* (London, 1661), 178, 183; Joseph Glanvill, *Plus Ultra* (London, 1668), preface.

81. See Webster, *The Great Instauration*.

82. John Webster, *Academiarum examen: or the Examination of Academies* (London, 1654), 19.

83. Sprat, *History*, 322f., 349f., 438.

84. Glanvill, *Plus Ultra*, 87; Joseph Glanvill, "The Usefulness of Real Philosophy to Religion," 39, in *Essays*.

85. Glanvill, *Scepsis Scientifica*, sigs. b3r–v; Cr–v.

86. John Ray, *Wisdom of God*, 116f.

87. For criticisms along these lines, see Casaubon, *A Letter*, 5f.; Stubbe, *Legends no Histories*, preface. For a response see Glanvill, *Plus Ultra*, 83f. See also Michael Hunter, *Science and Society in Restoration England* (Cambridge: Cambridge University Press, 1981), 111.

88. Jean le Rond d'Alembert, "Discours préliminaire," in *L'Encyclopédie, ou Dictionnaire raisonné des sciences, des arts et des métiers*, 17 vols., 11 vols. of plates (Paris, 1751–72), vol. 1, xvii.

89. Bacon, "Great Instauration," in *The Works of Francis Bacon*, vol. 4, 30f. Cf. "De augmentis," in *The Works of Francis Bacon*, vol. 4, 295.

90. Robert Plot, *The Natural History of Oxfordshire* (Oxford, 1677), To the Reader; Ray and Willoughby, *Ornithology*, preface; Ray, *Wisdom of God*, preface.

91. Claude Perrault, *Memoir's [sic] for a Natural History of Animals. Containing the Anatomical Descriptions of Several Creatures Dissected by the Royal Academy of Sciences at Paris* (London, 1688), tr. of *Memoires pour servir a l'histoire naturelle des animaux* (Paris, 1671–6), sig. A2v.

92. Bacon, "De augmentis," in *The Works of Francis Bacon*, vol. 4, 295, 362; Bacon, "Great Instauration," in *The Works of Francis Bacon*, vol. 4, 28f.; Bacon, "Novum Organum," in *The Works of Francis Bacon*, vol. 4, 71.

93. See Richard Yeo, "Classifying the Sciences," in *The Cambridge History of Science*, ed. Roy Porter (Cambridge: Cambridge University Press, 2003), vol. 4, 241–66; Peter Anstey, "Locke, Bacon, and Natural History," *Early Science and Medicine* (2002) **7**: 65–92.

94. *Encyclopaedia Britannica* (1788–97), 3rd ed., s.v. "Philosophy."

95. John Gascoigne, "The Eighteenth-Century Scientific Community: A Prospographical Study," *Social Studies of Science* (1995) **25**: 575–81 (577f).

96. Quoted in John Gascoigne, "The Study of Nature," in *The Cambridge History of Eighteenth Century Philosophy*, ed. K. Haakonssen (Cambridge: Cambridge University Press, 2006), 860. Gascoigne provides an excellent account of the relations between natural history and natural philosophy.

97. J. Llana, "Natural History and the *Encyclopédie*," *Journal of the History of Biology* (2000) **33**: 1–25.

Chapter Two

1. Renato G. Mazzolini and Shirley A. Roe, eds., *Science Against the Unbelievers: The Correspondence of Bonnet and Needham, 1760–1780*, Studies on Voltaire and the Eighteenth Century, no. 243 (Oxford: Voltaire Foundation, 1986), 309; Letter from Needham to Bonnet, 28 October (September) 1779. See also Shirley A. Roe, "John Turberville Needham and the Generation of Living Organisms," *Isis* (1983) **74**: 159–84.

2. René Descartes, *Treatise of Man*, French text with translation and commentary by Thomas Steele Hall (Cambridge: Harvard University Press, 1972).

3. René Descartes, *L'homme et un traitté de la formation du foetus* [Man and a treatise on the formation of the fetus] (Paris: Charles Angot, 1664), 146. Unless otherwise noted, all translations are my own.

4. Nicolas Malebranche, *Oeuvres complètes* [Complete works], ed. André Robinet, 20 vols. (Paris: Librairie Philosophique J. Vrin, 1958–67), vol. 12, 264, 252, 253.

5. Marcello Malpighi, *Dissertatio epistolica de formation pulli in ovo* [A dissertation on the formation of the chicken in the egg] (London: J. Martyn, 1673); Jan Swammerdam, *Miraculum naturae, sive uteri muliebris fabrica* [A wonder of nature, or the fabric of the female uterus] (Leiden: S. Matthaei, 1672); Jacques Roger, *The Life Sciences in Eighteenth-Century French Thought*, ed. Keith R. Benson, transl. Robert Ellrich (Stanford: Stanford University Press, 1997); Peter J. Bowler, "Preformation and Pre-existence in the Seventeenth Century: A Brief Analysis," *Journal of the History of Biology* (1971) **4**: 221–44; Shirley A. Roe, *Matter, Life, and Generation: Eighteenth-Century Embryology and the Haller-Wolff Debate* (Cambridge: Cambridge University Press, 1981), 2–9; Edward Reustow, *The Microscope in the Dutch Republic* (Cambridge: Cambridge University Press, 1996; and Carla Pinto-Correia, *The Ovary of Eve: Egg and Sperm and Preformation* (Chicago: University of Chicago Press, 1997).

6. Roger, *Life Sciences*, **262**. See also Elizabeth Gasking, *Investigations into Generation, 1651–1828* (Baltimore: Johns Hopkins University Press, 1967).

7. Claude Perrault, "La mechanique des animaux" (1680), in *Oeuvres diverses de physique et de mechanique* [Diverse works on physics and mechanics] (Leiden: Pierre Vander Aa, 1721), 481.

8. Pierre-Louis Moreau de Maupertuis, *Vénus physique* (n.p., 1745); quotation from *The Earthly Venus*, transl. Simone Brangier Boas, Sources of Science, no. 29 (New York: Johnson Reprint Corporation, 1966), 55–6. See also Mary Terrall, "Salon, Academy, and Boudoir: Generation and Desire in Maupertuis' Science of Life," *Isis* (1996) **87**: 217–29; Mary Terrall, *The Man Who Flattened the Earth: Maupertuis and the Sciences in the Enlightenment* (Chicago: University of Chicago Press, 2002); David Beeson, *Maupertuis: An Intellectual Biography*, Studies on Voltaire and the Eighteenth Century, no. 299 (Oxford: Voltaire Foundation, 1992).

9. Aram Vartanian, "Review of *Abraham Trembley of Geneva: Scientist and Philosopher* by John R. Baker," *Isis* (1953) **44**: 388. See also Virginia P. Dawson, *Nature's Enigma: The Problem of the Polyp in the Letters of Bonnet, Trembley and Réaumur* (Philadelphia: American Philosophical Society, 1987).

10. See Roe, *Isis* (1983) **74**: 159–84; Mazzolini and Roe, *Science Against the Unbelievers*, 12–5.

11. Georges-Louis Leclerc, Comte de Buffon, *Histoire naturelle, générale et particulière, avec la description du cabinet du roy* [Natural history, general and particular, with a description of the king's cabinet] (Paris: De l'Imprimerie Royale, 1749–89), vol. 2, 17. See also Jacques Roger, *Buffon: A Life in Natural History*, transl. Sarah Lucille Bonnefoi, ed. L. Pearce Williams (Ithaca, NY: Cornell University Press, 1997).

12. See Shirley A. Roe, "Buffon and Needham: Diverging Views on Life and Matter," in *Buffon 88: Actes du Colloque international* [Buffon 88: Proceedings from an international colloquium], ed. Jean Gayon (Paris: Librairie philosophique J. Vrin, 1992), 439–50. Needham's theory was presented in his "A Summary of some late Observations upon the Generation, Composition, and Decomposition of Animal and Vegetable Substances; Communicated in a Letter to Martin Folkes Esq; President of the Royal Society," *Philosophical Transactions* (1748) **45**: 615–66.

13. Denis Diderot, *Lettre sur les aveugles à l'usage de ceux qui voyent* [Letter on blindness for the use of those who have their sight] (London, 1749).

14. See Shirley A. Roe, "The Development of Albrecht von Haller's Views on Embryology," *Journal of the History of Biology* (1975) **8**: 167–90; and Roe, *Matter, Life, and Generation*, 21–44.

15. Charles Bonnet, *Considérations sur les corps organisés* [Considerations on organized bodies] (Amsterdam: Marc-Michel Rey, 1762).

16. Raymond Savioz, ed., *Mémoires autobiographiques de Charles Bonnet* [Autobiographical memoires of Charles Bonnet] (Paris: Librairie philosophique J. Vrin, 1948), 210.

17. René Antoine Ferchault de Réaumur, *Art de faire éclorre et d'élever en toute saison des oiseaux domestique de toutes espèces* [The art of hatching and bringing

up domestic fowls by means of heat], 2 vols. (Paris: L'Imprimerie Royale, 1749), vol. 2, 362–3.

18. Albrecht von Haller, *Réflexions sur le systême de la génération, de M. de Buffon* [Reflections on the system of generation of Mr. Buffon] (Geneva: Barrillot, 1751), xi.

19. François Marie Arouet de Voltaire, *Éléments de la philosophy de Newton* [Elements of Newton's philosophy] (London [Paris], 1738).

20. Voltaire, *La Métaphysique de Neuton, ou, Parallèle des sentimens de Neuton et de Leibnitz* [The metaphysics of Newton, or a comparison between the views of Newton and Leibniz] (Amsterdam: Étienne Ledet, 1740).

21. Voltaire, *Oeuvres complètes* [Complete works], ed. Louis Moland, 52 vols. (Paris: Garnier-Frères, 1877–85), vol. 23, 573. For a more complete treatment of Voltaire's biological views, see Shirley A. Roe, "Voltaire versus Needham: Atheism, Materialism, and the Generation of Life," *Journal of the History of Ideas* (1985) **46**: 65–87; and Mazzolini and Roe, *Science Against the Unbelievers*, 77–92.

22. See René Pomeau, *La religion de Voltaire* [Voltaire's religion] (Paris: Librairie Nizet, 1969), 393–4. For Diderot's letter to Damilaville at Ferney, see Denis Diderot, *Correspondance*, ed. Georges Roth, 16 vols. (Paris: Éditions de Minuit, 1955–79), vol. 5, 117–21.

23. Voltaire, *Oeuvres*, vol. 25, 394, 393–4.

24. Voltaire, *Oeuvres*, vol. 27, 159.

25. [Paul-Henri Thiry d'Holbach], *Système de la nature, ou des loix du monde physique et du monde moral, par M. Mirabaud* [System of nature, or the laws of the physical and moral world by M. Mirabaud] (London [Amsterdam]: 1770), 23, 23n5.

26. [Voltaire], *Questions sur l'encyclopédie, par des amateurs* [Questions on the encyclopedia, by amateurs] 9 vols. ([Geneva], 1770–2), vol. 4 (1771), 285–96.

27. Voltaire, *Correspondence and Related Documents*, definitive edition by Theodore Besterman, vols. 85–135 of *The Complete Works of Voltaire* (Geneva and Oxford: Voltaire Foundation, 1968–77), D15189, vol. 118, 33, 34; Letter to Phillipe Charles François Joseph de Pavée, marquis de Villevielle, 26 August 1768.

28. Voltaire, *Correspondence*, D16736, vol. 121, p. 58; Letter to Louis François Armand du Plessis, duc de Richelieu 1 November 1770. See also Theodore Besterman, "Voltaire, Absolute Monarchy, and the Enlightened Monarch," *Studies on Voltaire and the Eighteenth Century* (1965) **32**: 7–21; Theodore Besterman, "Voltaire's God," *Studies on Voltaire and the Eighteenth Century* (1967) **55**: 23–41.

29. *Traité des trois imposteurs* [Treatise on the three imposters] ([Amsterdam], 1765).

30. *Encyclopédie, ou dictionnaire raisonné des sciences, des arts et des métiers, par une société des gens de lettres* [Encyclopedia, or a systematic dictionary of the sciences, the arts and the crafts, by a society of learned gentlemen], ed. Denis

Diderot and Jean le Rond d'Alembert, 17 vols. (Paris: Briasson, David, Le Breton, Durand, 1751–65); 11 vols. of plates (1762–72). D'Alembert's co-editorship ended in 1758.

31. Diderot, *Pensées philosophique* [Philosophical thoughts] (La Haye [Paris], 1746).

32. Diderot, *Correspondance*, vol. 8, 234; Letter to Sophie Volland, 22 November 1768. For Diderot's life, see P. N. Furbank, *Diderot: A Critical Biography* (New York: Alfred A. Knopf, 1992); Otis Fellows, *Diderot* (Boston: Twayne Publishers, 1989); and Arthur Wilson, *Diderot* (New York: Oxford University Press, 1972). For Diderot's relationship with d'Holbach, see Alan Charles Kors, *D'Holbach's Coterie: An Enlightenment in Paris* (Princeton, NJ: Princeton University Press, 1976).

33. Diderot, *Oeuvres philosophiques* [Philosophical works], ed. Paul Vernière (Paris: Éditions Garnier Frères, 1964), 17. See also Aram Vartanian, "From Deist to Atheist: Diderot's Philosophical Orientation 1746–1749," *Diderot Studies* (1949) **1**: 46–63.

34. Diderot, *Pensées sur l'interpretation de la nature* [Thoughts on the interpretation of nature] (London [Paris], 1754).

35. Diderot, *Correspondance*, vol. 2, 283, Letter to Sophie Volland, 15 October 1759. See also Diderot's letter to Duclos in *Correspondance*, vol. 5, 141; Letter to M. Duclos [10 October 1765].

36. Diderot, *Correspondance*, vol. 9, 126–7, Letter to Sophie Volland [31 August 1769].

37. Diderot, *Rêve de d'Alembert* [Dream of d'Alembert] in *Oeuvres philosophiques*, 299, 303. See also Furbank, *Diderot*, 327–42; Wilda Anderson, *Diderot's Dream* (Baltimore: Johns Hopkins University Press, 1990).

38. For Diderot's writings on morality, see John Hope Mason, *The Irresistible Diderot* (London: Quartet Books, 1982), 282–306. For d'Holbach's, see Virgil W. Topazio, *D'Holbach's Moral Philosophy: Its Background and Development* (Geneva: Institute et Musée Voltaire, 1956), and Everett C. Ladd, "Helvétius and d'Holbach: 'La moralisation de la politique,'" *Journal of the History of Ideas* (1962) **23**: 221–38.

39. For Diderot's politics see especially Mason, *Irresistible Diderot*, 320–58; and Anthony Strugnell, *Diderot's Politics: A Study of the Evolution of Diderot's Political Thought after the Encyclopédie* (The Hague: Martinus Nijhoff, 1973).

40. *Arrêt du Conseil d'État du Roi . . . du 7 févier 1752* [Decree from the King's Council of State], in Denis Diderot, *Oeuvres complètes* [Complete works], ed. Herbert Diekmann, Jean Fabre, and Jacques Proust, 25 vols. (Paris: Hermann, 1975–2004), vol. 5, 37.

41. Quoted in John Lough, *The Encyclopédie* (New York: David McKay, 1971), 23.

42. *Arrests de la Cour de Parlement, portant condamnation de plusieurs Livres & autres Ouvrages imprimés* [Decree from the court of Parlement, condemning several books and other printed works], 23 January 1759, 1–2, 3.

43. *Encyclopédie*, vol. 6, 55A.

44. *Encyclopédie*, vol. 4, 278B.

45. *Arrêt du Conseil d'État du Roi . . . du 8 Mars 1759* [Decree from the King's Council of State . . . from 8 March 1759], in Diderot, *Oeuvres complètes*, vol. 5, 44.

46. Wilson, *Diderot*, 339.

47. [Odet Joseph de Vaux de Giry de Saint-Cyr], "Avis utile" [Useful warning], *Mercure de France*, October 1757, 15–9. Reprinted in *L'Affaire des Cacouacs: Trois pamphlets contre les Philosophes des Lumières* [The Cacouac affair: Three pamphlets against the philosophers of the Enlightenment], ed. Gerhardt Stenger (Saint-Étienne: Publications de l'Université de Saint-Étienne, 2004).

48. [Jacob Nicolas Moreau], *Nouveau mémoire pour servir à l'histoire des Cacouacs* [A new memoire contributing to the story of the Cacouacs] (Amsterdam, 1757), in Stenger, *L'Affaire des Cacouacs*, 42, 58. To the questions asked the young man Moreau added footnote references to Diderot's *Pensées sur l'interprétation de la nature*.

49. [Giry de Saint-Cyr], *Catéchisme de décisions de cas de conscience à l'usage des Cacouacs* [Catechism of decisions in matters of conscience for the use of Cacouacs] (Cacopolis, 1758), in Stenger, *L'Affaire des Cacouacs*, 77, 79.

50. Diderot, *Correspondance*, vol. 11, 19–20, Letter to La Princesse Dashkiff [3 April 1771].

51. Mazzolini and Roe, *Science Against the Unbelievers*, 257; Letter from Needham to Bonnet, 25 August 1768.

Chapter Three

1. Johann Gottfried Herder, *Ideen zur Philosophie der Geschichte der Menschheit* [Ideas on the philosophy of the history of mankind], in *Herders sämmtliche Werke*, ed. Bernhard Suphan, 33 vols. (Berlin, 1877–1913; hereafter cited as *SWS*), vol. 13, 7.

2. *SWS*, vol. 14, 145.

3. For skepticism in the late eighteenth century, see Johan van der Zande and Richard Popkin, eds., *The Skeptical Tradition around 1800* (Dordrecht: Kluwer, 1998).

4. The following summarizes in brief some of the points I make in my book *Vitalizing Nature in the Enlightenment* (Berkeley and Los Angeles: University of California Press, 2005).

5. Georges-Louis Leclerc, Comte de Buffon, *Histoire naturelle, générale et particulière*, 36 vols. (Paris: L'Imprimerie royale, 1749–78), vol. 1, 4.

6. Adam Smith, *The Theory of Moral Sentiments*, ed. D. D. Raphael and A. L. Macfie (Oxford: Clarendon Press, 1976; hereafter cited as *TMS*), VI.i.11 (215).

7. *Wilhelm von Humboldts Briefe an Karl Gustav von Brinkmann* [Wilhelm von Humboldt's letter to Karl Gustav von Brinkmann], ed. Albert Leitzmann (Leipzig: Karl Heirsemann, 1939), 60–1.

8. Isaiah Berlin, "The Counter-Enlightenment," in *Against the Current: Essays in the History of Ideas*, ed. Henry Hardy (New York: Viking, 1980), 1–24.

9. Since the 1970s a concerted effort has been made to place Smith back in his eighteenth-century context by investigating his relation to themes such as republicanism, natural rights, moral philosophy, and jurisprudence. Amongst the most important books and articles in this regard are: Istvan Hont and Michael Ignatieff, eds., *Wealth and Virtue: The Shaping of Political Economy in the Scottish Enlightenment* (Cambridge: Cambridge University Press, 1983); Roy Hutchenson Campbell and Andrew S. Skinner, *Adam Smith* (London and Canberra: Helm, 1982); Donald Winch, *Adam Smith's Politics: An Essay in Historiographic Revision* (Cambridge: Cambridge University Press, 1978); Stephen Copley and Kathryn Sutherland, eds., *Adam Smith's Wealth of Nations: New Interdisciplinary Essays* (Manchester: Manchester University Press, 1995); Vivienne Brown, *Adam Smith's Discourse: Canonicity, Commerce, and Conscience* (London: Routledge, 1994); Jennifer Pitts, *A Turn to Empire: The Rise of Imperial Liberalism in Britain and France* (Princeton, NJ: Princeton University Press, 2005). An excellent article by Catherine Packham has clearly demonstrated Smith's debt to eighteenth-century physiology: "The Physiology of Political Economy: Vitalism and Adam Smith's *Wealth of Nations*," *Journal of the History of Ideas* (2005) **63**: 465–81.

10. Herder's struggle with his role as outsider was clearly posed in his fascinating *Journal meiner Reise im Jahr 1769*, written for himself and published only after his death. Herder, *Journal meiner Reise im Jahr 1769: Historisch-Kritische Ausgabe* [Journal of my journey in 1769: critical historical edition], ed. Katherina Mommsen (Stuttgart: Philipp Reclam, 1976).

11. *SWS*, vol. 14, 145.

12. *SWS*, vol. 13, 20.

13. *SWS*, vol. 13, 20.

14. Dugald Stewart, "Account of the Life and Writings of Adam Smith," in *Adam Smith, Essays on Philosophical Subjects with Dugald Stewart's Account of Adam Smith*, ed. W. P. D. Wightman and J. C. Bryce (Oxford: Clarendon Pess, 1980; hereafter cited as *EPS*), 270.

15. Smith, "The Principles which Lead and Direct Philosophical Enquiries: Illustrated by the History of Astronomy," in *EPS*, 33–105; hereafter cited as *Astronomy*.

16. Catherine Packham argues that Smith probably acquired his knowledge of Robert Whytt's physiological system through his participation in the activities of the Philosophical Society: Packham, *J. Hist. Ideas* (2005) **63**: 474.

17. Campbell and Skinner, *Adam Smith*, 220.

18. The catalog of Smith's library was first assembled by James Bonar, *A Catalogue of the Library of Adam Smith*, 2nd ed. (London: Macmillan, 1932), and expanded by Hiroshi Mizuta, *Adam Smith's Library: A Supplement to Bonar's Catalogue with a Checklist of the Whole Library* (Cambridge: Cambridge University Press, 1967). It reveals that Smith's collection of scientific works was very extensive. Amongst the leaders of vitalist thought, Smith held works by Bonnet, Buffon, Cullen, Diderot, John and William Hunter, James Hutton, Pierre Macquer, Maupertuis, Joseph Priestley, J. B. Robinet, and the abbé Spallanzani.

19. *EPS*, vol. 14, 26 (322).

20. Adam Smith, *An Inquiry into the Nature and Causes of the Wealth of Nations*, ed. R. H. Campbell and A. S. Skinner, 2 vols. (Oxford: Clarendon Press, 1976; hereafter cited as *WN*), L.i.9 (21).

21. *Astronomy*, vol. 2, 12 (45).

22. *Astronomy*, vol. 2, 12 (46–7).

23. *Astronomy*, vol. 4, 27 (71).

24. *Astronomy*, vol. 4, 33(75).

25. *SWS*, vol. 14, 92–105.

26. *SWS*, vol. 13, 9.

27. *WN*, IV.vii.c (605–6).

28. *WN*, IV.vi.c (606).

29. *WN*, IV.ix.28 (673–4).

30. *WN*, I.xi.p (265).

31. *WN*, I.v.17 (54).

32. *WN*, I.ii.2 (26).

33. *WN*, I.ii.2 (26).

34. *WN*, IV.vii.c (630).

35. *SWS*, vol. 8, 169.

36. *SWS*, vol. 13, 213.

37. *SWS*, vol. 13, 8.

38. *SWS*, vol. 14, 144.

39. *SWS*, vol. 13, 213.

40. *SWS*, vol. 13, 173. "*Bildung* [*formation*] is an effect of internal powers, acting upon a mass prepared by nature for development." Herder borrowed this idea from Johann Friedrich Blumenbach.

41. Blumenbach as a proponent of epigenesis also drew a distinction between dead and living matter, denying in the process the Great Chain of Being. Johann Friedrich Blumenbach, *Ueber den Bildungstrieb* [On the formative drive], 3rd ed. (Göttingen, 1791), 79–80.

42. *WN*, I.vii.3 (80).

43. *Adam Smith, Lectures on Jurisprudence*, ed. R. L. Meek, D. D. Raphael, and P. G. Stein (Oxford: Clarendon Press, 1978), A.I.27 (14).

44. *WN*, II.iii.36 (345).
45. *WN*, III.iv.17 (422).
46. *WN*, I.viii.43 (99).
47. *WN*, IV.vii.c (608).
48. *WN*, IV.ii.43 (471).
49. *WN*, I.xi.l (239).
50. *WN*, III.ii.4 (383).
51. *WN*, IV.vi.c (606).

52. Smith hardly used the term "invisible hand" in his analysis. He used it only once in the *Wealth of Nations*. Certainly it plays no central role in his discussion of his system and its dynamics, a point made in Packham, *J. Hist. Ideas* (2005) **63**: 480.

53. *SWS*, XIII, 8.
54. *SWS*, XIV, 230.
55. *SWS*, XIV, 208.
56. *WN*, III.i.3 (377).
57. *WN*, II.iii.28 (341).
58. *WN*, II.iii.28 (341).
59. *WN*, II.ii.86 (320).
60. *WN*, II.iii.31 (343).
61. *EPS*, 250.
62. *WN*, I.vii.15 (75).
63. Campbell and Skinner, *Adam Smith*, 179.
64. *WN*, I.x.a (116).
65. *SWS*, VIII, 169.
66. *SWS*, XIII, 129.
67. *SWS*, VIII, 170.
68. *SWS*, VIII, 170.
69. *Astronomy*, II, 8 (41–2).
70. *Astronomy*, II.12 (47).
71. *SWS*, XIV, 98–100.
72. *WN*, II.ii.86 (320).
73. *WN*, I.v.7 (51).
74. *WN*, IV.v. a (510).

75. *Astronomy*, IV, 33 (75). As an example, Smith evaluated the rise in prices in a manner opposed to generally held opinion, seeing their rise not as a "calamity" but "the necessary forerunner and attendant of the greatest of all publick advantages." *WN*, I.xi.i (245).

76. *SWS*, XIII, 123.
77. *TMS*, I.i.1.3 (10).
78. *TMS*, I.i.4.7 (22).
79. Campbell and Skinner, *Adam Smith*, 105.

Chapter Four

1. William Paley, *Natural Theology; or, Evidences of the Existence and Attributes of the Deity, Collected from the Appearances of Nature* (London: R. Faulder, 1802), 19; Richard Dawkins, *The Blind Watchmaker* (Harmondsworth: Penguin, 2006), 5; Charles Darwin, *On the Origin of Species by Means of Natural Selection, or the Preservation of Favoured Races in the Struggle for Life* (London: John Murray, 1859).

2. Nora Barlow, ed., *The Autobiography of Charles Darwin, 1809–1822, with Original Omissions Restored* (London: Collins, 1958), 87.

3. Francis Darwin, ed., *More Letters of Charles Darwin: A Record of His Work in a Series of Hitherto Unpublished Letters*, 2 vols. (London: John Murray, 1908), vol. 1, 154.

4. John Hedley Brooke, *Science and Religion: Some Historical Perspectives* (Cambridge: Cambridge University Press, 1991), 192–203; John Hedley Brooke, "Why Did the English Mix Their Science and Religion?" in *Science and the Imagination in Eighteenth-Century British Culture*, ed. S. Rossi (Milan: Edizioni Unicopli, 1987), 57–78.

5. Brooke, *Science and Religion*, 209–19 (on the critiques of Kant and Hume, see also 181–9 and 203–9); John Hedley Brooke, "The Natural Theology of the Geologists: Some Theological Strata," in *Images of the Earth: Essays in the History of the Environmental Sciences*, ed. L. J. Jordanova and Roy Porter, 2nd ed. ([n.p.]: British Society for the History of Science, 1997), 53–74. For the dubious claim that natural theology had ceased to have any importance for the practice or cognitive content of science by 1859, see Neal C. Gillespie, "Preparing for Darwin: Conchology and Natural Theology in Anglo-American Natural History," *Studies in History of Biology* (1983) **7**: 93–145.

6. William Kirby, *On The Power, Wisdom and Goodness of God as Manifested in the Creation of Animals and in Their History, Habits and Instincts*, 2 vols. (London: William Pickering, 1835), vol. 1, 181.

7. Aileen Fyfe, *Science and Salvation: Evangelical Popular Science Publishing in Victorian Britain* (Chicago and London: University of Chicago Press, 2004).

8. Pietro Corsi, *Science and Religion: Baden Powell and the Anglican Debate, 1800–1860* (Cambridge: Cambridge University Press, 1988); Jonathan R. Topham, "Science, Natural Theology, and the Practice of Christian Piety in Early Nineteenth-Century Religious Magazines," in *Science Serialized: Representations of the Sciences in Nineteenth-Century Periodicals*, ed. Geoffrey Cantor and Sally Shuttleworth (Cambridge, MA: MIT Press, 2004), 37–66.

9. Aileen Fyfe, "Publishing and the Classics: Paley's *Natural Theology* and the Nineteenth-Century Scientific Canon," *Studies in History and Philosophy of Science* (2002) **33**: 733–55 (735).

10. Jonathan R. Topham, "Beyond the 'Common Context': The Production and Reading of the Bridgewater Treatises," *Isis* (1998) **89**: 233–62.

11. Robert M. Young, "Natural Theology, Victorian Periodicals, and the Fragmentation of a Common Context," in *Darwin's Metaphor: Nature's Place in Victorian Culture* (Cambridge: Cambridge University Press, 1985), 126–63, 265–71 (127–8).

12. Paley, *Natural Theology*, vii–viii; Aileen Fyfe, "The Reception of William Paley's *Natural Theology* in the University of Cambridge," *British Journal for the History of Science* (1997) **30**: 321–35; William Paley, *The Principles of Moral and Political Philosophy* (London: R. Faulder, 1785); Paley, *Horæ Paulinæ; or, The Truth of the Scripture History of St. Paul Evinced, by a Comparison of the Epistles which Bear His Name, with the Acts of the Apostles, and with One Another* (London: R. Faulder, 1790); Paley, *A View of the Evidences of Christianity: in Three Parts*, 3 vols. (London: R. Faulder, 1794).

13. Fyfe, *Stud. Hist. Phil. Sci.* (2002) **33**: 736.

14. Paley, *Natural Theology*, 18.

15. John Brooke and Geoffrey Cantor, *Reconstructing Nature: The Engagement of Science and Religion* (Edinburgh: T. & T. Clark, 1998), 182.

16. Matthew D. Eddy, "The Rhetoric and Science of Paley's *Natural Theology*," *Literature and Theology* (2004) **18**: 1–22 (7); Paley, *Natural Theology*, 81–3.

17. Paley, *Natural Theology*, 409.

18. James Keill, *The Anatomy of the Humane Body Abridged; or, A Short and Full View of All the Parts of the Body. Together with Their Several Uses Drawn from Their Compositions and Structures*, 15th ed. (London: William Keblewhite, 1771); Oliver Goldsmith, *An History of the Earth, and Animated Nature*, 8 vols. (London: J. Nourse, 1774).

19. Fyfe, *Stud. Hist. Phil. Sci.* (2002) **33**: 735; Joseph John Gurney to Whewell, 31 October 1833, Trinity College, Cambridge, Add.MS.a.205[63], quoted by kind permission of the Master and Fellows, Trinity College, Cambridge.

20. Charles Bell, *The Hand: Its Mechanism and Vital Endowments as Evincing Design* (London: William Pickering, 1833); *Examiner*, 9 June 1833, 357a; Charles Bell and Henry Peter Brougham, eds., *Paley's Natural Theology, with Illustrative Notes*, 2 vols. (London: Charles Knight, 1836), vol. 1, 1n–2n.

21. William Buckland, *Geology and Mineralogy Considered with Reference to Natural Theology*, 2 vols. (London: William Pickering, 1836).

22. Peter Mark Roget, *Animal and Vegetable Physiology Considered with Reference to Natural Theology*, 2 vols. (London: William Pickering, 1834), vol. 1, ix.

23. William Whewell, *Astronomy and General Physics Considered with Reference to Natural Theology* (London: William Pickering, 1833), 3; Charles Babbage, *The Ninth Bridgewater Treatise: A Fragment* (London: John Murray, 1837). See also William Prout, *Chemistry, Meteorology and the Function of Digestion Considered with Reference to Natural Theology* (London: William Pickering, 1834).

24. Kirby, *On the Power*, vol. 1, xviii.

25. Thomas Chalmers, *On the Power, Wisdom and Goodness of God as Manifested in the Adaptation of External Nature to the Moral and Intellectual Constitution*

of Man, 2 vols. (London: William Pickering, 1833), vol. 2, 281; Whewell, *Astronomy*, 166, 41, vi. See also Jonathan R. Topham, "Evangelicals, Science, and Natural Theology in Early Nineteenth-Century Britain: Thomas Chalmers and the *Evidence* Controversy," in *Evangelicals and Science in Historical Perspective*, ed. Daryl Hart, David Livingstone, and Mark Noll (New York: Oxford University Press, 1998), 142–74; and John Hedley Brooke, "Indications of a Creator: Whewell as Apologist and Priest," in *William Whewell: A Composite Portrait*, ed. Menachem Fisch and Simon Schaffer (Oxford: Clarendon Press, 1991), 149–73.

26. John Kidd, *On the Adaptation of External Nature to the Physical Nature of Man, Principally with Reference to the Supply of His Wants and the Exercise of His Intellectual Faculties* (London: William Pickering, 1833), 339; Bell, *The Hand*, xi.

27. Topham, "Science."

28. Bell, *The Hand*, 18, 21, 23, 79. For a helpful introduction to Cuvier's work see Peter Bowler, *Evolution: The History of an Idea*, 3rd ed. (Berkeley: University of California Press, 2003), 108–15.

29. Anon., "Infidelity in Disguise—Geology," *Church of England Quarterly Review* (1837) **2**: 450–91 (465); Buckland, *Geology*, vol. 1, vii–viii.

30. Anon., "Geology and Mineralogy Considered with Reference to Natural Theology," *Spectator* (1836) **9**: 946–7 (947).

31. Buckland, *Geology*, vol. 1, 21, 586.

32. Dov Ospovat, *The Development of Darwin's Theory: Natural History, Natural Theology, and Natural Selection, 1838–1859* (Cambridge: Cambridge University Press, 1981), 33–4.

33. Roget, *Animal and Vegetable Physiology*, vol. 1, 48.

34. Roget, *Animal and Vegetable Physiology*, vol. 1, 49–52.

35. Adrian Desmond, *The Politics of Evolution: Morphology, Medicine, and Reform in Radical London* (Chicago and London: University of Chicago Press, 1989).

36. Philip F. Rehbock, *The Philosophical Naturalists* (Madison and London: University of Wisconsin Press, 1983), 26–30; Kirby, *On the Power*, vol. 2, 40–1; Muriel Blaisdell, "Natural Theology and Nature's Disguises," *Journal of the History of Biology* (1982) **15**: 163–89.

37. C. Carter Blake, "The Life of Dr. Knox," *Journal of Anthropology* (1870–1) **1**: 332–8 (334).

38. Chalmers, *On the Power*, vol. 1, 15–31; see also Crosbie Smith, "From Design to Dissolution: Thomas Chalmers' Debt to John Robison," *British Journal for the History of Science* (1979) **12**: 59–70.

39. Whewell, *Astronomy*, 360–1, 11–2.

40. Chalmers, *On the Power*, vol. 1, 18; Whewell, *Astronomy*, 184, 349.

41. Buckland, *Geology*, vol. 1, 40n, 580–1.

42. [William Whewell] "Lyell—*Principles of Geology*," *British Critic* (1831) **9**: 180–206 (194), quoted in Buckland, *Geology*, vol. 1, 586; M. J. S. Hodge, "The

History of the Earth, Life, and Man: Whewell and Palaetiological Science," in *William Whewell: A Composite Portrait*, ed. by Menachem Fisch and Simon Schaffer (Oxford: Clarendon Press, 1990), 255–88.

43. Babbage, *The Ninth Bridgewater*, 44–6.

44. [Robert Chambers], *Vestiges of the Natural History of Creation* (London: John Churchill, 1844), 157–8; John H. Brooke, "The Relations between Darwin's Science and Religion," in *Darwinism and Divinity: Essays on Evolution and Religious Belief*, ed. John Durant (London: Blackwell, 1985), 40–75; John H. Brooke, "Darwin and Victorian Christianity," in *The Cambridge Companion to Darwin*, ed. Jonathan Hodge and Gregory Radick (Cambridge: Cambridge University Press, 2003), 192–213 (197); Darwin, *Origin of Species*, [ii]; James R. Moore, *The Post-Darwinian Controversies: A Study of the Protestant Struggle to Come to Terms with Darwin in Great Britain and America, 1870–1900* (Cambridge: Cambridge University Press, 1979), chap. 11.

45. Ospovat, *The Development*, esp. chap. 1.

46. Walter F. Cannon, "The Bases of Darwin's Achievement: A Revaluation," *Victorian Studies* (1961) **5**: 109–34; Brooke, "The Relations"; George Bentham, "Anniversary Address," *Journal of the Proceedings of the Linnean Society of London (Botany)* (1862) **6**: lxvi–lxxxiii (lxviii, lxxxi); Darwin to John Murray, 21 September [1861], in *Correspondence of Charles Darwin*, ed. Frederick Burkhardt et al., 15 vols. (Cambridge: Cambridge University Press, 1985–), vol. 9, ed. Frederick Burkhardt, Janet Browne, Duncan M. Porter, and Marsha Richmond, 272.

Chapter Five

1. This chapter is a synthesis of some recent work on race and science, in what is becoming an increasingly fragmented historiography, located in history of science, intellectual history, imperial history and area studies.

2. The details of these executions are taken from the following sources: Henry Marshall, *Ceylon, A General Description of the Island and its Inhabitants* (first published London, 1846; reprinted Dehiwela, Sri Lanka: Tisara Prakasakayo, 1982), Appendix X; P. E. Pieris, *Sinhale and the Patriots, 1815–1818* (first published, Colombo, 1950; republished Delhi: Navrang, 1995), 418–20 and corresponding footnotes.

3. Marshall, *Ceylon*, 198

4. For more on phrenology, see Roger Cooter, *The Cultural Meaning of Popular Science: Phrenology and the Organization of Consent in Nineteenth-Century Britain* (Cambridge: Cambridge University Press, 1984).

5. This paragraph and the next draw from the recent book by Colin Kidd, *The Forging of Races: Race and Scripture in the Protestant Atlantic World, 1600–2000* (Cambridge: Cambridge University Press, 2006). See also David N. Livingstone,

Adam's Ancestors: Race, Religion, and the Politics of Human Origins (Baltimore: Johns Hopkins University Press, 2008).

6. William F. Bynum, "The Great Chain of Being After Forty Years: An Appraisal," *History of Science* (1975) **13**: 1–28; and the classic, A. O. Lovejoy, *The Great Chain of Being: The Study of the History of an Idea* (Cambridge: Cambridge University Press, 1936).

7. Cited in Nancy Stepan, *The Idea of Race in Science: Great Britain, 1800–1960* (London: Macmillan, 1982), 7.

8. This paragraph draws on Stepan, *The Idea of Race in Science*, chap. 1, and also David Bindman, *Ape to Apollo: Aesthetics and the Idea of Race in the Eighteenth-Century* (London: Reaktion, 2002), chap. 4.

9. J. F. Blumenbach, *Anthropological Treaties,* transl. T. Bendyshe (London: Longman, 1865), 269.

10. M. Meijer, *Race and Aesthetics in the Anthropology of Petrus Camper (1722–1789)* (Amsterdam: Rodopi, 1999). Also Bindman, *Ape to Apollo*, 201ff.

11. Meijer, *Race and Aesthetics*, 191.

12. Jacques Roger, *Buffon: A Life in Natural History*, transl. Sarah Lucille Bonnefoi, ed. L. Pearce Williams (Ithaca, NY: Cornell University Press, 1997), 180.

13. See Bindman, *Ape to Apollo,* 62–6. For the history of the use of the term "race" by naturalists, see Nicholas Hudson, "From 'Nation' to 'Race': The Origin of Racial Classification in Eighteenth-Century Thought," *Eighteenth-Century Studies* (1996) **29**: 247–64.

14. Londa Schiebinger, "The Anatomy of Difference: Sex and Race in Eighteenth-Century Science," *Eighteenth-Century Studies* (1990) **23**: 387–405 (393).

15. Cited in Bindman, *Ape to Apollo,* 68.

16. For discussion of this footnote and how it was revised, see John Immerwahr, "Hume's Revised Racism," *Journal of the History of Ideas* (1992) **53**: 481–6 (the two versions of the footnote appear on 481 and 483).

17. Kidd, *The Forging of Races*, 96–100.

18. For more on the early shows of London, see Richard Altick, *The Shows of London* (Cambridge, MA: Belknap Press, 1978).

19. See for instance, Anne Maxwell, *Colonial Photography and Exhibitions: Representations of the 'Native' and the Making of European Identities* (London: Leicester University Press, 1999), chap. 1.

20. Sadiah Qureshi, "Displaying Sara Baartman, the 'Hottentot Venus,'" *History of Science* (2004) **42**: 233–57.

21. Qureshi, *Hist. Sci.* (2004) **42**: 241–2.

22. Seymour Drescher, "The Ending of the Slave Trade and the Evolution of European Scientific Racism," *Social Science History* (1990) **14**: 415–50.

23. Citation from Drescher, *Soc. Sci. Hist.* (1990) **14**: 422–23. For an extended discussion of Long's work, see Roxanne Wheeler, *The Complexion of Race: Cate-*

gories of Difference in Eighteenth-Century British Culture (Philadelphia: University of Pennsylvania Press, 2000).

24. Citation from Stephen J. Gould, "The Great Physiologist of Heidelberg—Friedrich Tiedemann," *Natural History* (1999) **108** (July/August): 26–29, 62–70.

25. This paragraph draws on George Stocking, "From Chronology to Ethnology: James Cowles Prichard and British Anthropology, 1800–1850," in James Cowles Prichard, *Researches into the Physical History of Man*, ed. George Stocking (Chicago: University of Chicago Press, 1973), ix–cx, and George Stocking, *Victorian Anthropology* (New York: Free Press, 1987) esp. 48–52, 240–44.

26. Prichard, *Researches*, 233.

27. Kidd, *The Forging of Races*, 139.

28. This argument is taken from Michael O'Brien, *Conjectures of Order: Intellectual Life in the American South, 1800–1860* (Chapel Hill: University of North Carolina Press, 2004), vol. 1, chap. 5.

29. This account draws from O'Brien, *Conjectures of Order*, vol.1, 240–47; Lester D. Stephens, *Science, Race, and Religion in the American South: John Bachman and the Charleston Circle of Naturalists, 1815–1895* (Chapel Hill: University of North Carolina Press, 2000), chap. 10; G. Blair Nelson, "'Men Before Adam!': American Debates about the Unity and Antiquity of Humanity," in *When Science and Christianity Meet*, ed. David C. Lindberg and Ronald L. Numbers (Chicago: University of Chicago Press, 2003), 161–81.

30. Cited in Stephens, *Science*, 198.

31. Cited in Kidd, *The Forging of Races*, 142.

32. Stepan, *The Idea of Race in Science*, 23–6.

33. George Combe, *On the Relation between Religion and Science* (Edinburgh: MacLachlan, Stewart & Co., 1847), 6.

34. Nancy Stepan, "Race and Gender: The Role of Analogy in Science," *Isis* (1986) **77**: 261–77.

35. Dorinda Outram, *Georges Cuvier: Vocation, Science, and Authority in Post-Revolutionary France* (Manchester: Manchester University Press, 1984).

36. Michael Banton, *Racial Theories* (Cambridge: Cambridge University Press, 1987), 48.

37. William Coleman, *Georges Cuvier Zoologist: A Study in the History of Evolution Theory* (Cambridge, MA: Harvard University Press, 1964) 165–8; Martin J. S. Rudwick, *Georges Cuvier, Fossil Bones and Geological Catastrophes* (Chicago: University of Chicago Press, 1997).

38. Toby A. Appel, *The Cuvier–Geoffroy Debate: French Biology in the Decades Before Darwin*, 2 vols. (Oxford: Oxford University Press, 1987).

39. The account of institutions in this paragraph and the following relies on Stocking, *Victorian Anthropology*, chap. 7.

40. Cited in Stocking, *Victorian Anthropology*, 249.

41. Cited in Stocking, *Victorian Anthropology*, 251.

42. See Felix Driver, *Geography Militant: Cultures of Exploration and Empire* (Oxford: Oxford University Press, 2001).

43. James Secord, *Victorian Sensation: The Extraordinary Publication, Reception, and Secret Authorship of Vestiges of the Natural History of Creation* (Chicago: University of Chicago Press, 2000). See also James Secord, "Introduction," in *Robert Chambers, Vestiges of the Natural History of Creation: and Other Evolutionary Writings*, ed. James Secord (Chicago: University of Chicago Press, 1994).

44. This account draws on Stuart McCook, "It May Be the Truth, But It Is Not Evidence: Paul Du Chaillu and the Legitimation of Evidence in the Field Sciences," *Osiris* (1996) **11**: 177–97. See also Nicolaas Rupke, *Richard Owen: Victorian Naturalist* (London: Yale University Press, 1994), 314–22.

45. Cited in McCook, *Osiris* (1996) **11**: 190

46. This follows the argument of Peter Mandler, "'Race' and 'Nation' in Mid-Victorian Thought," in *History, Religion, and Culture: British Intellectual History 1750–1950*, ed. S. Collini, R. Whatmore, and B. Young (Cambridge: Cambridge University Press, 2000), 224–44.

47. For the role of the Pacific oceans in opening up new vistas for science, see Bernard Smith, *European Vision and the South Pacific, 1768–1850: A Study in the History of Art and Ideas* (Oxford: Clarendon Press, 1960). For the sense of the Pacific as a laboratory, see for instance Roy MacLeod and Philip F. Rehbock, eds., *Darwin's Laboratory: Evolutionary Theory and Natural History in the Pacific* (Honolulu: University of Hawaii Press, 1994).

48. Smith, *European Vision*, 7.

49. Cited in Smith, *European Vision*, 5.

50. Cited in Harry Liebersohn, *The Travelers' World: Europe to the Pacific* (Cambridge, MA: Harvard University Press, 2006), 30.

51. Nicholas Thomas, "The Force of Ethnology: Origins and Significance of the Melanesia/Polynesia Division," *Current Anthropology* (1989) **30**: 27–41, citing p. 29.

52. Liebersohn, *The Travelers' World*, 197ff.

53. Thomas, *Curr. Anthropol.* (1989) **30**: 30ff; citation from d'Urville below is from 30.

54. I am drawing here on: Jane Samson, "Ethnology and Theology: Nineteenth-century Mission Dilemmas in the South Pacific," in *Christian Missions and the Enlightenment*, ed. Brian Stanley (Grand Rapids, MI: William B. Eerdmans, 2001), 99–123; Sujit Sivasundaram, *Nature and the Godly Empire: Science and Evangelical Mission in the Pacific, 1795–1850* (Cambridge: Cambridge University Press, 2005).

55. Cited in Samson, "Ethnology," 113.

56. Cited in Samson, "Ethnology," 119.

57. The account of Indian Orientalism draws from Michael Dodson, *Orientalism, Empire and National Culture, 1770–1880* (Basingstoke, Hants: Palgrave 2007), and of particular relevance here, Tony Ballantyne, *Orientalism and Race: Aryanism in the British Empire* (New York: Palgrave, 2002). Also Thomas Trautmann, *Aryans and British India* (Berkeley: University of California Press, 1997).

58. See Dodson, *Orientalism*, 24–33; Ballantyne, *Orientalism*, 26–30.

59. Cited in Ballantyne, *Orientalism*, 43.

60. Thomas R. Metcalf, *Ideologies of the Raj* (Cambridge: Cambridge University Press, 1994, reprinted 2005), 82. See also Trautmann, *Aryans and British India*, chaps. 6 and 7.

61. Citations from Metcalf, *Ideologies*, 119.

62. The following account of phrenology is a summary of Shruti Kapila, "Race Matters: Orientalism and Religion, India and Beyond, c. 1770–1880," *Modern Asian Studies* (2007) **41**: 471–513.

63. This draws on Mark Harrison, "'The Tender Frame of Man': Disease, Climate, and Racial Difference in India and the West Indies, 1760–1860," *Bulletin of the History of Medicine* (1996) **70**: 68–93. See also Mark Harrison, *Climates and Constitutions: Health, Race, Environment, and British Imperialism in India, 1600–1850* (Oxford: Oxford University Press, 1999).

64. For more on Buchanan, see Maria Vicziany, "Imperialism, Botany, and Statistics in Early-Nineteenth Century India," *Modern Asian Studies* (1986) **20**: 625–60.

65. This summarizes Sujit Sivasundaram, "Trading Knowledge: The East India Company's Elephants in India and Britain," *Historical Journal* (2005) **48**: 27–63. See also Mahesh Rangarajan, *India's Wildlife History* (Delhi: Permanent Black, 2001).

66. Alison Winter, *Mesmerized: Powers of Mind in Victorian Britain* (Chicago: University of Chicago Press, 1998), chap. 8.

67. Janet Browne, "Missionaries and the Human Mind: Charles Darwin and Robert Fitzroy," in MacLeod and Rehbock, *Darwin's Laboratory*, 263–82.

68. Niel Gunson, "British Missionaries and Their Contribution to Science in the Pacific Islands," in MacLeod and Rehbock, *Darwin's Laboratory*, 283–316.

69. For the account of the Fuegians, see Janet Browne, *Charles Darwin Voyaging* (Princeton, NJ: Princeton University Press, 1995), chap. 10.

70. Cited in Gillian Beer, "Travelling the Other Way," in *Cultures of Natural History*, ed. Nicholas Jardine, James A. Secord and E. C. Spary (Cambridge: Cambridge University Press, 1996), 332.

71. Jim Moore, "Socialising Darwinism: Historiography and the Fortunes of a Phrase," in *Science as Politics*, ed. Les Levidow (London: Free Association Books, 1986), 38–80.

Chapter Six

1. Uwe Hossfeld and Rainer Brömer, eds., *Darwinismus und/als Ideologie* [Darwinism and/as ideology] (Berlin: Verlag für Wissenschaft und Bildung, 2001).

2. Several of Young's seminal papers were collected in Robert M. Young, *Darwin's Metaphor: Nature's Place in Victorian Culture* (Cambridge: Cambridge University Press, 1985).

3. Adrian Desmond and James Moore, *Darwin* (London: Joseph, 1991); Janet Browne, *Charles Darwin*, 2 vols. (New York: Knopf, 1995, 2002).

4. For example, see James Moore, "Wallace's Malthusian Moment: The Common Context Revisited," in *Victorian Science in Context*, ed. Bernard Lightman (Chicago and London: University of Chicago Press, 1997), 290–311.

5. David N. Livingstone, "Science, Region, and Religion: The Reception of Darwinism in Princeton, Belfast, and Edinburgh," in *Disseminating Darwinism: The Role of Place, Race, Religion, and Gender*, ed. Ronald L. Numbers and John Stenhouse (Cambridge: Cambridge University Press, 1999), 7–38; Ronald L. Numbers, *Darwinism Comes to America* (Cambridge, MA, and London: Harvard University Press, 1998). See also Nicolaas A. Rupke, "Zu einer Taxonomie der Darwin-Literatur nach ideologischen Merkmalen" [Toward a taxonomy of the Darwin literature based on ideological characteristics], in *Evolutionsbiologie von Darwin bis heute*, ed. Rainer Brömer, Uwe Hossfeld, and Nicolaas Rupke (Berlin: Verlag für Wissenschaft und Bildung, 1999), 59–68.

6. To cite just one of several instances: Mario A. Di Gregorio, *From Here to Eternity: Ernst Haeckel and Scientific Faith* (Göttingen: Vandenhoeck and Ruprecht, 2005).

7. James Moore, "Deconstructing Darwinism: The Politics of Evolution in the 1860s," *Journal of the History of Biology* (1991) **24**: 353–408. This paper was written for and revised in the light of the conference Ideology in the Life Sciences at Harvard University in April 1989, which meeting represented a pinnacle of interest in the science-and-ideology topic as it had developed through the 1970s and '80s. The point was then made that talking about ideological science is pleonastic and that science-and-ideology is a tautology. James Moore, personal communication, 10 March 2008.

8. Charles C. Gillispie, *Essays and Reviews in History and History of Science* (Philadelphia, PA: American Philosophical Society, 2007), 408.

9. Ernst Mayr, *One Long Argument: Charles Darwin and the Genesis of Modern Evolutionary Thought* (Cambridge, MA: Harvard University Press, 1999), 96.

10. Michael Ruse, *Mystery of Mysteries: Is Evolution a Social Construction?* (Cambridge, MA, and London: Harvard University Press, 1999).

11. See, for example, Bentley Glass, "Heredity and Variation in the Eighteenth Century Concept of the Species," in *Forerunners of Darwin: 1745–1859*, ed. Bentley Glass, Owsei Temkin, and William L. Straus, Jr. (Baltimore: Johns Hopkins Uni-

versity Press, 1959), 144–51. See also Nicolaas A. Rupke, "The Origin of Species from Linnaeus to Darwin," in *Aurora Torealis: Studies in the History of Science and Ideas in Honor of Tore Frängsmyr*, ed. Marco Beretta, Karl Grandin, and Svante Lindqvist (Sagamore Beach, MA: Science History Publications, 2008), 73–87.

12. Nicolaas A. Rupke, "Biology without Darwin," MS in preparation.

13. Heinrich Czolbe, *Neue Darstellung des Sensualismus* [A new presentation of sensualism] (Leipzig: Costenoble, 1855), 161–85; Rudolf Virchow, "Alter und neuer Vitalismus," *Archiv für pathologische Anatomie und Physiologie und für klinische Medizin* (1856) **9**: 3–55 (23–5).

14. From among the several monographs on the subject, see Otto Taschenberg, *Die Lehre von der Urzeugung sonst und jetzt* [The theory of spontaneous generation, past and present] (Halle: Niemeyer, 1882); Edmund O. von Lippmann, *Urzeugung und Lebenskraft* [Spontaneous generation and life force] (Berlin: Springer, 1933); John Farley, *The Spontaneous Generation Controversy from Descartes to Oparin* (Baltimore and London: Johns Hopkins University Press, 1974); Iris Fry, *The Emergence of Life on Earth: Historical and Scientific Overview* (New Brunswick, NJ, and London: Rutgers University Press, 2000); James Strick, *Sparks of Life: Darwinism and the Victorian Debates over Spontaneous Generation* (Cambridge, MA: Harvard University Press, 2000).

15. See, for example, Heinrich Georg Bronn, *Handbuch einer Geschichte der Natur* [Textbook of a history of nature] (Stuttgart: Schweizerbart, 1843), vol. 2, 30.

16. Johann Friedrich Blumenbach, *Handbuch der Naturgeschichte* [Textbook of natural history] (Göttingen: Dieterich, 1830), 2–3; *Beyträge zur Naturgeschichte* [Contributions to natural history], 2nd ed. (Göttingen: Dieterich, 1806), pt. I, 19–20.

17. Gottfried Reinhold Treviranus, *Biologie, oder Philosophie der lebenden Natur für Naturforscher und Aerzte* [Biology, or philosophy of living nature], 6 vols. (Göttingen: Röwer, 1802–22), vol. 2, 77.

18. Treviranus, *Biologie*, vol. 2, 378.

19. For example, Rudolf Leuckart, "Urerzeugung" [Spontaneous generation], *Handwörterbuch der Physiologie mit Rücksicht auf physiologische Pathologie* (Braunschweig: Vieweg, 1853), vol. 4, 991–1000.

20. Karl Friedrich Burdach, *Anthropologie für das gebildete Publicum* [Anthropology for the general public] (Stuttgart: Baltz, 1837), 726–7, 742–4.

21. Johannes Müller, *Handbuch der Physiologie des Menschen* [Textbook of human physiology] (Coblenz: Hölscher, 1844), 12, 17, 23–4; Rudolf Virchow, "Alter und neuer Vitalismus [Old and new vitalism]," *Archiv für pathologische Anatomie und Physiologie und für klinische Medizin* (1856) **9b**: 3–55 (25–6).

22. Blumenbach, *Beyträge zur Naturgeschichte*, 19–20.

23. K. E. A. von Hoff, *Erinnerung an Blumenbach's Verdienste um die Geologie* [Commemorating Blumenbach's contributions to geology] (Gotha: Engelhard und Reyher, 1826), 16–7.

24. Jean-Claude de Lamétherie, *Considérations sur les êtres organisés* (Paris: Courcier, 1804), vol. 2, 438.

25. Lamétherie, *Théorie de la terre*, 3 vols. (Paris: Maradan 1795), vol. 3, 161–2.

26. Lamétherie, *Théorie de la terre*, vol. 3, 164–5.

27. Hermann Burmeister, *Geschichte der Schöpfung* [History of creation] (Leipzig: Wigand, 1872), 314.

28. Nicolaas A. Rupke, *Richard Owen: Victorian Naturalist* (New Haven, CT, and London: Yale University Press, 1984), 225–30.

29. Alphonse de Candolle, *Introduction à l'étude de la botanique* [Introduction to the study of botany] (Paris: Roret, 1835), vol. 2, 313.

30. Burdach, *Anthropologie*, 742.

31. Karl Asmund Rudolphi, *Grundriss der Physiologie* [Outline of physiology] (Berlin: Ferdinand Dümmler, 1821), vol. 1, 50–7.

32. Carl Gustav Carus, *System der Physiologie für Naturforscher und Ärzte* [Compendium of physiology for scientists and doctors] (Dresden and Leipzig: Fleischer, 1838), pt. I, 112–3.

33. Carl Gustav Carus, *Denkschrift zum hundertjährigen Geburtsfeste Goethe's* [Commemorative publication on the occasion of the centenary of Goethe's birth] (Leipzig: Brockhaus, 1849).

34. Rupke, *Richard Owen*, 259–68.

35. In addition to the above cited Lamétherie, there were J.-B. Fray, *Essai sur l'origine des corps organisés et inorganisés* [An essay on organized and disorganized bodies] (Paris: Courcier, 1817); and Frédéric Gérard, "Génération spontanée ou primitive" [Spontaneous or primitive generation], *Dictionnaire universel d'histoire naturelle* (Paris: Renard, Martinet and Co., 1845), 53–71.

36. Charles Lyell, *Principles of Geology, being an Attempt to Explain the Former Changes of the Earth's Surface, by Reference to Causes now in Operation*, 3 vols. (London: Murray, 1830–3), vol. 2, 182–3; Edward Forbes, "On the Connexion Between the Distribution of the Existing Fauna and Flora of the British Isles, and the Geological Changes which have Affected their Area, Especially During the Epoch of the Northern Drift," *Memoirs of the Geological Society of Great Britain* (1846) 1: 336–432. See also Forbes, "Abstract of the theory of specific centres," in *Essays on the Spirit of the Inductive Philosophy*, ed. Baden Powell (London: Longman, Brown, Green, Longmans, & Roberts, 1855), 498–500.

37. Heinrich G. Bronn, *Untersuchungen über die Entwickelungs-Gesetze der organischen Welt* [Inquiries into the laws of development of the organic world] (Stuttgart: Schweizerbart, 1858), 78; Carl Vogt, *Altes und Neues aus Thier- und Menschenleben* [Animal and human life: old and new], 2 vols. (Frankfurt am Main: Literarische Anstalt, 1859), vol. 1, 366–8.

38. Lyell, *Principles of Geology*, vol. 2, 182–3.

39. Thomas Henry Huxley, "Vestiges of the Natural History of Creation," *Brit-

ish and Foreign Medico-Chirurgical Review (1854) **26**: 425–39; Alphonse de Candolle, *Géographie botanique raisonnée* [A rational geography of plants], 2 vols. (Paris: Masson; Geneva: Kessmann, 1855), vol. 2, 151–7, 248–56.

40. Rupke, *Annals Hist. Philos. Biol.* (2005) **10**: 163.

41. John C. Greene, *Science, Ideology, and World View* (Berkeley and Los Angeles: University of California Press, 1981), 7.

42. It is to be included in Rupke, "Biology without Darwin," MS in preparation.

43. Charles Darwin, *The Origin of Species by Means of Natural Selection*, 1st ed. (London: John Murray, 1859), 483.

44. Charles Darwin, *The Origin of Species by Means of Natural Selection*, 4th ed. (London: John Murray, 1866), xiii.

45. Bronn, *Untersuchungen.*

46. Pieter Harting, *Leerboek van de grondbeginselen der dierkunde in haren geheelen omvang* [Textbook of the principles of all of zoology], 3 vols. (Tiel: Campagne, 1862), vol. 1, 286–313.

47. Charles Darwin, *The Origin of Species by Means of Natural Selection*, 6th ed. (London: John Murray, 1876), 424.

48. For examples of autogenists who have been reclassified as evolutionists, see Nicolaas A. Rupke, "Neither Creation Nor Evolution: The Third Way in Mid-Nineteenth Century Thinking about the Origin of Species," *Annals of the History and Philosophy of Biology* (2005) **10**: 143–72 (165).

49. Bentley Glass, preface, in *Forerunners of Darwin: 1745–1859*, ed. Bentley Glass, Owsei Temkin, and William L. Straus, Jr. (Baltimore: Johns Hopkins University Press, 1959), v–vi.

50. Carl Vogt, *Vorlesungen über den Menschen, seine Stellung in der Schöpfung und in der Geschichte der Erde* [Lectures on man, his position in nature, and in the history of the earth], 2 vols. (Giessen: Ricker, 1863), vol. 2, 259.

51. Roy Porter, "Charles Lyell and the Principles of the History of Geology," *British Journal for the History of Science* (1976) **9**: 91–103 (93).

52. Porter, *Brit. J. Hist. Sci.* (1976) **9**: 94–5.

53. Martin J. S. Rudwick, "The Strategy of Lyell's *Principles of Geology*," *Isis* (1970) **61**: 4–33.

54. James Moore, personal communication, 18 March 2008.

55. Darwin, *Origin of Species*, 1st ed., 484, 490. On Darwin's contrasting his theory with the belief that God independently created each species yet invoking special creation for the origin of life, see Chris Cosans, "Was Darwin a Creationist?," *Perspectives in Biology and Medicine* (2005) **48**: 362–71.

56. Neal C. Gillespie, *Charles Darwin and the Problem of Creation* (Chicago and London: University of Chicago Press, 1979), 130.

57. See, for example, Alexander Graham Cairns-Smith, *Seven Clues to the Origin of Life* (Cambridge: Cambridge University Press, 1985).

58. Darwin, *Origin of Species*, 1st ed., 483.

59. On the *Bathybius* affair, see Philip F. Rehbock, "Huxley, Haeckel, and the Oceanographers: The Case of *Bathybius haeckelii*," *Isis* (1975) **66**: 504–33; Nicolaas A. Rupke, "*Bathybius Haeckelii* and the Psychology of Scientific Discovery," *Studies in the History and Philosophy of Science* (1976) **7**: 53–62.

60. Rupke, *Richard Owen*, 251.

61. Daniel G. Gibson et al., "Complete Chemical Synthesis, Assembly, and Cloning of a *Mycoplasma genitalium* Genome," *Science* (2008) 319: 1215–20.

62. Erwin Schrödinger, "Ist die Naturwissenschaft milieubedingt?" [Is science conditioned by context?] in *Quantenmechanik und Weimarer Republik*, ed. Karl von Meyenn (Brunswick and Wiesbaden: Vieweg), 295–332 (310).

Chapter Seven

1. Francis Galton to Charles Darwin, 24 December 1869, in *The Life, Letters and Labours of Francis Galton*, ed. Karl Pearson (Cambridge: Cambridge University Press, 1914), vol. 1, pl. 2.

2. For example, Francis Galton, *Hereditary Genius: An Inquiry into Its Laws and Consequences* (New York: Appleton, 1869), 330–9; Francis Galton, "Hereditary Talent and Character," *Macmillan's Magazine* (1869) **14**: 157–66, 314–27 (325–6).

3. For historical analysis of the socially constructed prejudices that Galton brought to his science, see Ruth Schwartz Cowan, *Francis Galton and the Study of Heredity in the Nineteenth Century* (New York: Garland, 1985), 64–9, 255–61; and Raymond E. Fancher, "Francis Galton's African Ethnography and Its Role in the Development of his Psychology," *British Journal for the History of Science* (1983) **16**: 67–79 (67–8, 79).

4. Galton, *Macmillan's Mag.* (1869) **14**: 322 (emphasis in original).

5. For expanded discussion of Galton's selectionism with further references, see Peter J. Bowler, *Evolution: The History of an Idea* (Berkeley: University of California Press, 1984), 240.

6. Bowler, *Evolution*, 437–9.

7. Bowler, *Evolution*, 241–2.

8. This sentence is compiled from two parallel passages in Galton's voluminous writings: Francis Galton, *Macmillan's Mag.* (1869) **14**: 163 (first quote); and Francis Galton, *Inquiries into Human Faculty and its Development* (London: Macmillan, 1883), 217 (second quote).

9. Galton, *Macmillan's Mag.* (1869) **14**: 165.

10. Francis Galton, *Memories of My Life* (London: Methuen, 1908), 315–6.

11. Galton, *Macmillan's Mag.* (1869) **14**: 319 ("weakly and incapable" quote); Francis Galton, "Hereditary Improvement," *Fraser's Magazine* (1873) **7**: 116–30 (125–8) (suggests compulsory eugenic segregation). Galton did not use the phrase

"negative eugenics," though it is generally used by historians to distinguish eugenic measures designed to discourage reproduction from those designed to encourage reproduction.

12. Galton, *Macmillan's Mag.* (1869) **14**: 319–20.

13. Galton, *Macmillan's Mag.* (1869) **14**: 165–6.

14. Galton, *Memories of My Life*, 323.

15. Galton, *Inquiries into Human Faculty*, 200.

16. Charles Darwin, *The Descent of Man and Selection in Relation to Sex* (New York: Appleton, 1871), vol. 1, 228–31 (in subchapter entitled "On the Extinction of the Races of Man").

17. R. L. Dugdale, *The Jukes: A Story in Crime, Pauperism, Disease, and Heredity*, 5th ed. (New York: Putnam, 1895), 7–15, 69–70.

18. Darwin, *Descent of Man*, vol. 1, 106–7.

19. Dugdale, *The Jukes*, 55, 57, 65.

20. This particular phrase was the motto of the eugenicist and birth-control advocate Margaret Sanger, but it was broadly representative of early twentieth-century eugenicists generally. For an example of Sanger's use of the phrase, see Margaret Sanger, "Why Not Birth Control Clinics in America?" *Birth Control Review*, May 1919: 10–11 (10).

21. Arthur H. Estabrook, *The Jukes in 1915* (Washington: Carnegie Institution, 1916), 85.

22. Henry Herbert Goddard, *The Kallikak Family: A Study in the Heredity of Febble-Mindedness* (New York: Macmillan, 1913), 60.

23. For a comparison of eugenics legislation in the USA and the United Kingdom, see Daniel J. Kevles, *In the Name of Eugenics: Genetics and the Uses of Human Heredity* (New York: Knopf, 1985), 101–7; and Edward J. Larson, "The Rhetoric of Eugenics: Expert Authority and the Mental Deficiency Bill," *British Journal for the History of Science* (1991) **24**: 45–60.

24. For a representative discussion of these four methods by leading members of the American eugenics movement, see Paul Popenoe and Roswell Hill Johnson, *Applied Eugenics* (New York: Macmillan, 1918), 184–96. The three areas of state action were singled out for analysis in the 1914 eugenics tract written by Stevenson Smith, Madge W. Wilkinson, and Louisa C. Wagoner, *A Summary of the Laws of the Several States* (Seattle: University of Washington, 1914).

25. Popenoe and Johnson, *Applied Eugenics*, 196.

26. Conn. Gen. Stat. sec. 1354 (1902).

27. Paul Popenoe and E. S. Gosney, *Twenty-eight Years of Sterilization in California* (Pasadena, CA: Human Betterment Foundation, 1938), 19–20. See also Popenoe and Johnson, *Applied Eugenics*, 197. Eugenicist Marian S. Olden also singled out these programs as models, and described the South Dakota effort in some detail, in Marian S. Olden, "Present Status of Sterilization Legislation in the United States," *Eugenical News* (1946) **31**: 1–10 (7–8).

28. For a listing and analysis of all eugenic marriage laws, see Charles B. Davenport, *State Laws Limiting Marriage Selection Examined in Light of Eugenics* (Cold Spring Harbor, NY: Eugenics Record Office, 1913), Tab. I; Mary Laack Oliver, "Eugenic Marriage Laws of the Forty-Eight States," M.A. thesis, University of Wisconsin, 1937, A1-A63; Smith, Wilkinson, and Wagoner, *Summary*, 4–12.

29. Davenport, *State Laws*, 27–36 (weighed advantages and disadvantages of interracial unions from a eugenics viewpoint, and suggested that only limited restrictions were justified, which were much less than imposed by the antimiscegenation statute in any Southern state). See also Mark H. Haller, *Eugenics: Hereditarian Attitudes in American Thought* (New Brunswick, NJ: Rutgers University Press, 1963), 158–9. For general discussions of racism and eugenics, see Kevles, *In the Name of Eugenics*, 74–6; and Haller, *Eugenics*, 50–7.

30. Henry H. Goddard, *Feeble-mindedness: Its Causes and Consequences* (New York: Macmillan, 1914), 565–6.

31. J. E. Wallace Wallin, *The Odyssey of a Psychologist: Pioneering Experiences in Special Education, Clinical Psychology, and Mental Hygiene* (Wilmington, IN: privately printed, 1955), 88.

32. Popenoe and Johnson, *Applied Eugenics*, 196.

33. Davenport, *State Laws*, 12.

34. Popenoe and Johnson, *Applied Eugenics*, 442.

35. Davenport, *State Laws*, 12.

36. Gerald N. Grob, *Mental Illness and American Society, 1875–1940* (Princeton, NJ: Princeton University Press, 1983), 7. See also Albert Deutsch, *The Mentally Ill in America: A History of Their Care and Treatment from Colonial Times* (New York: Columbia University Press, 1949, revised ed.), 158–245; and Daniel J. Rothman, *The Discovery of the Asylum: Social Order and Disorder in the New Republic*, revised ed. (Boston: Little, Brown, 1990), 130–54.

37. For example, Charles B. Davenport, preface, in Estabrook, *Jukes in 1915*, iv.

38. Deutsch, *Mentally Ill*, 367. See also Royal Commission on the Care and Control of the Feeble-Minded, *Report upon Their Visit to American Institutions* (London: HMSO, 1908), vol. 8, 132–3.

39. Alexander Johnson, quoted in Deutsch, *Mentally Ill*, 368. See also Alexander Johnson, "Report of Committee on Colony for Segregation of Defectives," in *Proceedings of the National Conference of Charities and Corrections, 1903*, ed. Isabel C. Barrow (n.p.: Herr Press, 1903), 248–9.

40. Henry H. Goddard, *Sterilization or Segregation* (New York: Russell Sage Foundation, 1913), 4.

41. Goddard, *Sterilization*, 5.

42. H. H. Goddard, "Four Hundred Feeble-Minded Children Classified by the Binet Method," *Journal of Psycho-Asthenics* (1910) **15**:17–30 (17, 26–7). An early description of Goddard's classification system as applied to residents of California's

Home for the Feebleminded is in F. W. Hatch, "Report of the General Superintendent," in *Ninth Biennial Report of the Commission in Lunacy of California* (Sacramento: State Printing, 1914), 15–6. For a general discussion of Goddard's role in early intelligence testing, see Haller, *Eugenics*, 95–100; and Kevles, *In the Name of Eugenics*, 77–84.

43. Goddard, *Sterilization*, 6–7.

44. Goddard, *Sterilization*, 6.

45. Goddard estimated that the number of feeble-minded persons in the USA would be cut by two-thirds in a generation. Deutsch, *Mentally Ill*, 369. Popenoe and Johnson were similarly optimistic. Popenoe and Johnson, *Applied Eugenics*, 186.

46. Goddard, *Sterilization*, 5.

47. E. R. Johnstone, "Waste Land Plus Waste Humanity," *Training School Bulletin* (1914) **11**: 60–3 (61–2).

48. Royal Commission, *Report*, 136.

49. Popenoe and Johnson, *Applied Eugenics*, 186.

50. American Eugenics Society, *A Eugenics Catechism* (New York: American Eugenics Society, 1926), 8–9.

51. For example, compare Popenoe and Johnson, *Applied Eugenics*, 195 ("There are cases where [sterilization] is advisable, in states too poor or niggardly to care adequately for their defectives and delinquents, but eugenicists should favor segregation as the main policy") with Popenoe and Johnson, *Applied Eugenics*, 2nd ed. (New York: Macmillan, 1935), 153 ("Sterilization is only one factor, though an important one, in a complete and well-balanced system of state care for the defective and mentally diseased"). See also Deutsch, *Mentally Ill*, 368–9.

52. For example, Popenoe and Johnson, *Applied Eugenics*, 185.

53. Haller, *Eugenics*, 40–7.

54. For example, John E. Purdon, "Social Selection: The Extirpation of Criminality and Hereditary Disease," *Transactions of the Medical Association of the State of Alabama* (1901), 459–66 (465) (reprint of address made on behalf of the Tri-State Medical Society of Alabama, Georgia, and Tennessee in which Purdon offered castration as an alternative to lynching).

55. Haller, *Eugenics*, 48; and Kenneth M. Ludmerer, *Genetics and American Society: A Historical Appraisal* (Baltimore: Johns Hopkins University Press, 1972), 91.

56. Haller, *Eugenics*, 48–9; Kevles, *In the Name of Eugenics*, 30–3; and Ludmerer, *Genetics*, 91.

57. A. J. Ochsner, "Surgical Treatment of Habitual Criminals," *Journal of the American Medical Association* (1899) **32**: 867–8.

58. H. C. Sharp, "The Severing of the Vasa Deferentia and Its Relation to the Neuropsychopathic Constitution," *New York Medical Journal* (1902) **75**: 411–4; and H. C. Sharp, "The Indiana Plan," in *Proceedings of the National Prison Association* (Pittsburgh: National Prison Association, 1909), 36.

59. 1907 Ind. Acts ch. 215.

60. Popenoe and Johnson, *Applied Eugenics*, 185. The progressive disposition of the states that initially enacted sterilization laws was noted at the time by a Southern eugenicist in A. B. Cooke, "Safeguarding Society from the Unfit," *Southern Medical Journal* (1910) **3**: 16–22 (16).

61. The decisions against these statutes were as follows: *Williams v. State*, 190 Ind. 526, 121 N.E.2 (1921) (Indiana—due process); *Davis v. Berry*, 216 Fed. 413 (S.D. Iowa, 1914) (Iowa—cruel and unusual punishment, and due process); *Haynes v. Lapeer*, 201 Mich. 138, 166 N.W. 938 (1918) (Michigan—equal protection); *Mickel v. Heinrichs*, 262 Fed. 688 (D.C. Nev. 1918) (Nevada—cruel or unusual punishment); *Smith v. Board of Examiners of Feeble-Minded*, 85 N.J. Law 46, 88 Atl. 963 (1913) (New Jersey—equal protection); *In re Thompson*, 169 N.Y.S. 638 (Sup. Ct. 1918), aff'd sub nom., *Osborn v. Thompson*, 169 N.Y.S. 638 (App. Div. 1918) (New York—equal protection); and *Cline v. State Board of Eugenics*, Marion County, Ore., Cir. Ct. (1921) (Oregon—due process). An eugenicist's analysis of these decisions is in Otis H. Castle, "The Law and Human Sterilization," in *Collected Papers On Eugenic Sterilization in California: A Critical Study of Results in 6000 Cases*, ed. E. S. Gosney (Pasadena, CA: Human Betterment Foundation, 1930), 558–64.

62. See Popenoe and Johnson, *Applied Eugenics*, 192; and Harry H. Laughlin, *The Legal, Legislative and Administrative Aspects of Sterilization* (Cold Spring Harbor, NY: Eugenics Record Office, 1914), 83. Five revised versions of the California statute are reprinted in *Sterilization Laws: Compilation of the Sterilization Laws of Twenty-Four States* (Des Moines, IA: Huston, [1928]), 2–3.

63. Popenoe and Johnson, *Applied Eugenics*, 191; and *Williams v. State*, 190 Ind. 526, 121 N.E.2 (1921).

64. *Buck v. Bell*, 274 U.S. 200, 207 (1927).

65. Kevles, *In the Name of Eugenics*, 111–2; and Philip R. Reilly, *The Surgical Solution: A History of Involuntary Sterilization in the United States* (Baltimore: Johns Hopkins University Press, 1991), 84–7.

66. Jonas Robitscher, ed., *Eugenic Sterilization* (Springfield, IL: Thomas, 1973), 123.

67. 1917 Cal. Stats. ch. 489. For an analysis of the California legislation and its impact, see Wendy Kline, *Building a Better Race: Gender, Sexuality, and Eugenics from the Turn of the Century to the Baby Boom* (Berkeley: University of California Press, 2001), 50–60.

68. For sterilization statistics by state, see Robitscher, *Eugenic Sterilization*, 118–9.

69. On British eugenicists, see Larson, *Brit. J. Hist. Sci.* (1991), **24**: 45–60. For Davenport, see, for example, Charles B. Davenport, H. H. Laughlin, David F. Weeks, E. R. Johnstone, and Henry H. Goddard, *The Study of Human Heredity* (Cold Spring Harbor, NY: Eugenics Record Office, 1911), 28; and Davenport,

State Laws, 36 ("This may be done by segregation during the reproductive period, or even, as a last resort, by sterilization"). See also Haller, *Eugenics*, 93, 124; and Kevles, *In the Name of Eugenics*, 47–8.

70. Goddard, *Sterilization*, 10–1.

71. Frances Hassencahl, "Harry H. Laughlin, 'Expert Eugenics Agent' for the House Committee on Immigration and Naturalization, 1921 to 1931," Ph.D. thesis, Case Western Reserve University, 1970, 89–160.

72. Laughlin, *Legal Aspects of Sterilization*, 144–5.

73. Laughlin, *Legal Aspects of Sterilization*, 142–3.

74. For Roosevelt and Coolidge, see, for example, Kevles, *In the Name of Eugenics*, 74, 85–8. As governor, in 1911, Wilson signed New Jersey's eugenic sterilization bill into law. See Reilly, *Surgical Solution*, 46.

75. Hamilton Cravens, *The Triumph of Evolution: The Heredity-Environment Controversy, 1900–1941* (reprint, Baltimore: Johns Hopkins University Press, 1988), 49–55, 77–85, and 138–47; Carl N. Degler, *In Search of Human Nature: The Decline and Revival of Darwinism in American Social Thought* (New York: Oxford University Press, 1991), 43–5; and Ludmerer, *Genetics*, 9.

76. Erwin Baur, quoted in Max Weinreich, *Hitler's Professors: The Part of Scholarship in Germany's Crimes Against the Jewish People* (New Haven, CT: Yale University Press, reprinted 1999), 31 (emphasis as in original).

77. E. Alec-Tweede, "Eugenics," *Fortnightly Review* (1912) **92**: 25–7.

78. George William Hunter, *A Civic Biology* (New York: American, 1914), 261–3. Regarding the popularity of this textbook, see Edward J. Larson, *Summer for the Gods: The Scopes Trial and America's Continuing Debate Over Science and Religion* (New York: Basic, 1997), 23.

79. The movie was first released in 1916 under the title *The Black Stork* but survives only in its 1927 form, *Are You Fit to Marry?*. All quotes are from a VHS tape of the movie, *Are You Fit to Marry?* (Quality Amusement Corp., 1927), John E. Allen Archives, Nebraska ETV Network, Lincoln, Nebraska. As the movie is silent, with all words printed as subtitles, the quotes can be read from the screen. The origins, contents, and reception of the movie are discussed in Martin S. Pernick, *The Black Stork: Eugenics and the Death of "Defective" Babies in American Medicine and Motion Pictures since 1915* (New York: Oxford University Press, 1996), 143–58. Pernick's book also contains many quotes from the movie.

80. Christine Rosen, *Preaching Eugenics: Religious Leaders and the American Eugenics Movement* (New York: Oxford University Press, 2004), 115–28.

81. Rosen, *Preaching Eugenics*, 133–4; Larson, *Brit. J. Hist. Sci.* (1991), **24**: 55–6; Kevles, *In the Name of Eugenics*, 68.

82. For a discussion of the position of the Roman Catholic Church and an extended excerpt from the 1930 encyclical, see Reilly, *Surgical Solution*, 118–22.

83. For an analysis of the role of Roman Catholic clergy and laity in defeating eugenic sterilization legislation in Louisiana, see Edward J. Larson, *Sex, Race, and*

Science: Eugenics in the Deep South (Baltimore: Johns Hopkins University Press, 1995), 107–13.

84. G. K. Chesterton, *Eugenics and Other Evils* (London: Cassell, 1922), 180.

85. Larson, *Brit. J. Hist. Sci.* (1991), **24**: 53–9.

86. Kevles, *In the Name of Eugenics*, 104–7.

87. H. J. Muller, "Dominance of Economics over Eugenics," in Third International Congress of Eugenics, *A Decade of Progress in Eugenics* (reprint, New York: Garland, 1984), 141–2.

88. H. J. Muller, *Out of the Night* (reprint, New York: Garland, 1984), 113.

89. Abraham Myerson, J. B. Ayer, T. J. Putnam, C. E. Keeler, and L. Alexander, *Eugenical Sterilization: A Reorientation of the Problem* (New York: Macmillan, 1936), 4, 177–83.

90. Diane Paul, *Controlling Human Heredity: 1865 to the Present* (Atlantic Highlands, NJ: Humanities Press, 1995), 119.

91. As late as 1930, the influential American geneticist Edwin G. Conklin still asserted that "all modern geneticists approve the segregation or sterilization of those who are know to have serious hereditary defects, such as hereditary feeble-mindedness, insanity, etc." Edwin G. Conklin, "The Purposive Improvement of the Human Race," in *Human Biology and Population Improvement*, ed. E. V. Cowdry (New York: Hoeber, 1930), 577.

92. For example, Garland E. Allen, "Is a New Eugenics Afoot?" *Science* **294** (2001): 59–61; and Edward J. Larson, "The Meaning of Human Gene Testing for Disability Rights," *Cincinnati Law Review*, **70** (2002): 1–26.

Chapter Eight

1. Paul Weindling, "Akteure in eigener Sache: Die Aussagen der Überlebvfenden und die Verfolgung der medizinischen Kriegsverbrechen nach 1945" [Participants in their own history: the statements of survivors and the pursuit of medical war crimes after 1945], in *Die Verbindung nach Auschwitz—Biowissenschaften und Menschenversuche an Kaiser-Wilhelm-Instituten*, ed. C. Sachse (Göttingen: Wallstein Verlag, 2004), 255–82; Paul Weindling, "Dissecting German Social Darwinism: Historicizing the Biology of the Organic State," *Science in Context* (1998) 11: 619–37.

2. Bernhard Strebel and Jens-Christian Wagner, *Zwangsarbeit für Forschungseinrichtungen der Kaiser-Wilhelm-Gesellschaft: Ein Überblick*. See http://www.mpiwg-berlin.mpg.de/KWG/Ergebnisse/Ergebnisse11.pdf (last accessed 30 December 2008).

3. Ute Deichmann, *Biologists under Hitler* (Cambridge, MA: Harvard University Press, 1996).

4. See the periodical *Der Biologe*.

5. Karl von Frisch, *Du und das Leben: Dr.-Goebbels-Spende für die deutsche Wehrmacht* [You and life: Dr. Goebbels's fund for the German army] (Berlin: Deutscher Verlag, 1936); Donald A. Dewsbury, "The 1973 Nobel Prize for Physiology or Medicine: Recognition for Behavioral Science?" *American Psychologist* (2003) **58**: 747–52.

6. Benedikt Fäger and Klaus Taschwer, *Die andere Seite des Spiegels: Konrad Lorenz und der Nationalsozialismus* [The other side of the mirror: Konrad Lorenz and National Socialism] (Vienna: Czernin, 2001).

7. Hannsjoachim Koch, *Der Sozialdarwinismus: Seine Genese und sein Einfluß auf das imperialistische Denken* [Social Darwinism: its genesis and influence on imperialist thinking] (Munich: Beck, 1973).

8. Nicholas Goodrick-Clarke, *The Occult Roots of Nazism* (Wellingborough, Northants.: Aquarian Press, 1985).

9. Phillip Gassert and D. W. Mattern, *The Hitler Library: A Bibliography* (Westport, CT: Greenwood Press, 2001)

10. Adolf Hitler, *Mein Kampf* [My struggle] (New York: Reynal and Hitchcock, 1941).

11. Richard Weikart, *From Darwin to Hitler: Evolutionary Ethics, Eugenics, and Racism in Germany* (New York: Palgrave, 2004).

12. Hitler, *Mein Kampf*, 614.

13. Weikart, *From Darwin to Hitler*.

14. Oskar Hertwig, *Zur Abwehr des ethischen, des socialen, des politischen Darwinismus* [Defense against ethical, social, and political Darwinism] (Jena: G. Fischer, 1918); Paul Weindling, *Darwinism and Social Darwinism in Imperial Germany: The Contribution of the Cell Biologist Oscar Hertwig (1849–1922)* (Stuttgart: G. Fischer, 1991).

15. Paul Weindling, "Ernst Haeckel, Darwinismus, and the Secularization of Nature," in *History, Humanity, and Evolution: Perspectives in the History of Evolutionary Naturalism*, ed. J. R. Moore (Cambridge: Cambridge University Press, 1989), 311–27.

16. Ernst Haeckel, *The History of Creation* (New York: Appleton, 1896), vol. 1, 152–3, 170, 289–91; Charles Darwin, *Descent of Man*, 2nd ed. (London: John Murray, 1879), 3, 595.

17. Freiburg University Library, Weismann papers Kopierbuch 1 Bl 739, 784, 803; D. Crook, *Benjamin Kidd* (Cambridge: Cambridge University Press, 1984); Paul Weindling, *Health, Race, and German Politics* (Cambridge: Cambridge University Press, 1989).

18. Benjamin Kidd, *Soziale Evolution mit einem Vorwort von Professor August Weismann* [Social evolution, with a foreword by Profesor August Weismann] (Jena: G. Fischer, 1895).

19. Richard Semon, *The Mneme* (London: George Allen & Unwin, 1921).

20. T. H. Huxley, prolegomena, in *Evolution and Ethics* (London: Macmillan, 1893); Zygmunt Bauman, *Modernity and the Holocaust* (Ithaca, NY: Cornell University Press, 1989).

21. Friedrich Hertz, *Race and Civilisation* (London: Kegan Paul, 1928).

22. Paul Weindling, "Central Europe Confronts German Racial Hygiene: Friedrich Hertz, Hugo Iltis, and Ignaz Zollschan as Critics of German Racial Hygiene," in *Blood and Homeland: Eugenics in Central Europe 1900–1940*, ed. Marius Turda and Paul Weindling (Budapest: Central University Press, 2006), 263–80.

23. For *Humani Generis Unitas* [The unity of the human race] (1938–40), see Georges Passelecq and Bernard Suchecky, *The Hidden Encyclical of Pius XI* (New York: Harcourt Brace, 1997).

24. Victoria Barnett, *For the Soul of the People: Protestant Protest Against Hitler* (New York: Oxford University Press, 1992).

25. Widukind Lenz to the author, remark made when discussing history of eugenics.

26. Wilhelm Schallmayer, *Über die drohende körperliche Entartung der Kulturmenschheit und die Verstaatlichung des ärztlichen Standes* [Concerning the threatening physical degeneration of cultural mankind and the nationalization of medical conditions] (Berlin: Heuser, 1891).

27. Weindling, *Health, Race and German Politics*, 426–7.

28. Benno Müller-Hill, *Murderous Science: Elimination by Scientific Selection of Jews, Gypsies, and Others, Germany 1933–1945*, transl. George Fraser (Cold Spring Harbor, NY: Cold Spring Harbor Laboratory Press, 1998).

29. Julian Huxley, A. C. Haddon, and A. M. Carr-Saunders, *We Europeans* (Harmondsworth, Middx.: Penguin Books, 1939). Pierre-André Taguieff, *La force du préjugé : essai sur le racisme et ses doubles* (Paris: Gallimard, 1987), 123, points out that the Germanist Henri Lichtenberger introduced the term "racist" in 1922.

30. "Congrès International de la Population, Paris, 1937," *Races et Racisme* (1937), no. 5.

31. Michael Burleigh and Wolfgang Wipperman, *The Racial State: Germany 1933-1945* (Cambridge: Cambridge University Press, 1991).

32. Michael H. Kater, *Das "Ahnenerbe" der SS 1935–1945: Ein Beitrag zur Kulturpolitik des Dritten Reiches* [The "ancestral heritage" of the SS, 1935–1945: a contribution to the cultural policy of the Third Reich] (Stuttgart: Deutsche Verlags-Anstalt, 1974); Gisela Bock, *Zwangssterilisation im Nationalsozialismus* [Compulsory sterilization under National Socialism] (Opladen: Westdeutscher Verlag, 1986); Claudia Koonz, "Ethical Dilemmas and Nazi Eugenics: Single-Issue Dissent in Religious Contexts," *Journal of Modern History* (1992) **64**: S8–S31; Michael Burleigh, *The Third Reich: A New History* (London: Macmillan, 2000), chap. 5, 343–404; Richard J. Evans, *The Third Reich in Power: 1933–1939* (New York: Penguin, 2005).

33. Cf. Jonathan Harwood, *Styles of Scientific Thought* (Chicago: University of Chicago Press, 1993).

34. U. Deichmann and B. Müller-Hill, "The fraud of Abderhalden's enzymes," *Nature* **393** (1998): 109–11.

35. Roger Griffin, *Modernism and Fascism: The Sense of a Beginning under Mussolini and Hitler* (London: Palgrave, 2007).

36. Paul Weindling, "The Medical Publisher," 159–70.

37. Thomas Junker and Uwe Hossfeld, "The Architects of the Evolutionary Synthesis in National Socialist Germany: Science and Politics," *Biology and Philosophy* (2002) **17**: 223–49; Thomas Junker, *Die zweite Darwinische Revolution: Geschichte des synthetischen Darwinismus in Deutschland 1924 bis 1950* [The second Darwinian revolution: a history of the founding of Darwinism in Germany 1924 to 1950] (Marburg: Basilisken-Presse, 2004).

38. Sara Jansen, *"Schädlinge": Geschichte eines wissenschaftlichen und politischen Konstrukts, 1840–1920* ["Parasites": a history of a scientific and political construct, 1840–1920] (Frankfurt am Main and New York: Campus-Verlag, 2000); Paul Weindling, *Epidemics and Genocide in Eastern Europe, 1890–1945* (New York: Oxford University Press, 2000).

39. Myklos Nyiszli, *Auschwitz: A Doctor's Eyewitness Account* (New York: Frederick Fell, 1960).

40. See website http://www.adl.org/PresRele/HolNa_52/5277_52.htm (last accessed 3 May 2008).

41. C. P. Blacker, *Eugenics: Galton and After* (London: Duckworth, 1952).

42. Paul Weindling, *Nazi Medicine and the Nuremberg Trials: From Medical War Crimes to Informed Consent* (Basingstoke: Palgrave-Macmillan, 2004).

Chapter Nine

1. Conway Zirkle, *Evolution, Marxian Biology, and the Social Scene* (Philadelphia: University of Pennsylvania Press, 1959), 6.

2. R. C. Cook, "Lysenko's Marxist Genetics: Science or Religion," *Journal of Heredity* (1949) **40** (7): 169–202; see also B. D. Wolfe, "Science Joins the Party," *Antioch Review* (1950) Spring: 47–60.

3. See, for example, A. Morton, *Soviet Genetics* (London: Lawrence and Wishart, 1951).

4. David Joravsky, *The Lysenko Affair* (Cambridge, MA: Harvard University Press, 1970), VII.

5. L. R. Graham, *Science and Philosophy in the Soviet Union* (New York: Vintage Books, 1974), 30, 206, 230.

6. R. Lewontin and R. Levins, "The Problem of Lysenkoism," in *The Radicalization of Science*, ed. H. Rose and S. Rose (London: Macmillan Press, 1976),

32–64; F. Belardelli, "The 'Lysenko Affair' in the Framework of the Relations Between Marxism and the Natural Sciences," *Scientia* (1977) **112**: 33–50; R. Young, "Getting Started on Lysenkoism," *Radical Science Journal* (1978) **6/7**: 80–105; I. Wainright and J. Fredman, "A Re-examination of the Lysenko Controversy," *Proletariat* (1979) **5**(1): 32–44; (2): 6–18.

7. J. A. Rogers, "Marxist and Russian Darwinism," *Jahrbücher für Geschichte Osteuropas* (1965) **13**: 199–211; M. W. Mikulak, "Darwinism, Soviet Genetics, and Marxism–Leninism," *Journal of the History of Ideas* (1970) **31**: 359–76; D. Lecourt, *Proletarian Science? The Case of Lysenko* (London: NBL, 1977); D. Buican, *L'éternel retour de Lyssenko* (Paris: Copernic, 1978); J. Kotek and D. Kotek, *L'affaire Lyssenko* (Bruxelles: Éditions complexe, 1986); R. Ladous, *Darwin, Marx, Engels, Lyssenko et les autres* (Lyons: Institut interdisciplinaire d'études epistémologiques, 1984).

8. Ethan Pollock, *Stalin and the Soviet Science Wars* (Princeton, NJ: Princeton University Press, 2006), 70.

9. Alexander Vucinich, *Darwin in Russian Thought* (Berkeley: University of California Press, 1988).

10. Compare, for instance, A. N. Beketov, "Darvinizm s tochki zreniia obshche-fizicheskikh nauk [Darwinism from the point of view of physical sciences]," *Trudy Sankt-Peterburgskogo obshchestva ispytatelei prirody* (1882) **13**(1): 353–63, and N. Ia. Danilevskii, *Darvinizm: Kriticheskoe issledovanie* [Darwinism: A critical investigation], 2 vols. (St. Petersburg: M. E. Komarov, 1885–9).

11. See Charl'z Darvin, *O proiskhozhdenii vidov v tsarstvakh zhivotnom i rastitel'nom putem estestvennogo podbora rodichei ili o sokhranenii usovershenst-vovannykh porod v bor'be za sushchestvovanie* [On the origin of species in animal and plant kingdoms by means of natural selection or on the preservation of favored races in the struggle for existence] (St. Petersburg: A. I. Glazunov, 1964).

12. On Timiriazev, see Abba E. Gaissinovitch, "Contradictory Appraisal by K. A. Timiriazev of Mendelian Principles and Its Subsequent Perception," *History and Philosophy of the Life Sciences* (1985) **7**: 257–86.

13. For a detailed analysis, see A. E. Gaissinovitch, "Problems of Variation and Heredity in Russian Biology in the Late Nineteenth Century," *Journal of the History of Biology* (1973) **6**: 97–123.

14. See Daniel Todes, *Darwin without Malthus* (New York: Oxford University Press, 1989), and Vucinich, *Darwin in Russian Thought*.

15. See Jan Sapp, *Evolution by Association: A History of Symbiosis* (New York: Oxford University Press, 1994).

16. Todes, *Darwin without Malthus*.

17. See M. M. Kovalevskii *et al.*, ed., *Pamiati Darvina* [In memory of Darwin] (Moscow: Nauchnoe Slovo, 1910).

18. See Albert Resis, "*Das Kapital* Comes to Russia," *Slavic Review* (1970) **29**: 219–37.

19. See Samuel H. Baron, "The First Decade of Russian Marxism," *American Slavic and East European Review* (1955) **14**: 315–30.

20. G. V. Plekhanov, *Izbrannye filosofskie proizvedeniia v 5-ti tomakh* [Selected philosophical writings in 5 volumes] (Moscow: Politizdat, 1956–8), vol. 2, 690. On Plekhanov's attitude to Darwinism, see Mark Bassin, "Geographical Determinism in Fin-de-siècle Marxism: Georgii Plekhanov and the Environmental Basis of Russian History," *Annals of the Association of American Geographers* (1992) **82**: 3–22.

21. M. Kovalevskii, "Darvinizm v sotsiologii [Darwinism in sociology]," in Kovalevskii *et al.*, *Pamiati Darvina*, 117–58.

22. A. Bogdanov, *Osnovnye elementy istoricheskogo vzgliada na prirodu. Priroda. Zhizn'. Psikhika. Obshchestvo* [Major elements of a historical outlook on nature: nature, life, psyche, society] (St. Petersburg: Seiatel', 1899).

23. On the history of these institutions, see Michael David-Fox, *Revolution of the Mind: Higher Learning among the Bolsheviks, 1918–1929* (Ithaca, NY: Cornell University Press, 1998).

24. See F. Engel's, *Ot obez'iany k cheloveku* [From ape to human] (Gomel': Gomel'skii rabochii, 1922); F. Engel's, "Dialektika prirody [The dialectics of nature]," *Arkhiv Marksa i Engel'sa* (Moscow–Leningrad: GIZ, 1925), vol. 2.

25. M. Ravich-Cherkasskii, "Predislovie k pervomu izdaniiu [Preface to the first edition]," in *Darvinizm i Marksizm*, ed. M. Ravich-Cherkasskii (Khar'kov: GIZ Ukrainy, 1925), 6.

26. Ravich-Cherkasskii, "Predislovie k pervomu izdaniiu," 9.

27. See G. Ia. Graf, *Ot biblii k Darvinu* [From the Bible to Darwin] (Leningrad: GIZ, 1925).

28. L. Trotskii, "K pervomu vserossiiskomu s"ezdu nauchnykh rabotnikov [To the first all-Russia congress of scientific workers]," *Izvestiia* (1923), 24 November: 1.

29. James T. Andrews, *Science for the Masses* (College Station: Texas A&M University Press, 2003).

30. See D. Joravsky, *Soviet Marxism and Natural Science, 1917–1932* (New York: Columbia University Press, 1961); David-Fox, *Revolution of the Mind*.

31. "Ot redaktsii [editorial]," *Pod Znamenem Marksizma* (hereafter called *PZM*) (1922) nos. 1–2: 3–4 (4).

32. V. I. Lenin, "O znachenii voinstvuiushchego materializma [On the significance of militant materialism]," *PZM* (1922) no. 3: 29.

33. Anonymous, "O politike partii v oblasti khudozhestvennoi literatury [On the politics of the party in the field of literary fiction]," *Zvezda* (1925) no. 4: 256–9 (257).

34. K. A. Timiriazev, "Ch. Darvin i K. Marks [Ch. Darwin and K. Marx]," *Proletarskaia Kul'tura* (1919) nos. 9–10: 20–5.

35. See K. A. Timiriazev, *Istoricheskii metod v biologii* [Historical method in biology] (Moscow: Granat, 1922).

36. See K. Timiriazev, *Charl'z Darvin i ego uchenie* [Charles Darwin and his theory] (Moscow: GIZ, 1919–21).

37. For a detailed analysis of this emerging symbiosis, see Nikolai Krementsov, *Stalinist Science* (Princeton, NJ: Princeton University Press, 1997).

38. See Peter J. Bowler, *The Eclipse of Darwinism* (Baltimore: Johns Hopkins University Press, 1983).

39. See Mark B. Adams, "Through the Looking Glass: The Evolution of Soviet Darwinism," in *New Perspectives on Evolution*, ed. Leonard Warren and Hillary Kopowski (New York: Wiley–Liss, 1991), 37–63.

40. N. I. Vavilov, *Zakon gomologicheskikh riadov v nasledstvennoi izmenchivosti* [The law of homologous series in hereditary variation] (Saratov: n.p., 1920).

41. Iu. Filipchenko, *Izmenchivost' i evoliutsiia* [Variability and evolution] (Petrograd and Berlin: Grzhebin, 1921).

42. Iu. Filipchenko, *Evoliutsionnaia ideia v biologii* [The evolutionary idea in biology] (Moscow: Sabashnikov, 1923).

43. N. Kol'tsov, "Obrazovanie novykh vidov i chislo khromosom [The formation of new species and the number of chromosomes]," *Uspekhi Eksperimental'noi Biologii* (1922) **1**(2): 181–96.

44. For a historical analysis of this campaign, see A. E. Gaissinovitch, "The Origins of Soviet Genetics and the Struggle with Lamarckism, 1922–1929," *Journal of the History of Biology* (1980) **13**: 1–51.

45. For a detailed analysis of Pavlov's involvement with Lamarckism, see Krementsov, *Stalinist Science*, 260–70.

46. I. P. Pavlov, "New Research on Conditioned Reflexes," *Science* (1923) **58**: 359–61 (360).

47. See "Doklad akademika Pavlova [Academician Pavlov's report]," *Izvestiia*, 22 April 1924: 4.

48. See Daniel Todes, "Pavlov and the Bolsheviks," *History and Philosophy of the Life Sciences* (1995) **17**: 379–418.

49. N. K. Kol'tsov, "Noveishie popytki dokazat' nasledstvennost' blagopriobretennykh priznakov [The latest attempts to prove the inheritance of acquired characteristics]," *Russkii Evgenicheskii Zhurnal* (1924) **3**: 159–67 (161).

50. See N. K. Kol'tsov, "I. P. Pavlov: Trud zhizni velikogo biologa [I. P. Pavlov: The life-work of the great biologist]," *Biologicheskii Zhurnal* (1936) **5**: 387–402.

51. Kol'tsov, *Russkii Evgenicheskii Zhurnal* (1924) **3**: 159–67.

52. T. H. Morgan and Iu. A. Filipchenko, *Nasledstvenny li priobretennye priznaki?* [Are acquired characteristics inheritable?] (Leningrad: Seiatel', 1925). For Morgan's original article, see T. H. Morgan, "Are Acquired Characteristics Inherited?" *Yale Review* (1924) **13**: 712–29.

53. See, for example, *Preformizm ili epigenezis?* [Preformation or epigenesis?] (Volodga: Severnyi pechatnik, 1926); P. A. Novikov, *Teoriia epigeneza v biologii: Istoriko-sistematicheskii ocherk* [The theory of epigenesis in biology: a systematic

historical survey] (Moscow: Izdatel'stvo komakademii, 1927); E. S. Smirnov, *Problema nasledovaniia priobretennykh priznakov: Kriticheskii obzor literatury* [The problem of the inheritance of acquired characteristics: a critical review of the literature] (Moscow: Izdatel'stvo komakademii, 1927).

54. On Kammerer's work, see Sander Gliboff, "The Case of Paul Kammerer: Evolution and Experimentation in the Early 20th Century," *Journal of the History of Biology* (2006) **39**: 525–63.

55. E. S. Smirnov, Iu. M. Vermel', and B. S. Kuzin, *Ocherki po teorii evoliutsii* [Essays on evolutionary theory] (Moscow: Krasnaia nov', 1924).

56. See P. Kammerer, *Obshchaia biologiia* [General biology] (Moscow and Leningrad: GIZ, 1925).

57. See V. Komarov, *Zh.-B. Lamark* [J.-B. Lamarck] (Moscow–Leningrad: GIZ, 1925).

58. Smirnov, *Problema nasledovaniia*.

59. See P. Kammerer, *Zagadka nasledstvennosti* [The enigma of heredity] (Leningrad: Priboi, 1927; Moscow: GIZ, 1927).

60. See A. Serebrovskii, "E. S. Smirnov, Iu. M. Vermel', B. S. Kuzin, "Ocherki po teorii evoliutsii [Essays on the theory of evolution]," *Pechat' i Revoliutsiia* (1925), no. 4: 257–8; F. G. Dobrzhanskii [Th. Dobzhansky], "K voprosu o nasledovanii priobretennykh priznakov [On the question of the inheritance of acquired characteristics]," in *Preformizm ili epigenezis?*, 27–47; F. G. Dobrzhanskii [Th. Dobzhansky], *Chto i kak nasleduetsia u zhivykh sushchestv?* [What and how do living beings inherit?] (Leningrad: GIZ, 1926).

61. S. S. Chetverikov, "O nekotorykh momentakh evolutsionnogo protsessa s tochki zreniia sovremennoi genetiki [On certain moments of evolutionary process from the viewpoint of modern genetics]," *Zhurnal Eksperimental'noi Biologii* (1926) **2**(1): 3–54.

62. For an early analysis of this debate, see David Joravsky, "Soviet Marxism and Biology before Lysenko," *Journal of the History of Ideas* (1959) **20**: 85–104.

63. See his commentaries in G. V. Plekhanov, *Osnovnye voprosy marksizma* [The basic questions of Marxism] (Moscow: GIZ, 1923), 120–2.

64. V. Sarab'ianov, "Nazrevshii vopros [The looming question]," *Sputnik Kommunista* (1923) no. 20: 215–34.

65. Compare, for instance, D. Gul'be, "Darvinizm i teoriia mutatsii [Darwinism and the theory of mutation]," *PZM* (1924) nos. 8–9: 157–66, and F. Duchinskii, "Darvinizm i mutatsionnaia theoriia [Darwinism and mutation theory]," *PZM* (1925) no. 3: 128–39.

66. See, for instance, an article by the founder of the society Leninism in Medicine, Petr Obukh, "Nauka i organizatsiia meditsiny [Science and the organization of medicine]," *Pravda*, 3 February 1923: 2; or the volume of proceedings issued by the Circle of Materialist Physicians, *Meditsina i dialekticheskii materialism* [Medicine and dialectical materialism] (Moscow: Moscow University, 1926), vol. 1.

67. See B. M. Zavadovskii, "Darvinizm i lamarkizm i problema nasledovaniia priobretennykh priznakov [Darwinism and Lamarckism and the problem of the inheritance of acquired characteristics]," *PZM* (1925) nos. 10–11: 79–114; B. M. Zavadovskii, *Darvinizm i marksizm* [Darwinism and Marxism] (Moscow: Gosudarstvennoe izdatel'stvo, 1926).

68. On Kammerer's life story, see Arthur Koestler, *The Case of the Midwife-Toad* (New York: Random House, 1971).

69. See A. S. Serebrovskii, "Teoriia nasledstvennosti Morgana i Mendelia i marksisty [Mendel's and Morgan's theory of heredity and Marxists]," *PZM* (1926) no. 3: 98–117.

70. See, for example, I. I. Agol, "Dialektika i metafizika v biologii [Dialectics and metaphysics in biology]," *PZM* (1926) no. 3: 118–50; M. M. Mestergazi, "Epigenezis i genetika [Epigenesis and genetics]," *Vestnik Kommunisticheskoi Akademii* (1927) no. 19: 187–252; F. Duchinskii, "Darvinizm, lamarkizm i neodarvinizm [Darwinism, Lamarckism and Neo-Darwinism]," *PZM* (1926) nos. 7–8: 95–122.

71. See M. Levin, "E. Smirnov, 'Problema nasledovaniia priobretennykh priznakov' [E. Smirnov, 'The problem of the inheritance of acquired characteristics']," *Pravda*, 13 May 1927: 4.

72. See H. J. Muller, "Artificial transmutation of the gene," *Science* (1927) **66**: 84–7.

73. A. Serebrovskii, "Chetyre stranitsy, kotorye vzvolnovali uchenyi mir [Four pages that shook the scientific world]," *Pravda* (1927), 11 September: 5. The title of the article alluded to the title of the famous book about the Bolshevik Revolution by John Reed, *Ten Days that Shook the World* (New York: Boni and Liveright, 1919).

74. See, for example, I. Agol, *Vitalizm, mekhanisticheskii materialism i marksizm* [Vitalism, mechanistical materialism, and Marxism] (Moscow: Moskovskii rabochii, 1928); I. Agol, *Dialekticheskii metod i evoliutsionnaia teoriia* [Dialectical method and evolutionary theory] (Moscow: Izdatel'stvo komakademii, 1930); S. Levit, "Evoliutsionnye teorii v biologii i marksizm [Evolutionary theories in biology and Marxism]," *Meditsina i dialekticheskii materializm* (1926) no. 1: 15–32; S. Levit, "Dialekticheskii materializm v meditsine [Dialectical materialism in medicine]," *Vestnik Sovremennoi Meditsiny* (1927) no. 23: 1481–90; Vas. Slepkov, "Vitalism, mekhanizm, i dialektika [Vitalism, mechanicism, and dialectics]," *PZM* (1926) no. 10: 89–107; Vas. Slepkov, *Biologiia i marksizm. Ocherki materialisticheskoi biologii* [Biology and Marxism: essays on materialistic biology] (Moscow–Leningrad: GIZ, 1928).

75. See Mark B. Adams, "Darwinism as Science and Ideology," in *Science and Culture in the Western Tradition: Sources and Interpretations*, ed. John G. Burke (Scottsdale, AZ: Gorsuch Scarisbrick, 1987), 199–208.

76. "Ot redaktsii [editorial]," in *Dialektika v prirode* [Dialectics in nature] (Vologda: Severnyi pechatnik, 1926), vol. 2: VIII.

77. Agol, *Dialekticheskii metod*, 150.

78. V. A. Vagner, *Vozniknovenie i razvitie psikhicheskikh sposobnostei* [The origin and development of psychic abilities] (Leningrad: Nachatki znaniia, 1927), no. 5: 29.

79. See A. Deborin, "Engel's i dialektika v biologii [Engels and dialectics in biology]," *PZM* (1926) nos. 1–2: 54–89; no. 3: 5–28; nos. 9–10: 5–25.

80. See N. S. Poniatskii, *Velikii uchenyi-revoliutsioner, Kliment Arkad'evich Timiriazev* [The great scientist-revolutionary Kliment Arkadievich Timiriazev] (Moscow–Leningrad: GIZ, 1923); F. Duchinskii, "K. A. Timiriazev kak darvinist [K. A. Timiriazev as a Darwinist]," *PZM* (1925) no. 7: 86–97; for a historical analysis of the Timiriazev's views on genetics and their uses by various interested parties, see Gaissinovitch, *Hist. Phil. Life Sci.* (1985) **7**: 257–86.

81. A. N. Severtsov, "Istoricheskoe napravlenie v zoologii [Historical direction in biology]," in *Nauka i tekhnika v SSSR, 1917–27* (Moscow: Rabotnik prosveshcheniia, 1928), 143–93.

82. See M. A. Menzbir, ed., *Polnoe sobranie sochinenii Charl'za Darvina* [A complete collection of works by Charles Darwin], 4/8 vols. (Moscow and Leningrad: GIZ, 1928–9).

83. See "Mendelizm i ego otnoshenie k darvinizmu [Mendelism and its relation to Darwinism]," in Menzbir, *Polnoe sobranie*, vol. 3, part 1, 3–19; M. A. Menzbir, "Teorii nasledstvennosti [Theories of heredity]," in Menzbir, *Polnoe sobranie*, vol. 4, pt. 2, 424–69.

84. O. Iu. Shmidt, "Zadachi marksistov v oblasti estestvoznaniia [The tasks of Marxists in the field of natural sciences]," *Nauchnoe Slovo* (1929) no. 5: 3–18 (4).

85. See Joravsky, *Soviet Marxism*; and especially Graham, *Science and Philosophy*.

86. See *Protiv mekhanitsizma i men'shevistvuiushchego idealizma v biologii* [Against mechanisticism and menshevizing idealism in biology] (Moscow: Kommunisticheskaia akademiia, 1931). On the details of the campaign "on two fronts" and its ramifications for Soviet science, see Joravsky, *Soviet Marxism*; Graham, *Science and Philosophy*; and Krementsov, *Stalinist Science*.

87. For details, see Krementsov, *Stalinist Science*.

88. I. V. Stalin, "Rech' na VIII S"ezde VLKSM [Speech at the 8th Congress of the Communist Youth]," in *Sobranie Sochinenii* (Moscow: Partizdat, 1947), vol. 11, 77.

89. See B. P. Tokin and M. I. Aizupet, eds., *K. Marks, F. Engels, V. Lenin o biologii* [K. Marks, F. Engels, and V. Lenin on biology] (Moscow: Partizdat, 1934).

90. See I. I. Prezent, ed., *Khrestomatiia po evoliutsionnomu ucheniiu* [Reader on evolutionary theory] (Leningrad: Kubuch, 1934).

91. *Trudy Chetvertogo Vsesoiuznogo S"ezda Zoologov, Anatomov i Gistologov* [Proceedings of the 4th all-Union congress of zoologists, anatomists, and histologists] (Kiev: Gosmedizdat USSR, 1931), 5–6.

92. "Ot redaktsii [editorial]," *Vestnik Kommunisticheskoi Akademii* (1932) nos. 4–5: 119–20.

93. *Izvestiia* (1932), 18 April: 1–3.

94. *Pravda* (1932), 19 April: 1–3.

95. See P. I. Valeskaln and B. P. Tokin, eds., *Uchenie Darvina i marksizm-leninizm* [Darwin's doctrine and Marxism-Leninism] (Moscow: Partizdat, 1932).

96. See M. Mitskevich and B. P. Tokin, "Darvinizm na sluzhbu sotsstroitel'stvu [Darwinism to the service of socialist construction]," *Izvestiia* (1932), 18 April: 2.

97. See I. I. Prezent, *Teoriia Darvina v svete dialekticheskogo materializma* [Darwin's theory in light of dialectical materialism] (Leningrad: Lenmedizdat, 1932).

98. See N. Bukharin, *Darvinizm i marksizm* [Darwinism and Marxism] (Leningrad: Akademiia nauk, 1932).

99. See *Pamiati K. A. Timiriazeva* [In memory of K. A. Timiriazev] (Moscow and Leningrad: Biomedgiz, 1936).

100. Bukharin, *Darvinizm i marksizm*, 30.

101. Prezent, *Teoriia Darvina*, 20.

102. On the history of Soviet eugenics, see Mark B. Adams, "Eugenics in Russia," in *The Wellborn Science*, ed. Mark B. Adams (New York: Oxford University Press, 1990), 153–201.

103. See *Materialy k Vsesoiuznoi konferentsii po planirovaniiu genetiko-selektsionnykh issledovanii* [Materials for the all-Union conference on the planning of genetic and breeding research] (Leningrad: VASKhNIL, 1932).

104. For agrobiologists' critique and geneticists' responses, see *Sbornik rabot po diskussionym problemam genetiki i selektsii* [A collection of work on issues in genetics and breeding] (Moscow: VASKhNIL, 1936).

105. For the materials of the discussion, see *Spornye voprosy genetiki i selektsi* [Issues in genetics and breeding] (Moscow: Sel'khozgiz, 1937).

106. For details, see Nikolai Krementsov, "Eugenics, *Rassenhygiene*, and Human Genetics in the late 1930s: The Case of the Seventh International Genetics Congress," in Susan G. Solomon, ed., *Doing Medicine Together: Germany and Russia between the Wars* (Toronto: University of Toronto Press, 2006), 369–404.

107. See Ia. Iakovlev, "O darvinizme i nekotorykh anti-darvinistakh [On Darwinism and certain anti-Darwinists]," *Pravda* (1937), 12 April: 1.

108. For details, see Krementsov, *Stalinist Science*, 54–82.

109. For details, see Nikolai Krementsov, "A 'Second Front' in Soviet Genetics: An International Dimension of the Lysenko Controversy, 1944–1947," *Journal of the History of Biology* (1996) **29**: 229–50.

110. For details, see Krementsov, *Stalinist Science*, 149–57.

111. For details, see Krementsov, *Stalinist Science*, 193–250.

112. For a detailed analysis of the "resurrection" of Soviet genetics, see Mark B. Adams, *Networks in Action* (Trondheim: Studies on East European Cultures & Societies, 2000).

Chapter Ten

1. Michael Ruse, *Darwinism and Its Discontents* (Cambridge: Cambridge University Press, 2006).

2. J. B. Bury, *The Idea of Progress: An Inquiry into Its Origin and Growth* (London: Macmillan, [1920] 1924).

3. A very extensive collection of Billy Graham's sermons can be found in the Billy Graham Archives, held at Wheaton College, Illinois. Many of them can be downloaded from the Internet. See, for example, the sermon "Christ's answer to the world," given at Charlotte, North Carolina, on 21 September 1958.

4. Erasmus Darwin, *The Temple of Nature* (London: J. Johnson, 1803), vol. 1, chap. 11, 295–314.

5. M. Hesse, *Revolutions and Reconstructions in the Philosophy of Science* (Bloomington: Indiana University Press, 1980); Ernan McMullin, "Values in Science," in *PSA 1982*, ed. P. D. Asquith and T. Nickles (East Lansing, MI: Philosophy of Science Association, 1983), 3–28.

6. Charles Darwin, *On the Origin of Species by Means of Natural Selection or the Preservation of Favoured Races in the Struggle for Life* (London: John Murray, 1859).

7. Charles Darwin's annotation to the flyleaf of [Robert Chambers], *Vestiges of the Natural History of Creation*, 5th ed. (London: Churchill, 1846).

8. B. Rosemary Grant and Peter R. Grant, *Evolutionary Dynamics of a Natural Population: The Large Cactus Finch of the Galapagos* (Chicago: University of Chicago Press, 1990).

9. Michael Ruse, *Monad to Man: The Concept of Progress in Evolutionary Biology* (Cambridge, MA: Harvard University Press, 1996), 486.

10. C. Darwin, *On the Origin of Species*.

11. The now standard discussion of Greek thinking about origins is David Sedley, *Creationism and Its Critics in Antiquity* (Berkeley: University of California Press, 2008). Aristotle, *Physics*, II 8, 198b29, criticizes Empedocles as ridiculous.

12. Denis Diderot, "On the Interpretation of Nature" (1754), in *Diderot: Interpreter of Nature*, ed. Jonathan Kemp, transl. Jean Stewart and Jonathan Kemp (New York: International Publishers, 1943), 48.

13. Denis Diderot, "Supplement to Bougainville's Voyage" (1772), in *Diderot: Interpreter of Nature*, ed. Jonathan Kemp, transl. Jean Stewart and Jonathan Kemp (New York: International Publishers, 1943), 152.

14. Erasmus Darwin, *A Plan for the Conduct of Female Education in Boarding Schools* (Derby: J. Johnson, 1797).

15. Erasmus Darwin, *Zoonomia; or The Laws of Organic Life* (London: J. Johnson, 1794), vol. 1, 509.

16. Jean-Baptiste Lamarck, *Recherches sur l'organisation des corps vivans* [Researches in to the organization of the living body] (Paris: Maillard, 1802), 38.

17. Jean-Baptiste Lamarck, *Philosophie zoologique* [Zoological philosophy] (Paris: Dentu, 1809).

18. Jean-Baptiste Lamarck, *Système analytique des connaissances positives de l'homme, restreintes à celles qui proviennent directement ou indirectement de l'observation* [Analytic system of the positive facts about man, restricted to those that are open directly or indirectly to observation.] (Paris: Belin, 1820), 92.

19. [Chambers], *Vestiges*, 400–2.

20. Étienne Geoffroy Saint-Hilaire, *Philosophie anatomique* [Anatomical philosophy] (Paris: Mequignon-Marvis, 1818).

21. E. S. Russell, *Form and Function: A Contribution to the History of Animal Morphology* (London: John Murray, 1916); R. J. Richards, *The Meaning of Evolution: The Morphological Construction and Ideological Reconstruction of Darwin's Theory* (Chicago: University of Chicago Press, 1992).

22. Erasmus Darwin, "The Love of the Plants," in *The Botanic Garden* (London: J. Johnson, 1879); G. Canning, H. Frere, and G. Ellis, "The Loves of the Triangles," *Anti-Jacobin* (1798), 16 April, 23 April, and 17 May.

23. "The Progress of Man," *Anti-Jacobin* (1798), 26 February.

24. Michael Ruse, *Darwin and Design: Does Evolution Have a Purpose?* (Cambridge, MA: Harvard University Press, 2003).

25. W. Coleman, *Georges Cuvier Zoologist: A Study in the History of Evolution Theory* (Cambridge, MA: Harvard University Press, 1964); T. A. Appel, *The Cuvier–Geoffroy Debate: French Biology in the Decades before Darwin* (New York: Oxford University Press, 1987).

26. Adam Sedgwick, "Vestiges," *Edinburgh Review* (1845) **82**: 1–85 (4).

27. J. W. Clark and T. M. Hughes, eds., *Life and Letters of the Reverend Adam Sedgwick*. (Cambridge: Cambridge University Press, 1890), vol. 2, 116.

28. Letter to Herbert Spencer, on being sent an evolutionary essay, 29 July 1853; Spencer Papers, Senate House, University of London.

29. Benjamin Franklin, et al., "Report of the Commissioners charged by the King to examine animal magnetism," in *Foundations of Hypnosis from Mesmer to Freud*, ed. M. N. Tinterow (Springfield, IL: Charles C. Thomas, 1970), 89–124 (124).

30. François Magendie, *An Elementary Treatise on Human Physiology: On the Basis of the Précis Élémentaire de Physiologie*, trans. John Revere (New York: Harper & Brothers, 1855.)

31. W. F. Cannon, "The Impact of Uniformitarianism. Two letters from John Herschel to Charles Lyell, 1836–1837," *Proceedings of the American Philosophical Society* (1961) **105**: 301–14.

32. C. Darwin, *Origin of Species*, 1.

33. Michael Ruse, *The Darwinian Revolution: Science Red in Tooth and Claw* (Chicago: University of Chicago Press, 1979).

34. John F. W. Herschel, *A Preliminary Discourse on the Study of Natural Philosophy* (London: Longman, Rees, Orme, Brown, Green, and Longman, 1830).

35. William Whewell, *The History of the Inductive Sciences* (London: Parker, 1837); John F. W. Herschel, "Review of Whewell's History of the Inductive Sciences and Philosophy of the Inductive Sciences," *Quarterly Review* (1841) **135**: 177–238.

36. Michael Ruse, "Darwin's Debt to Philosophy: An Examination of the Influence of the Philosophical Ideas of John F. W. Herschel and William Whewell on the Development of Charles Darwin's Theory of Evolution," *Studies in History and Philosophy of Science* (1975) **6**: 159–81.

37. Herschel, *Preliminary Discourse*.

38. Whewell, *History*; William Whewell, *The Philosophy of the Inductive Sciences* (London: Parker, 1840).

39. C. Darwin, *Origin of Species*, 63.

40. C. Darwin, *Origin of Species*, 80–1.

41. C. Darwin, *Origin of Species*, 446.

42. Charles Darwin, "Letter of 22 May 1863," in *The Correspondence of Charles Darwin*, ed. Frederick Burkhardt and Sydney Smith (Cambridge: Cambridge University Press, 1985), vol. 1, 433.

43. William Paley, *Natural Theology, or Evidences of the Existence and Attributes of the Deity collected from the Appearances of Nature* (1802), in *Collected Works* (London: Rivington, 1819), vol. IV.

44. William Whewell, *Astronomy and General Physics*, Bridgewater Treatise number 3 (London: Pickering, 1833).

45. H. W. Bates, "Contributions to an Insect Fauna of the Amazon Valley" [1862], in *Collected Papers of Charles Darwin*, ed. P. H. Barrett (Chicago: University of Chicago Press, 1977), 87–92.

46. C. Darwin, *Origin of Species*, 489.

47. M. Peckham, ed., *The Origin of Species by Charles Darwin: A Variorum Text* (Philadelphia, University of Pennsylvania Press, 1959), 222.

48. For example, Thomas Kuhn, *The Structure of Scientific Revolutions* (Chicago: University of Chicago Press, 1962).

49. Ruse, *Monad to Man*; Michael Ruse, *The Evolution-Creation Struggle*. (Cambridge, MA: Harvard University Press, 2005).

50. T. H. Huxley, "Evolution and Ethics," in *Evolution and Ethics* (London: Macmillan, 1893), 46–111.

51. Ernst Haeckel, *Generelle Morphologie der Organismen* (Berlin: Georg Reimer, 1866).

52. Julian Huxley, *Memories* (New York: Harper and Row, 1970), 150.

53. See Ruse, *Monad to Man*.

54. Maude Royden, Address at Queen's Hall, London, 16 July 1917, reported in *The Times*, 17 July 1917.

55. From Cecil F. Alexander, "All Things Bright and Beautiful," in *Hymns for Little Children* (London: J. Masters, 1848).

56. T. H. Huxley, *Lay Sermons, Addresses and Reviews* (London: Macmillan, 1870).

57. The *Lay Sermons* are a good place to start. I discuss Huxley's views on social Progress and biological progress in Ruse, *Monad to Man*, 210-4.

58. Charles Darwin, *The Descent of Man* (London: John Murray, 1871).

59. Herbert Spencer, "Progress: Its Law and Cause." *Westminster Review* (1857) LXVII: 244–67.

60. J. B. S. Haldane, *The Causes of Evolution* (New York: Longmans Green, 1932); Ronald A. Fisher, *The Genetical Theory of Natural Selection* (Oxford: Oxford University Press, 1930); Sewall Wright, "Evolution in Mendelian Populations," *Genetics* (1931) **16**: 97–159; Sewall Wright, "The Roles of Mutation, Inbreeding, Crossbreeding and Selection in Evolution," *Proceedings of the Sixth International Congress of Genetics* (1932) **1**: 356–66.

61. Ruse, *Monad to Man*, 339–43.

62. Theodosius Dobzhansky, *Genetics and the Origin of Species* (New York: Columbia University Press, 1937); Ernst Mayr, *Systematics and the Origin of Species* (New York: Columbia University Press, 1942); George Gaylord Simpson, *Tempo and Mode in Evolution* (New York: Columbia University Press, 1944); G. Ledyard Stebbins, *Variation and Evolution in Plants* (New York: Columbia University Press, 1950).

63. A. J. Cain, *Animal Species and Their Evolution* (London: Hutchinson, 1954); P. M. Sheppard, *Natural Selection and Heredity* (London: Hutchinson, 1958). H. B. D. Kettlewell, "Selection Experiments on Industrial Melanism in the Lepidoptera," *Heredity* (1955) **9**: 323–42; H. B. D. Kettlewell, *The Evolution of Melanism* (Oxford: Clarendon Press, 1973).

64. Some of the clearest evidence for this claim, apart from the heavy Spencerian language in Wright's seminal paper, "Evolution in Mendelian Populations," *Genetics* (1931) **16**: 97–159, is in letters he wrote in the second decade of the twentieth century to his brother Quincy Wright. Not only is there high praise for Henderson but also there are all sorts of claims about "change in equilibrium" and so forth. (The Quincy Wright Papers are held at the University of Chicago. The phrase just used is from a letter of 27 February 1916.) These and other influences are discussed in Ruse, *Monad to Man*, 380–4. Henderson's key pro-Spencer books are *The Fitness of the Environment* (New York: Macmillan, 1913); and *The Order of Nature* (Cambridge, MA: Harvard University Press, 1917).

65. Simpson, *The Meaning of Evolution: A Study of the History of Life and of Its Significance for Man* (New Haven, CT: Yale University Press, 1949), see especially pt. 3, "Evolution, Humanity, and Ethics."

66. Comments to Michael Ruse, taped Spring 1993.

67. Karl R. Popper, *Objective Knowledge* (Oxford: Oxford University Press, 1972).

68. In his biography of Sewall Wright, *Sewall Wright and Evolutionary Biology* (Chicago: University of Chicago Press, 1986), W. B. Provine quotes Dobzhansky on Wright: "He has a lot of extremely abstruse, in fact almost esoteric, mathematics. Mathematics, incidentally, of a kind which I certainly do not claim to understand. I am not a mathematician at all. My way of reading Sewall Wright's papers, which I still think is perfectly defensible, is to examine the biological assumptions the man is making, and to read the conclusions he arrives at, and hope to goodness that what comes in between is correct. 'Papa knows best' is a reasonable assumption, because if the mathematics were incorrect, some mathematician would have found it out." Provine, *Sewall Wright*, 346.

69. "Letter from Mayr to W. A. Gosline, 5 March 1948," *Evolution Papers*. This is the collection of Ernst Mayr's correspondence for the early years of the journal *Evolution*, They are held at the American Philosophical Society in Philadelphia.

70. "Letter from Mayr to A. Conquist, 14 November 1949," *Evolution Papers*.

71. "Letter from Mayr to R. H. Flower, 23 January 1948," *Evolution Papers*.

72. "Letter from Mayr to G. G. Ferris, 29 March 1948," *Evolution Papers*.

73. Julian S. Huxley, *Evolution: The Modern Synthesis*. (London: Allen and Unwin, 1942).

74. J. S. Huxley, *Evolution*, 569.

75. *The Captive Shrew and Other Poems of a Biologist* (Oxford: Blackwell, 1932).

76. Peter Medawar, review of *The Phenomenon of Man*, in *The Art of the Soluble*, ed. Peter Medawar (London: Methuen and Co. Ltd, [1961] 1967), 71–81.

77. This was from a review of Teilhard de Chardin's 1955 masterwork, *Le phénomène humain* [The human phenomenon], for which Huxley had written a laudatory preface. (*The Phenomenon of Man*, transl. Bernard Wall (London: Collins, 1959)). It is not entirely clear who is the target of Medawar's attack, Teilhard or Huxley, but really that was the whole point.

78. Simpson, *Tempo and Mode*; Simpson, *The Meaning of Evolution*; George Gaylord Simpson, *The Major Features of Evolution* (New York: Columbia University Press, 1953).

79. W. Warren Wagar, *Good Tidings: The Belief in Progress from Darwin to Marcuse* (Bloomington: Indiana University Press, 1972).

80. Stephen Jay Gould, "On Replacing the Idea of Progress with an Operational Notion of Directionality," in *Evolutionary Progress*, ed. M. H. Nitecki (Chicago: University of Chicago Press, 1988), 319–38 (319).

81. Interview with Michael Ruse, April 1988.

82. Stephen Jay Gould, *The Mismeasure of Man* (New York: Norton, 1981); Stephen Jay Gould, *Wonderful Life: The Burgess Shale and the Nature of History* (New York: W. W. Norton Co., 1989).

83. E. O. Wilson, *Sociobiology: The New Synthesis* (Cambridge, MA: Harvard University Press, 1975).

84. Wilson, *Sociobiology*, 379.

85. Wilson, *Sociobiology*, 382.

86. E. O. Wilson, *On Human Nature* (Cambridge: Cambridge University Press, 1978).

Chapter Eleven

1. Although historians and social scientists today almost universally recognize the socially constructed nature of race, historically many biologists have claimed race as another example of stable biological variation within humans. For example, see Michelle Brattain, "Race, Racism, and Antiracism: UNESCO and the Politics of Presenting Science to the Postwar Public," *American Historical Review* (2007) **112**: 1386–413; Charles Darwin, *The Descent of Man, and Selection in Relation to Sex, with an Introduction by James Moore and Adrian Desmond*, 2nd ed. (London: Penguin Books, [1879] 2004), 652–75.

2. I do not mean to suggest that "male" and "female" are immutable categories, but simply that both male and female individuals are produced in each generation; see Joan Roughgarden, *Evolution's Rainbow: Diversity, Gender, and Sexuality in Nature and People* (Berkeley: University of California Press, 2004).

3. Carolyn Merchant, *The Death of Nature: Women, Ecology, and the Scientific Revolution* (San Francisco: Harper & Row, 1980); Londa Schiebinger, *Nature's Body: Gender in the Making of Modern Science* (New York: Beacon Press, 1993), reissued with a new preface (New York: Rutgers University Press, 2004); Patricia Fara, *Sex, Botany, and Empire: The Story of Carl Linnaeus and Joseph Banks* (New York: Columbia University Press, 2004).

4. I use *gender* here as a category describing the social roles of men and women in human society. As the material in the chapter covers biologists' research on presumed universal differences between male and female animals and humans, I prefer to use the actor's category, *sex*. Anne Fausto-Sterling, *Sexing the Body: Gender Politics and the Construction of Sexuality* (New York: Basic Books, 2000). Additionally, throughout this chapter I refer to all non-human animals simply as "animals."

5. Donna Haraway, *Primate Visions: Gender, Race, and Nature in the World of Modern Science* (New York: Routledge, 1989).

6. Cynthia Eagle Russett, *Sexual Science: Victorian Construction of Womanhood* (Cambridge, MA: Harvard University Press, 1989).

7. Stevi Jackson and Amanda Rees, "The Appalling Appeal of Nature: The Popular Influence of Evolutionary Psychology as a Problem for Sociology," *Sociology* (2007) **41**: 917–30.

8. Some Internet dating sites allow members to vet hopeful applicants on the basis of their photo alone: http://www.darwindating.com/ and http://us.beautifulpeople.net/

9. Charles Darwin, *On the Origin of Species by Means of Natural Selection or the Preservation of Favoured Races in the Struggle for Life* (London: John Murray, 1859); Darwin, *The Descent of Man*.

10. James Moore and Adrian Desmond, "Introduction," in Charles Darwin, *The Descent of Man, and Selection in Relation to Sex*, 2nd ed. (London: Penguin Books, 2004), xi–lviii.

11. Eveleen Richards, "Darwin and the Descent of Woman," in *The Wider Domain of Evolutionary Thought*, ed. David Roger Oldroyd and Ian Langham (Dordrecht: D. Reidel, 1983), 57–111.

12. Michael Ruse, *Monad to Man: The Concept of Progress in Evolutionary Biology* (Cambridge, MA: Harvard University Press, 1997).

13. Russett, *Sexual Science*, 14; Nancy Leys Stepan, "Race and Gender: The Role of Analogy in Science," *Isis* (1986) **77**: 261–77; Francis Galton, *Hereditary Genius: An Inquiry into Its Laws and Consequences* (London: Macmillan and Co., 1869).

14. Russett, *Sexual Science*, 83–4; Darwin, *The Descent of Man*, 2nd ed., 654.

15. In *Sexual Science*, Russett addresses each of these issues in turn: the biogenetic law (chapter 2), laws of thermodynamics (chapter 4), and the physiological division of labor (chapter 5).

16. Russett, *Sexual Science*, 54–63; Ernst Haeckel, *Generelle Morphologie der Organismen* [General biology of organisms] (Berlin: Georg Reimer, 1866); Karl Ernst von Baer, *Entwicklungsgeschichte der Thiere: Beobachtung und Reflexion* [Developmental history of the animals: observation and reflection] (Königsberg: Bornträger, 1828); G. Stanley Hall, *Adolescence*, 2 vols. (New York: D. Appleton, 1904). For an excellent and thorough discussion of the history of evolutionary and developmental theory, see Stephen Jay Gould, *Ontogeny and Phylogeny* (Cambridge, MA: Harvard University Press, 1977).

17. On Wallace, Darwin, and sexual selection, see Helena Cronin, *The Ant and the Peacock: Altruism and Sexual Selection from Darwin to Today* (Cambridge: Cambridge University Press, 1991). On Geddes and Thompson, see Russett, *Sexual Science*, 90–1, 108–21; Patrick Geddes and J. Arthur Thompson, *The Evolution of Sex* (London: Walter Scott, 1889).

18. This belief certainly pre-dates the late nineteenth century, but was seized upon with renewed vigor by nineteenth-century scientists. See Merchant, *Death of Nature*; Londa Schiebinger, *The Mind Has No Sex? Women and the Origins of Modern Science* (Cambridge, MA: Harvard University Press, 1989); Lisbet Koerner, *Linnaeus: Nature and Nation* (Cambridge, MA: Harvard University Press, 1999); Barbara Taylor and Sarah Knott, eds., *Women, Gender, and Enlightenment* (New York: Palgrave Macmillan, 2005); and Ann Shteir and Bernard Lightman,

eds., *Figuring It Out: Science, Gender, and Visual Cultural* (Dartmouth, MA: Dartmouth College Press, 2006).

19. Russett, *Sexual Science*, 136–7.

20. Sally Gregory Kohlstedt and Mark R. Jorgensen, "'The Irrepressible Woman Question': Women's Responses to Evolutionary Ideology," in *Disseminating Darwinism: The Role of Place, Race, Religion, and Gender*, ed. Ronald L. Numbers and John Stenhouse (Cambridge: Cambridge University Press, 1999), 267–94; George Robb, "Eugenics, Spirituality, and Sex Differentiation in Edwardian England: The Case of Frances Swiney," *Journal of Women's History* (1998) **10**: 97–117.

21. Antoinette Brown Blackwell, *The Sexes Throughout Nature* (New York: G. P. Putnam's Sons, 1895); Charlotte Perkins Gilman, *Women and Economics: A Study of the Economic Relation between Men and Women as a Factor in Social Evolution* (Boston, MA: Small, Maynard & Company, 1898); Olive Schreiner, *Woman and Labour* (reprint, London: Virago, 1978); Eliza Burt Gamble, *The Evolution of Woman: An Inquiry into the Dogma of her Inferiority to Man* (New York: G. P. Putnam's Sons, 1894); Frances Swiney, *The Awakening of Women or Woman's Part in Evolution*, 3rd ed. (London: William Reeves, 1908).

22. Diane Paul, *Controlling Human Heredity: 1865 to the Present* (Atlantic Highlands, NJ: Humanities, 1995); Diane Paul, *The Politics of Heredity: Essays on Eugenics, Biomedicine, and the Nature-Nurture Debate* (Albany: SUNY Press, 1998); Penelope Deutscher, "The Descent of Man and the Evolution of Woman," *Hypatia* (2004) **19**(2), 35–55.

23. Alice Dreger, *Hermaphrodites and the Medical Invention of Sex* (Cambridge, MA: Harvard University Press, 1998).

24. Daniel Kevles, *In the Name of Eugenics: Genetics and the Uses of Human Heredity* (New York: Alfred A. Knopf, 1985); Wendy Kline, *Building a Better Race: Gender, Sexuality, and Eugenics from the Turn of the Century to the Baby Boom* (Berkeley: University of California Press, 2001); Paul, *Controlling Human Heredity*; Susan Marie Rensing, "Feminist Eugenics in America: From Free Love to Birth Control, 1880–1930," Ph.D. thesis, University of Minnesota, 2006; George Robb, "The Way of All Flesh: Degeneration, Eugenics, and the Gospel of Free Love," *Journal of the History of Sexuality* (1996) **6**: 589–603.

25. Kline, *Building a Better Race*.

26. For British fiction, see Angelique Richardson, *Love and Eugenics in the Late Nineteenth Century: Rational Reproduction and the New Woman* (Oxford: Oxford University Press, 2003). For American fiction, see Bert Bender, *The Descent of Love: Darwin and the Theory of Sexual Selection in American Fiction, 1871–1926* (Philadelphia: University of Pennsylvania Press, 1996); Bert Bender, *Evolution and "the Sex Problem": American Narratives during the Eclipse of Darwinism* (Kent, OH: Kent State University Press, 2004).

27. Alfred Russel Wallace, "Human Selection," *Fortnightly Review* (1890) **48**: 325–37; Alfred Russel Wallace, *Social Environment and Moral Progress* (New York: Cassell and Company, 1913), 140–1.

28. Ronald A. Fisher, "Evolution of Sex Preference," *Eugenics Review* (1915) **7**: 184–92 (191).

29. Fisher, *Eugenics Rev*. (1915) **7**: 189; Ronald A. Fisher, "Positive Eugenics," *Eugenics Review* (1917) **9**: 206-12.

30. Ronald Fisher, "Sexual Reproduction and Sexual Selection," in *The Genetical Theory of Natural Selection* (Oxford: Clarendon Press, 1930), chap. 6, 121–45.

31. Julian S. Huxley, "Courtship Habits of the Great Crested Grebe," *Proceedings of the Zoological Society of London* (1914) **35**: 92.

32. Julian S. Huxley, "Darwin's Theory of Sexual Selection and the Data Subsumed by it, in the Light of Recent Research," *American Naturalist* (1938) **72**: 416–33; Julian S. Huxley, "The present standing of the theory of sexual selection," in *Evolution: Essays on Aspects of Evolutionary Theory*, ed. Gavin de Beer (London: Oxford University Press, 1938), 11–42; Julian S. Huxley, Ludwig Koch, and Ylla, *Animal Language* (New York: Grosset & Dunlap, 1938; reprint, 1964).

33. Peter Laipson, "'Kiss without Shame, for She Desires It': Sexual Foreplay in American Marital Advice Literature, 1900–1925," *Journal of Social History* (1996) **29**: 507–25.

34. Marie Carmichael Stopes, "Mutual Adjustment," in *Married Love, or Love in Marriage*, with preface and notes by William J. Robinson, M.D. (New York: Critic and Guide Company, 1918), chap. 5.

35. Julian B. Carter, *The Heart of Whiteness: Normal Sexuality and Race in America, 1880–1940* (Durham, NC: Duke University Press, 2007).

36. For a thorough description of the influences on Huxley's approach to sexual selection and the history of animal behavior research in the early to mid-twentieth century, see Richard W. Burkhardt, Jr., *Patterns of Behavior: Konrad Lorenz, Niko Tinbergen, and the Founding of Ethology* (Chicago: University of Chicago Press, 2005).

37. Rebecca Lemov, *World as Laboratory: Experiments with Mice, Mazes, and Men* (New York: Hill and Wang, 2005).

38. Edward Steichen, *The Family of Man* (New York: published for the Museum of Modern Art by Simon and Schuster, 1955).

39. Haraway, *Primate Visions*.

40. For example, see Harry Harlow's research on the maternal instinct. Deborah Blum, *Love at Goon Park: Harry Harlow and the Science of Affection* (Cambridge, MA: Perseus Books, 2002).

41. Konrad Lorenz, *King Solomon's Ring: New Light on Animal Ways* (New York: Cromwell, 1952); Konrad Lorenz, *On Aggression*, translated by Marjorie Latzke (London: Methuen, 1966; original German edition, 1963). For a critique of Lorenz's popular writings, see Nikolaas Tinbergen, "On War and Peace in Animals and Men: An Ethologist's Approach to the Biology of Aggression," *Science* (1968) **160**: 1411–18.

42. On the history of the NRC–CRPS, see Sophie D. Aberle and George W. Corner, *25 Years of Sex Research; History of the National Research Council*

Committee for Research in Problems of Sex, 1922–1947 (New York: Saunders, 1953); Vern Bullough, *Science in the Bedroom: A History of Sex Research* (New York: Basic Books, 1994); Adele E. Clarke, *Disciplining Reproduction: Modernity, American Life Sciences, and the 'Problems of Sex'* (Berkeley: University of California Press, 1998). The research funded by the committee was published in numerous articles and edited volumes, in particular see Frank Ambrose Beach, ed., *Sex and Behavior* (New York: Wiley, 1965); Clellan S. Ford and Frank Ambrose Beach, eds., *Patterns of Sexual Behavior* (New York: Harper, 1951). For biographical information on Kinsey, see James H. Jones, *Alfred C. Kinsey: A Public/Private Life* (New York: W. W. Norton & Co, Inc., 1997); Jonathan Gathorne-Hardy, *Sex the Measure of All Things: A Life of Alfred C. Kinsey* (Bloomington: Indiana University Press, 2000).

43. Alfred Kinsey, *Sexual Behavior in the Human Male* (Philadelphia: Saunders, 1948); Alfred Kinsey, *Sexual Behavior in the Human Female* (Philadelphia: Saunders, 1953). For a thorough analysis of the reception of Kinsey's research by the American public, see Miriam Reumann, *American Sexual Character: Sex, Gender, and National Identity in the Kinsey Reports* (Berkeley: University of California Press, 2005).

44. Lemov, *World as Laboratory*.

45. For the influence of behaviorism on Kinsey's work, see Joanne Meyerowitz, "Sex Research at the Borders of Gender: Transvestites, Transsexuals, and Alfred C. Kinsey," *Bulletin of the History of Medicine* (2001) **75b**: 72–90. For the influence of psychobiologists, especially Frank Beach (another researcher funded by NRC–CRPS), see Joshua Levens, "Sex, Neurosis and Animal Behavior: The Emergence of American Psychobiology and the Research of W. Horsley Gantt and Frank A. Beach," Ph.D. thesis, Johns Hopkins University, 2005, 278–84.

46. Reumann, *American Sexual Character*.

47. For a fascinating and detailed description of ethology as a scientific and cultural enterprise, see Burkhardt, *Patterns of Behavior*; Richard W. Burkhardt, Jr., "Ethology, Natural History, the Life Sciences, and the Problem of Place," *Journal of the History of Biology* (1999) **32**: 489–508; Konrad Lorenz, "The Companion in the Bird's World," *Auk* (1937) **54**: 245–73; Konrad Lorenz, "The Comparative Method in Studying Innate Behaviour Patterns," *Symposia of the Society of Experimental Biology* (1950) **4**, 221–63; Nikolaas Tinbergen, *The Study of Instinct* (Oxford: Clarendon Press, 1951); Nikolaas Tinbergen, "On Aims and Methods of Ethology," *Zeitschrift für Tierpsychologie* (1963) **20**: 410–33.

48. On ethology's four questions, see Nikolaas Tinbergen and Robert Hinde, "The Comparative Study of Species-Specific Behavior," in *Behavior and Evolution,* ed. Anne Roe and George Gaylord Simpson (New Haven, CT: Yale University Press), 251–68. On the mechanomorphic linguistic representation of animals, see Eileen Crist, "The Ethological Constitution of Animals as Natural Objects:

The Technical Writings of Konrad Lorenz and Nikolaas Tinbergen," *Biology and Philosophy* (1998) **13**: 61–102.

49. David Blest, "The Concept of Ritualisation," in *Current Problems in Animal Behaviour*, ed. William Homan Thorpe and Oliver Louis Zangwill (Cambridge: Cambridge University Press, 1961), 102–24.

50. Julian S. Huxley, "Introduction: A Discussion on Ritualization of Behaviour in Animals and Man," *Philosophical Transactions of the Royal Society of London, Series B Biological* (1966) **251**: 249–71 (250).

51. Nikolaas Tinbergen, "'Derived' Activities: Their Causation, Biological Significance, Origin, and Emancipation During Evolution," *Quarterly Review of Biology* (1952) **27**: 1–32.

52. Lorenz, *On Aggression*, 23–48.

53. Alan John Marshall, *Bower Birds: Their Displays and Breeding Cycles, a Preliminary Statement* (Oxford: Clarendon Press, 1954). Although Marshall was not an ethologist by training, he developed his ideas on behavioral evolution in conversation with Julian Huxley during his years at Oxford University.

54. Critiques of ethological analyses of human behavior often referred to their language as "zoomorphic" or obsessed with understanding the animal in the human. For example, see the reviews of the English translation of Irenäus Eibl-Eibesfeldt, *Love and Hate: The Natural History of Behavior Patterns*, transl. Geoffrey Strachan (New York: Aldine de Gruyter, 1970); Irenäus Eibl-Eibesfeldt et al., "[Review] *Love and Hate*," *Current Anthropology* (1975) **16**: 151–62.

55. In addition to Eibl-Eibesfeldt's *Love and Hate*, and Lorenz's *On Aggression*, see Desmond Morris, *The Naked Ape: A Zoologist's Study of the Human Animal* (London: Jonathan Cape, 1967); Desmond Morris, *The Human Zoo* (New York: McGraw-Hill Book Company, 1969); and the anthropologically-oriented, Robert Ardrey, *African Genesis: A Personal Investigation into the Animal Origins and Nature of Man* (New York: Atheneum, 1961).

56. Lorenz, *On Aggression*, 236–75.

57. Several species of ants were known to wage wars of annihilation on other ant species, and to steal the eggs of other species to raise them as slaves, but within mammals humans seemed unique in their capacity to kill members of their own species. See Charlotte Sleigh, *Six Legs Better: A Cultural History of Myrmecology* (Baltimore: Johns Hopkins University Press, 2007). Jane Goodall's later observations of a four-year war (1974–78) of annihilation between groups of chimpanzees would change this perspective. Jane Goodall, *The Chimpanzees of Gombe: Patterns of Behavior* (Cambridge, MA: Harvard University Press, 1986).

58. Tinbergen, *Science* (1968) **160**: 1414.

59. Ashley Montagu, ed., *Man and Aggression* (Oxford: Oxford University Press, 1973): xi–xix.

60. Eibl-Eibesfeldt, *Love and Hate*, 48.

61. Eibl-Eibesfeldt, *Love and Hate*, 46–51.

62. Crist, "The Ethological Constitution of Animals as Natural Objects."

63. Edward O. Wilson, *Sociobiology: The New Synthesis* (Cambridge, MA: Belknap Press of Harvard University Press, 1975, 2000); Richard Dawkins, *The Selfish Gene* (Oxford: Oxford University Press, 1976).

64. Wilson, *Sociobiology*, 562.

65. Ullica Segerstråle, *Defenders of the Truth: The Battle for Science in the Sociobiology Debate and Beyond* (Oxford: Oxford University Press, 2002), 6–9.

66. Conrad Hal Waddington, "Mindless Societies," *New York Review of Books* (1975) **22**(13).

67. J. B. S. Haldane, *The Causes of Evolution*, with a new introduction and afterword by Egbert G. Leigh, Jr. (Princeton, NJ: Princeton University Press, [1932] 1990); William D. Hamilton, "The Evolution of Altruistic Behavior," *American Naturalist* (1963) **7**: 354–6; William D. Hamilton, "The Genetical Evolution of Social Behaviour, I–II," *Journal of Theoretical Biology* (1964) **7**: 1–52.

68. The attack: Anthony Leeds, Barbara Beckwith, Chuck Madansky, David Culver, Elizabeth Allen, Herb Schreier, Hiroshi Inouye, Jon Beckwith, Larry Miller, Margaret Duncan, Miriam Rosenthal, Reed Pyeritz, Richard C. Lewontin, Ruth Hubbard, Steven Chorover, and Stephen Jay Gould, "Against 'Sociobiology,'" *New York Review of Books* **22**(18), 13 November 1975. Wilson's response: Edward O. Wilson, "For 'Sociobiology,'" *New York Review of Books* **22**(20), 11 December 1975.

69. Wilson, *Sociobiology*, chap. 15.

70. Dawkins, *Selfish Gene*, 2.

71. Dawkins, *Selfish Gene*, 141–2.

72. Dawkins, *Selfish Gene*, 300–1.

73. Dawkins, *Selfish Gene*, 165.

74. Ruth Bleier, "Sociobiology, Biological Determinism, and Human Behavior," in *Science and Gender: A Critique of Biology and Its Theories on Women* (New York: Pergamon Press, 1984), chap. 2; Anne Fausto-Sterling, *Myths of Gender: Biological Theories about Men and Women* (New York: Basic Books, 1985); Haraway, *Primate Visions*; Sarah Blaffer Hrdy and George C. Williams, "Behavioral Biology and the Double Standard," in *Social Behavior of Female Vertebrates* (New York: Academic Press, 1983), 3–17; Ruth Hubbard, "Have Only Men Evolved?," in *Women Look at Biology Looking at Women: A Collection of Feminist Critiques*, ed. Ruth Hubbard, Mary Sue Henifen, and Barbara Fried (Cambridge, MA: Schenkman, 1979), 7–36; M. Lowe and Ruth Hubbard, "Sociobiology and Biosociology: Can Science Prove the Biological Basis of Sex Differences in Behavior?," in *Genes and Gender II*, ed. R. Hubbard and Marian Lowe (New York: Gordian Press, 1979), 91–111; and J. Sayers, *Biological Politics: Feminist and Anti-Feminist Perspectives* (New York: Tavistock, 1982).

75. Betty Friedan, *The Feminine Mystique* (New York: Norton, 1963), 15.

76. William Masters and Virginia Johnson, *Human Sexual Response* (Boston, MA: Little, Brown & Co., 1966).

77. Acknowledgments in Wilson, *Sociobiology*.

78. Sarah Blaffer Hrdy, *The Woman That Never Evolved* (Cambridge, MA: Harvard University Press, [1981] 1999).

79. Sarah Blaffer Hrdy, "Empathy, Polyandry, and the Myth of the Coy Female," in *Feminist Approaches to Science*, ed. Ruth Bleier (New York: Pergamon, 1986): 119; A. J. Bateman, "Intra-sexual Selection in *Drosophila*," *Heredity* (1948) **2**: 349–68.

80. Robert L. Trivers, "Parent-offspring Conflict," *American Zoologist* (1974) **14**: 249-64; Robert L. Trivers, "Parental Investment and Sexual Selection," in *Sexual Selection and the Descent of Man, 1871–1971*, ed. Bernard Clark Campbell (Chicago: Aldine, 1972): 136–79.

81. Hrdy identified the most cited paper in sociobiology as Hamilton's 1964 paper on kin selection (Hamilton, "The Genetical Evolution of Social Behavior"). Hrdy, "Empathy, Polyandry, and the Myth of the Coy Female," 119.

82. Haraway, *Primate Visions*, 279–367.

83. Hrdy, *The Woman That Never Evolved*, "preface, 1999: On Raising Darwin's Consciousness," xiii–xxxii.

84. Jeanne Altman, *Baboon Mothers and Infants* (Chicago: University of Chicago Press, [1980] 2001).

85. Barbara Smuts, *Sex and Friendship in Baboons* (New York: Aldine, 1985).

86. Sarah Blaffer Hrdy, *Mother Nature: A History of Mothers, Infants, and Natural Selection* (New York: Pantheon, 1999).

87. A. J. Moore, Patricia A. Gowaty, and P. J. Moore, "Females Avoid Manipulative Males and Live Longer," *Journal of Evolutionary Biology* (2003) **16b**: 523–30; Lee C. Drickamer, Patricia A. Gowaty, and Christopher M. Holmes, "Free Female Mate Choice in House Mice Affects Reproductive Success and Offspring Viability and Performance," *Animal Behaviour* (2000) **59**: 371–8.

88. Amanda Rees, "Reflections on the Field: Primatology, Popular Science and the Politics of Personhood," *Social Studies of Science* **37** (2007): 881–907. For the complicated relationship between feminism and primatology, see Shirley Carol Strum, Linda Marie Fedigan, *Primate Encounters: Models of Science, Gender, and Society* (Chicago: University of Chicago Press, 2000).

89. Frans B. M. de Waal, "The end of nature versus nurture," *Scientific American*, (1999) 281: 94–9; Frans B. M. de Waal, "Bonobo: Sex & Society," *Scientific American* (2006) 16: 14-21; Sue Savage-Rumbaugh and Roger Lewin, *Kanzi: The Ape at the Brink of the Human Mind* (New York: John Wiley & Sons, 1995); Sue Savage-Rumbaugh, Stuart Shanker, and Talbot Taylor, *Apes, Language and the Human Mind: Philosophical Primatology* (Oxford: Oxford University Press, 1999).

90. Randy Thornhill and Craig T. Palmer, *A Natural History of Rape: Biological Bases of Sexual Coercion* (Cambridge, MA: MIT Press, 2000); Elisabeth

Lloyd, "Science Gone Astray: Evolution and Rape" [a review of *A Natural History of Rape: Biological Bases of Sexual Coercion* by Randy Thornhill and Craig T. Palmer], *Michigan Law Review* (2001) **99**: 1536–59.

91. Stephen Jay Gould, "The Modern Synthesis as a Limited Consensus," in *Structure of Evolutionary Thought* (Cambridge, MA: Harvard University Press, 2004), chap. 7.

Chapter Twelve

1. This chapter is extracted in large part from Ronald L. Numbers, *The Creationists: From Scientific Creationism to Intelligent Design*, expanded ed. (Cambridge, MA: Harvard University Press, 2006). Charles Darwin, *The Descent of Man, and Selection in Relation to Sex*, 2 vols. (London: John Murray, 1871), vol. 1, 152–3. See also Charles Darwin to Asa Gray, 11 May 1863, quoted in Francis Darwin, ed., *The Life and Letters of Charles Darwin*, 2 vols. (New York: D. Appleton, 1896), vol. 2, 163–4.

2. Charles Darwin, *On the Origin of Species*, a facsimile of the first edition, with an introduction by Ernst Mayr (Cambridge, MA: Harvard University Press, 1966), 484; A. Hunter Dupree, *Asa Gray, 1810–1888* (Cambridge, MA: Harvard University Press, 1959), 300–1, 339; Ronald L. Numbers, *Darwinism Comes to America* (Cambridge, MA: Harvard University Press, 1998), 27 (Gray); Ronald L. Numbers, "George Frederick Wright: From Christian Darwinist to Fundamentalist," *Isis* (1988) **79**: 624–45.

3. Charles Darwin, *The Variation of Animals and Plants under Domestication*, 2 vols. (London: John Murray, 1868), vol. 2, 432.

4. Charles Hodge, *What Is Darwinism?* (New York: Scribner, Armstrong, 1874), 177. See also Ronald L. Numbers, "Charles Hodge and the Beauties and Deformities of Science," in *Charles Hodge Revisited: A Critical Appraisal of His Life and Work*, ed. John W. Stewart and James H. Moorhead (Grand Rapids, MI: William B. Eerdmans, 2002), 77–101.

5. B. B. Warfield, *Evolution, Science, and Scripture: Selected Writings*, ed. Mark A. Noll and David N. Livingstone (Grand Rapids, MI: Baker Books, 2000), 26 (purest water), 125 (thoroughgoing evolutionism), 128 (*interfering* hand), 130–1 (Eve, upshot), 234 (modification); David N. Livingstone and Mark A. Noll, "B. B. Warfield (1851–1921): A Biblical Inerrantist As Evolutionist," *Isis* (2000) **91**: 283–304. On Woodrow, see Ronald L. Numbers and Lester D. Stephens, "Darwinism in the American South," in *Disseminating Darwinism: The Role of Place, Race, Religion, and Gender*, ed. Ronald L. Numbers and John Stenhouse (Cambridge: Cambridge University Press, 1999), 123–43 (especially 126). On the variety of Calvinist responses to Darwinism, see David N. Livingstone, "Science, Region, and

Religion: The Reception of Darwinism in Princeton, Belfast, and Edinburgh," in Numbers and Stenhouse, *Disseminating Darwinism*, 7–38.

6. James R. Moore, *The Post-Darwinian Controversies: A Study of the Protestant Struggle to Come to Terms with Darwin in Great Britain and America, 1870–1900* (Cambridge: Cambridge University Press, 1979), 220 (Temple), 259–69 (Moore).

7. On American opposition to Darwinism, see Jon H. Roberts, *Darwinism and the Divine in America: Protestant Intellectuals and Organic Evolution, 1859–1900* (Madison: University of Wisconsin Press, 1988), chap. 8. On harmonizing Genesis and geology, see Nicolaas Rupke, *The Great Chain of History: William Buckland and the English School of Geology (1814–1849)* (Oxford: Clarendon Press, 1983); Rodney Lee Stiling, "The Diminishing Deluge: Noah's Flood in Nineteenth-Century American Thought," Ph.D. dissertation, University of Wisconsin-Madison, 1991.

8. Edward Hitchcock, "The Law of Nature's Constancy Subordinate to the Higher Law of Change," *Bibliotheca Sacra* (1863) **20**: 520–5; Edward Hitchcock, *Elementary Geology,* new ed. (New York: Ivison, Phinney, 1862), 373–4, 377–93.

9. H. L. Hastings, *Was Moses Mistaken? or, Creation and Evolution,* Anti-Infidel Library no. 36 (Boston: H. L. Hastings, 1896).

10. H. L. Hastings, "Preface to the Second Edition (1889)," in *The Errors of Evolution: An Examination of the Nebular Theory, Geological Evolution, the Origin of Life, and Darwinism,* 3rd ed., by Robert Patterson (Boston: Scriptural Tract Repository, 1893), iv (premillennialist); William B. Riley, *The Finality of the Higher Criticism; or, The Theory of Evolution and False Theology* (Minneapolis: n.p., 1909), 73 (Baptist cleric); Maynard M. Metcalf, *An Outline of the Theory of Organic Evolution* (New York: Macmillan, 1904), xix–xx (last stronghold).

11. Eberhard Dennert, *At the Deathbed of Darwinism,* transl. E. V. O'Harra and John H. Peschges (Burlington, IA: German Literary Board, 1904; originally published in Stuttgart in 1903 as *Vom Sterbelager des Darwinismus*); L. T. Townsend, *Collapse of Evolution* (New York: American Bible League, 1905), 48–53. See also Ronald L. Numbers, "Darwinism and the Dogma of Separate Creations: The Responses of American Naturalists to Evolution," in *Darwinism Comes to America* (Cambridge, MA: Harvard University Press, 1998), chap. 1.

12. Vernon L. Kellogg, *Darwinism To-Day* (New York: Henry Holt, 1907), 1–9; G. L. Young, "Relation of Evolution and Darwinism to the Question of Origins," *Bible Student and Teacher* (1909) **11**: 41.

13. W. J. Bryan to Johnnie Baldwin, 27 March 1923, Box 37, W. J. Bryan Papers, Library of Congress; Lawrence W. Levine, *Defender of the Faith: William Jennings Bryan: The Last Decade, 1915–1925* (New York: Oxford University Press, 1965), 261–5; Gregg Mitman, "Evolution as Gospel: William Patten, the Language of Democracy, and the Great War," *Isis* (1990) **81**: 446–63. See also Michael Kazin, *A Godly Hero: The Life of William Jennings Bryan* (New York: Alfred A. Knopf, 2006).

14. Levine, *Defender of the Faith,* 266–7. Mrs. Bryan's statement appears in Wayne C. Williams, *William Jennings Bryan* (New York: G. P. Putnam, 1936), 448. The Bryan Papers contain numerous letters reporting the insidious effects of evolution. On the Leopold-Loeb murder, see Simon Baatz, *For the Thrill of It: Leopold, Loeb, and the Murder That Shocked Chicago* (New York: HarperCollins, 2008); W. B. Riley and Henry B. Smith, *Should Evolution Be Taught in Tax Supported Schools? Riley–Smith Debate* (n.p., n.d.), 8, 25. A copy of the Riley–Smith debate can be found in Ronald L. Numbers, ed., *Creationism in Twentieth-Century America: A Ten-Volume Anthology of Documents, 1903–1961* (New York: Garland, 1995), vol. 2.

15. William Jennings Bryan, *In His Image* (New York: Fleming H. Revell, 1922), 94, 97–8. "The Menace of Darwinism" appears in this work as chap. 4, "The Origin of Man." For the apparent source of Bryan's illustration, see Alexander Patterson, *The Other Side of Evolution: An Examination of Its Evidences,* 3rd ed. (Chicago: Bible Institute Colportage Association, 1912; first published in 1903), 32–3. Bryan recommended Patterson's book in his popular lecture "The Bible and Its Enemies," reprinted in William Jennings Bryan, *The Dawn of Humanity* (Chicago: Altruist Foundation, 1925), 69.

16. Charles Darwin, *On the Origin of Species* (London: John Murray, 1859), 186–8; Asa Gray to Charles Darwin, 23 January 1860, and Darwin to Gray, February [?] 1860, in Francis Darwin, ed., *The Life and Letters of Charles Darwin,* 2 vols. (New York: D. Appleton, 1889), vol. 2, 66–7. See also Darwin's warning to himself in 1838 to avoid discussing the origin of the eye: Paul H. Barrett, P. J. Gautrey, S. Herbert, D. Kohn, and S. Smith, eds., *Charles Darwin's Notebooks, 1836–1844* (Ithaca, NY: Cornell University Press, 1987), 337. On the eye and atheism, see William Paley, *Natural Theology* (New York: Evert Duyckinck, 1820), 22.

17. T. T. Martin, *Hell and the High Schools: Christ or Evolution, Which?* (Kansas City: Western Baptist Publishing Co., 1923), 10, 72, 164; T. T. Martin, *Viewing Life's Sunset from Pikes Peak: Life Story of T. T. Martin* (Louisville: A. D. Muse, n.d.), 78, 80.

18. [William B. Riley], "The Evolution Controversy," *Christian Fundamentals in School and Church* (1922) **4**(April–June): 5; Bryan, *In His Image,* 94; L. L. Pickett, *God or the Guessers: Some Strictures on Present Day Infidelity* (Louisville, KY: Pentecostal Publishing Co., 1926), 11 (dictionaries); L. S. K[eyser], "No War against Science—Never!" *Bible Champion* (1925) **31**: 413. On using the dictionary to define science, see, for example, Riley, *The Finality of the Higher Criticism,* 76.

19. Townsend, *Collapse of Evolution* 47–53; T. T. M[artin], "The Three False Teachings of President Poteat of Wake Forest," *Western Recorder* (1920) **95** (5 February): 5, from a copy in Fld. 260, William Louis Poteat Papers, Baptist Historical Collection, Wake Forest University Library; William Louis Poteat, "Evolution," *Biblical Recorder* (1922) **87**(19 April): 3. In *Hell and the High Schools* (131), Martin responded to Poteat's critique with the queries: "Does the fact that they

are dead prove that they were not great scientists? Does death prove that they lied?" For another early list of alleged antievolutionists, see Patterson, *The Other Side of Evolution*, 7–11.

20. Albert Fleischmann, *Die Descendenztheorie* [The theory of descent] (Leipzig: Verlag von Arthur Georgi, 1901); Kellogg, *Darwinism To-Day*, 8; Albert Fleischmann, "The Doctrine of Organic Evolution in the Light of Modern Research," *Journal of the Transactions of the Victoria Institute* (1933) **65**: 194–214 (196, 205–6); Douglas Dewar to [name deleted], 2 November 1931, from a copy in the George McCready Price Papers, Adventist Heritage Center, Andrews University. For bibliographical information, see Georg Uschmann, "Albert Fleischmann," *Neue Deutsche Biographia* (Berlin: Duncker & Humblot, 1960), vol. 5, 234–5.

21. Townsend, *Collapse of Evolution*, 48; Patterson, *The Other Side of Evolution*, 9; Sidney F. Harmer to James H. Snowdon, 25 July 1922, quoted in W. C. Curtis, "Three Letters Bearing upon the Controversy over Evolution," *Science* (1925) **61**: 647–8 (648). One of the fullest versions of Etheridge's statement appears in Martin, *Hell and the High Schools*, 112.

22. William Bateson, "Evolutionary Faith and Modern Doubts," *Science* (1922) **55**: 55–61; G. M. Price to W. J. Bryan, 9 February 1922 (boost), Price Papers; *The Summarized Proceedings of the American Association for the Advancement of Science, 1921–1925* (Washington, DC, 1925), 66–7, quoted in Willard B. Gatewood, Jr., *Preachers, Pedagogues & Politicians: The Evolution Controversy in North Carolina, 1920–1927* (Chapel Hill: University of North Carolina Press, 1966), 169–70; Henry Fairfield Osborn, *Evolution and Religion in Education: Polemics of the Fundamentalist Controversy of 1922 to 1926* (New York: Charles Scribner's Sons, 1926), 29. See also E. G. Conklin to C. B. Davenport, 6 May 1922, Charles B. Davenport Papers, Library of the American Philosophical Society; Conklin headed up the AAAS Committee on Evolution, set up to combat antievolution propaganda. For Bateson's reaction to the flap, see his letter to W. C. Curtis, 11 December 1922, quoted in Curtis, *Science* (1925) **61**: 647.

23. Anonymous, "Big Scientist Rejects Darwin," *Christian Fundamentalist* (1929) **2**: 59; Austin H. Clark, *The New Evolution: Zoogenesis* (Baltimore: Williams & Wilkins, 1930); Ben F. Allen to G. M. Price, 12 June 1929, Price Papers. See also Austin H. Clark to G. M. Price, 23 March 1929, Price Papers; and Theodore Graebner, *God and the Cosmos: A Critical Analysis of Atheism* (Grand Rapids, MI: William B. Eerdmans, 1932), 287–9.

24. Numbers, *The Creationists*, 71–2.

25. Numbers, *The Creationists*, 72–5.

26. Numbers, *The Creationists*, 76–87.

27. [G. M. Price], "Letter to the Editor of Science from the Principal Scientific Authority of the Fundamentalists," *Science* (1926) **63**: 259.

28. Numbers, *The Creationists*, 88–119; G. M. Price to W. J. Bryan, 13 April 1922, Box 35, William Jennings Bryan Papers, Library of Congress; Geo. E.

McCready Price, *Outlines of Modern Christianity and Modern Science* (Oakland, CA: Pacific Press, 1902), 200; George McCready Price, *The Phantom of Organic Evolution* (New York: Fleming H. Revell, 1924), 19–20, 38–9. On White and Adventism, see Ronald L. Numbers, *Prophetess of Health: A Study of Ellen G. White* (New York: Harper & Row, 1976); Numbers, " 'Sciences of Satanic Origin': Adventist Attitudes toward Evolutionary Biology and Geology," in *Darwinism Comes to America*, chap. 5.

29. George McCready Price, *Q. E. D.; or, New Light on the Doctrine of Creation* (New York: Fleming H. Revell, 1917), 68–77; Price, *Outlines,* 199; George McCready Price, "Dear Fellow Science Teachers," *Watchman Magazine* (1925) **34** (January): 18.

30. Numbers, *The Creationists,* 101–2; Price, *The Phantom of Organic Evolution,* 91–112; George McCready Price, *A Textbook of General Science for Secondary Schools* (Mountain View, CA: Pacific Press, [1917]), 500–10; George McCready Price, *The Geological-Ages Hoax: A Plea for Logic in Theoretical Geology* (New York: Fleming H. Revell, 1931), 105–6.

31. Price, *Outlines,* 68; George McCready Price, *Why I Am Not an Evolutionist,* Bible Truth Series no. 52 (Mountain View, CA: Pacific Press, n.d.), 1; George McCready Price, *Poisoning Democracy: A Study of the Moral and Religious Aspects of Socialism* (New York: Fleming H. Revell, 1921), 25; George McCready Price and Robert B. Thurber, *Socialism in the Test-Tube* (Nashville: Southern Publishing Assn., 1921).

32. D. J. Whitney to B. C. Nelson, 3 and 18 June 1928, courtesy of Paul Nelson; Harold W. Clark, *Genes and Genesis* (Mountain View, CA: Pacific Press, 1940), 43, 143. On Clark, see Numbers, *The Creationists,* 142–8. On creationist debates over the fixity of species, see Ronald L. Numbers, "Ironic Heresy: How Young-Earth Creationists Came to Embrace Rapid Microevolution by Means of Natural Selection," in *Darwinian Heresies,* ed. Abigail J. Lustig, Robert J. Richards, and Michael Ruse (Cambridge: Cambridge University Press, 2004), 84–100.

33. Numbers, *The Creationists,* 148–53.

34. Frank Lewis Marsh, *Fundamental Biology* (Lincoln, NE: The Author, 1941), iii, 6, 11, 48, 56, 63, 100 (baramins).

35. Ernst Mayr to F. L. Marsh, 13 March 1945, Price Papers; T. Dobzhansky to F. L. Marsh, 29 November 1944, F. L. Marsh Papers, Adventist Heritage Center, Andrews University; Theodosius Dobzhansky, *Genetics and the Origin of Species* (New York: Columbia University Press, 1937), 8; Theodosius Dobzhansky, "Review of *Evolution, Creation, and Science,* by F. L. Marsh," *American Naturalist* (1945) **79**: 73–5. See also T. Dobzhansky to F. L. Marsh, 15 November 1944, Marsh Papers.

36. F. L. Marsh to T. Dobzhansky, 19 November 1944 (delight); F. L. Marsh to T. Dobzhansky, 4 December 1944 (varieties and natural selection); T. Dobzhansky to F. L. Marsh, 7 December 1944 (only antievolutionist); all in the Marsh Papers.

37. T. Dobzhansky to F. L. Marsh, 22 December 1944 (geological time); F. L. Marsh to T. Dobzhansky, 12 January 1945 (real proof); T. Dobzhansky to F. L. Marsh, 5 February 1945 (mouse); all in the Marsh Papers.

38. F. L. Marsh to T. Dobzhansky, 12 January 1945; T. Dobzhansky to F. L. Marsh, 22 December 1944; both in the Marsh Papers.

39. F. L. Marsh to T. Dobzhansky, 21 February 1945, Marsh Papers; Theodosius Dobzhansky, *Genetics and the Origin of Species,* 3rd ed. (New York: Columbia University Press, 1951), 11. Marsh quotes Dobzhansky's mention of him in his *Evolution or Special Creation?* (Washington: Review and Herald Publishing Assn., 1963), 46.

40. D. Dewar to G. M. Price, 25 July 1931, Price Papers; Douglas Dewar, *Difficulties of the Evolution Theory* (London: Edward Arnold, 1931), 5, 158. On Dewar, see Numbers, *The Creationists,* 166–73.

41. D. Dewar to G. M. Price, 25 July 1931, Price Papers (publishers); Douglas Dewar, "The Limitations of Organic Evolution," *Journal of the Transactions of the Victoria Institute* (1932) **64**: 142; Arnold Lunn and J. B. S. Haldane, *Science and the Supernatural: A Correspondence* (London: Erye and Spottiswoode, 1935), 181–2, 206, 218, 243, 324.

42. John C. Whitcomb, Jr., and Henry M. Morris, *The Genesis Flood: The Biblical Record and Its Scientific Implications* (Philadelphia: Presbyterian and Reformed Publishing Co., 1961); Numbers, *The Creationists,* 240–2 (Lammerts), 254 (CRS founders).

43. Henry M. Morris, letter to the editor, *Creation Research Society Quarterly* (1974–5) **11**: 173–5; Walter E. Lammerts, letter to the editor, *Creation Research Society Quarterly* (1974–5) **12**: 75–7.

44. W. E. Lammerts to H. H. Hartzler, 9 December 1965, Walter E. Lammerts Papers, Bancroft Library, University of California, Berkeley (creation research); Henry M. Morris, *A History of Modern Creationism,* (San Diego: Master Book Publishers, 1984), 251–4 (types of research); Walter E. Lammerts, "Planned Induction of Commercially Desirable Variation in Rose by Neutron Radiation," *Creation Research Society Quarterly* (1965–6) **2**: 39–48 (43). See also Walter E. Lammerts, "Mutations Reveal the Glory of God's Handiwork," *Creation Research Society Quarterly* (1967–8) **4**: 35–41; and Walter E. Lammerts, "Does the Science of Genetic and Molecular Biology Really Give Evidence for Evolution?" *Creation Research Society Quarterly* (1969–70) **6**: 5–12.

45. L. G. Butler to H. M. Morris, undated draft of letter [July 1971?] (image of respectability, psychopaths, and fanciful); Larry G. Butler, "A Critique of Creationist Research," undated MS [January 1972] (sensational), L. G. Butler to V. L. Bates, January 26, 1975 (lunatic fringe); all courtesy of Larry Butler. On Butler, see Numbers, *The Creationists,* 283–5.

46. "Creation Science, Sex Education, Abortion, and the ERA," *Creation-Science Report* (1980) no. 4: unpaginated; Nell J. Segraves, *The Creation Report* (San Diego: Creation-Science Research Center, 1977), 17 (widespread breakdown).

47. W. D. Burrowes, "Comment on the State of the Art in North America," North American Creation Movement, *Newsletter* (1988) no. 43(30 March): 10; David A. Kaufmann, "Feminism, Humanism and Evolution," *Creation Research Society Quarterly* (1988–9) **25**: 69–72 (72).

48. Henry M. Morris, ed., *Scientific Creationism*, public school ed. (San Diego: Creation-Life Publishers, 1974), 8–16. See also Numbers, *The Creationists*, 268–70, 369.

49. Larry Vardiman, "Scientific Naturalism as Science," *Impact* 293, inserted in *Acts & Facts* (1997) **26** (November): no pagination (reclaim science); William A. Dembski, "What Every Theologian Should Know about Creation, Evolution, and Design," *Transactions* (1995) **3**(May/June): 1–8 (ground rules). *Transactions* was published by the Center for Interdisciplinary Studies in Princeton, NJ. On the history of intelligent design, see Numbers, "Intelligent Design" in *The Creationists*, chap. 17; Barbara Forrest and Paul R. Gross, *Creationism's Trojan Horse: The Wedge of Intelligent Design* (New York: Oxford University Press, 2004).

50. Numbers, *The Creationists*, 376–7.

51. Michael J. Behe, *Darwin's Black Box: The Biochemical Challenge to Evolution* (New York: Free Press, 1996), 15 (ultimate black box), 39 (irreducibly complex), 193 (scientific data), 232–3 (greatest achievements).

52. Dembski, *Transactions* (1995) **3**(May/June): 3, 5.

53. J. W. Haas, Jr., "On Intelligent Design, Irreducible Complexity, and Theistic Science," *Perspectives on Science and Christian Faith* (1997) **49** (March): 1.

54. Stephen C. Meyer, "The Origin of Biological Information and the Higher Taxonomic Categories," *Proceedings of the Biological Society of Washington* (2004) **117**: 213–39 (234). Regarding the ensuing controversy, see David Klinghoffer, "The Branding of a Heretic," *Wall Street Journal* (2005), 28 January; David Klinghoffer, "Intelligent Design and the Smithsonian," *New York Times* (2005), 20 August: A26.

55. Decision in *Tammy Kitzmiller, et al. v. Dover Area School District*, available at http://ncseweb.org/webfm_send/73. See also Margaret Talbot, "Darwin in the Dock," *New Yorker* (2005), 5 December; and Matthew Chapman, "God or Gorilla: A Darwin Descendant at the Dover Monkey Trial," *Harper's Magazine* (2006) **312** (February): 54–63.

56. In 1999 the president of the International Council for Science estimated that there were "perhaps 3 to 10 million active scientists worldwide"; see Werner Arber, "World Conference on Science," *Science International* (1999), cover. I conservatively used 3 million in my calculation. George W. Gilchrist, "The Elusive Scientific Basis of Intelligent Design Theory," *NCSE Reports* (1997) **17** (May/June): 14–5, reported that a search of approximately 6,000 journals in the life sciences indexed by BIOSIS and published between 1991 and 1997 turned up only one reference to intelligent design. The cover of *Time* (2005), 15 August, announced the "Evolution Wars."

Chapter Thirteen

1. See especially Sam Harris, *The End of Faith: Religion, Terror, and the Future of Reason* (New York: W. W. Norton & Co., 2004); Daniel C. Dennett, *Breaking the Spell: Religion as a Natural Phenomenon* (New York: Viking, 2006); Richard Dawkins, *The God Delusion* (Boston: Houghton Mifflin Co., 2006); Christopher Hitchens, *God Is Not Great: How Religion Poisons Everything* (New York: Twelve, 2007).

2. For the background to this and similar events, see Christoph Reuter, *My Life Is a Weapon: A Modern History of Suicide Bombing* (Princeton, NJ: Princeton University Press, 2004).

3. See especially Richard Dawkins, *The Blind Watchmaker: Why Evidence of Evolution Reveals a Universe without Design* (New York: Norton, 1996).

4. See the huge literature, too complex to be summarized here, spawned by Edward O. Wilson, *Sociobiology: The New Synthesis* (Cambridge, MA: Harvard University Press, 1975).

5. It was widely agreed that challenges to existing understandings of divine causality seemed necessary: Note especially Muriel Blaisdell, "Natural Theology and Nature's Disguises," *Journal of the History of Biology* (1982) 15: 163–89 (especially 165): "The battles over evolution and natural selection were both heated and prolonged, because the Designer posited by natural theologians was replaced by the mechanistic secondary cause of natural selection, which was not even necessarily directional in its action."

6. Charles C. Gillispie, *Genesis and Geology: A Study in the Relations of Scientific Thought, Natural Theology and Social Opinion in Great Britain, 1790–1850* (Cambridge, MA: Harvard University Press, 1996).

7. See here Joan F. Box, *R. A. Fisher: The Life of a Scientist* (New York: Wiley, 1978).

8. For studies of the popular reception of Darwinism at earlier stages in its development, see Thomas F. Glick, *The Comparative Reception of Darwinism.* (Austin: University of Texas Press, 1972); Yvette Conry, *L'introduction du Darwinisme en France au XIXe siècle* (Paris: J. Vrin, 1974); Alfred Kelly, *The Descent of Darwin: The Popularization of Darwinism in Germany, 1860–1914* (Chapel Hill: North Carolina University Press, 1981).

9. See, for example, the approach of Victor J. Stenger, *God: The Failed Hypothesis: How Science Shows That God Does Not Exist* (Amherst, NY: Prometheus Books, 2007).

10. See the points made in Frank Miller Turner, "The Victorian Conflict between Science and Religion: A Professional Dimension," *Isis* (1978) 69: 356–76; Colin A. Russell, "The Conflict Metaphor and Its Social Origins," *Science and Christian Faith* (1989) 1: 3–26.

11. For this development, see John Durant, "Evolution, Ideology, and World View: Darwinian Religion in the Twentieth Century," in *History, Humanity, and*

Evolution: Essays for John C. Greene, ed. James R. Moore (Cambridge: Cambridge University Press, 1989), 355–73.

12. Examples of such critics include Richard C. Lewontin, *Biology as Ideology: The Doctrine of DNA* (New York: HarperPerennial, 1992); Mary Midgley, *Evolution as a Religion: Strange Hopes and Stranger Fears,* 2nd ed. (London: Routledge, 2002).

13. Thomas Dixon, "Scientific Atheism as a Faith Tradition," *Studies in History and Philosophy of Science C* (2002) 33: 337–59.

14. For useful accounts of Dawkins' work and influence, see Alan Grafen and Mark Ridley, eds., *Richard Dawkins: How a Scientist Changed the Way We Think: Reflections by Scientists, Writers, and Philosophers* (Oxford: Oxford University Press, 2006).

15. The key paper was published in two parts in 1964: William Hamilton, "The Genetical Evolution of Social Behaviour," *Journal of Theoretical Biology* (1964) 7: 1–16, 17–52.

16. Richard Dawkins, *The Selfish Gene* (Oxford: Oxford University Press, 1976); Dawkins, *Blind Watchmaker.*

17. On this general issue, see Donald Symons, "On the Use and Misuse of Darwinism in the Study of Human Behavior," in *The Adapted Mind*, ed. J. H. Barkow, L. Cosmides, and J. Tooby (Oxford: Oxford University Press, 1992), 137–59.

18. For a critical account of his understanding of the relation between religious belief and evolutionary biology, see Alister E. McGrath, *Dawkins' God: Genes, Memes, and the Meaning of Life* (Oxford: Blackwell Publishing, 2004).

19. Richard Dawkins, "A Survival Machine," in *The Third Culture*, ed. John Brockman (New York: Simon & Schuster, 1996), 75–95. For reflections on the historical interaction of these three options, see David N. Livingstone, "Natural Theology and Neo-Lamarckism: The Changing Context of Nineteenth-Century Geography in the United States and Great Britain," *Annals of the Association of American Geographers* (1984) 74: 9–28.

20. For a vigorous statement of this position, see Dawkins' essay "Darwin Triumphant: Darwinism as Universal Truth," in Richard Dawkins, *A Devil's Chaplain* (London: Weidenfield & Nicholson, 2003), 78–90.

21. On which see the interesting collection of essays gathered together in Noretta Koertge, ed., *A House Built on Sand: Exposing Postmodernist Myths About Science* (New York: Oxford University Press, 1998).

22. For the difference between the two approaches, see Karl-Otto Apel, "The Erklären-Verstehen Controversy in the Philosophy of the Natural and Human Sciences," in *Contemporary Philosophy: A New Survey*, ed. G. Floistad (The Hague: Nijhoff, 1982), 19–49.

23. For an illuminating example, see Luke Davidson, "Fragilities of Scientism: Richard Dawkins and the Paranoid Idealization of Science," *Science as Culture* (2000) 9: 167–99. For a more general postmodern response to scientific criticism of

cultural accounts of science, see Brian Martin, "Social Construction of an 'Attack on Science,'" *Social Studies of Science* (1996) 26: 161–73.

24. See especially Gerald M. Edelman, *Neural Darwinism: The Theory of Neuronal Group Selection* (New York: Basic Books, 1987).

25. Daniel C. Dennett, *Kinds of Minds: Toward an Understanding of Consciousness* (New York: Basic Books, 1996).

26. Daniel C. Dennett, *Freedom Evolves* (New York: Viking, 2003).

27. Dennett, *Breaking the Spell*.

28. Daniel C. Dennett, *Darwin's Dangerous Idea: Evolution and the Meaning of Life* (New York: Simon & Schuster, 1995), 11.

29. Dennett, *Darwin's Dangerous Idea*, 21.

30. Dennett, *Darwin's Dangerous Idea*, 82.

31. This has been a major theme in the writings of John C. Greene: see especially his *Darwin and the Modern World View* (Baton Rouge: Louisiana State University Press, 1961).

32. See the 1992 debate at Southern Methodist University between evolutionists and supporters of the intelligent design movement over whether Darwinism carries with it "an a priori commitment to metaphysical naturalism": John Buell and Virginia Hearn, eds., *Darwinism: Science or Philosophy?* (Richardson, TX: Foundation for Thought and Ethics, 1994).

33. Dennett, *Darwin's Dangerous Idea*, 82.

34. Dawkins, "Darwin Triumphant," 81.

35. Timothy Shanahan, "Methodological and Contextual Factors in the Dawkins/Gould Dispute over Evolutionary Progress," *Studies in History and Philosophy of Science* (2001) 31: 127–51.

36. For example, see David Sloan Wilson, *Darwin's Cathedral: Evolution, Religion, and the Nature of Society* (Chicago: University of Chicago Press, 2002). See also Dawkins, *God Delusion*, 172–9.

37. Thomas H. Huxley, *Lay Sermons, Addresses, and Reviews* (London: Macmillan, 1870), 330.

38. See David Grumett, "Teilhard de Chardin's Evolutionary Natural Theology," *Zygon* (2007) 42: 519–34; Mary Ann Gillies, *Henri Bergson and British Modernism* (Montreal: McGill–Queen's University Press, 1996), 14–15. The idea of "entelechy," as formulated by Hans Driesch (1867–1941), should also be noted here: Horst H. Freyhofer, *The Vitalism of Hans Driesch: The Success and Decline of a Scientific Theory* (Frankfurt am Main: Peter Lang, 1982). The clearest statement of Driesch's approach is to be found in his "Kausalität und Vitalismus [Causality and vitalism]," *Jahrbuch der Schopenhauer-Gesellschaft* (1939) 26: 3–79.

39. On Monod, see Horace F. Judson, *The Eighth Day of Creation: Makers of the Revolution in Biology* (Cold Spring Harbor, NY: Cold Spring Harbor Laboratory Press, 1996).

40. Colin S. Pittendrigh, "Adaptation, Natural Selection, and Behavior," in *Behavior and Evolution*, ed. Anne Roe and George Gaylord Simpson (New Haven, CT: Yale University Press, 1958), 390–416 (394).

41. Richard Dawkins, *River Out of Eden: A Darwinian View of Life* (London: Phoenix, 1995), 133.

42. Ernst Mayr, *Toward a New Philosophy of Biology: Observations of an Evolutionist* (Cambridge, MA: Harvard University Press, 1988), 38–66; Ernst Mayr, *What Makes Biology Unique? Considerations on the Autonomy of a Scientific Discipline* (Cambridge: Cambridge University Press, 2007), 39–66.

43. Important representative contributions from historical and biological perspectives include Francis J. Ayala, "Teleological Explanation in Evolutionary Biology," *Philosophy of Science* (1970) 37: 1–15; Ernst Nagel, "Teleology Revisited: Goal Directed Processes in Biology," *Journal of Philosophy* (1977) 74: 261–301; David Kohn, "Darwin's Ambiguity: The Secularization of Biological Meaning," *British Journal for the History of Science* (1989) 22: 215–39; Philippe Huneman, "Naturalising Purpose: From Comparative Anatomy to the 'Adventure of Reason,'" *Studies in History and Philosophy of Science* (2006) 37: 649–74.

44. M. R. Bennett and P. M. S. Hacker, *Philosophical Foundations of Neuroscience* (Oxford: Blackwell, 2003), 374: "It is wrong-headed to suppose that the only forms of explanation are scientific." The entire section dealing with reductionism (355–77) merits close study.

45. For example, see William E. Carroll, "At the Mercy of Chance? Evolution and the Catholic Tradition," *Revue des questions scientifiques* (2006) 177: 179–204.

46. Peter B. Medawar, *The Limits of Science* (Oxford: Oxford University Press, 1985), 66.

47. Dennett, *Darwin's Dangerous Idea*, 394. For his distinction between "good" and "greedy" reductionism, see 81–2.

48. Dennett, *Darwin's Dangerous Idea*, 83.

49. These criticisms are developed especially in Richard Dawkins, *Unweaving the Rainbow: Science, Delusion, and the Appetite for Wonder* (London: Penguin Books, 1998).

50. A much more satisfactory account may be found in John Hedley Brooke, "The Relations between Darwin's Science and His Religion," in *Darwinism and Divinity*, ed. John Durant (Oxford: Blackwell, 1985), 40–75.

51. Edward Aveling, *The Religious Views of Charles Darwin* (London: Freethought, 1883).

52. For a thoughtful account of this, see James R. Moore, "Of Love and Death: Why Darwin 'Gave Up on Christianity,'" in *History, Humanity, and Evolution: Essays for John C. Greene*, ed. James R. Moore (Cambridge: Cambridge University Press, 1987), 195–229. For an account of the tragedy in question, see Randal Keynes, *Annie's Box: Charles Darwin, His Daughter, and Human Evolution* (Lon-

don: Fourth Estate, 2001). Note also the richly documented analysis of Janet Browne, *Charles Darwin: Voyaging* (New York: Knopf, 1995), 503, who argues that Annie's death marked the beginning of Darwin's long process of gradual dissociation from traditional Christianity.

53. There is a useful account in Frank Burch Brown, *The Evolution of Darwin's Religious Views* (Macon, GA: Mercer University Press, 1986).

54. See the idea of "generic divinity" found in the writings of Benjamin Franklin: David T. Morgan, "Benjamin Franklin: Champion of Generic Religion." *Historian* (2000) 62: 723–9. While having difficulties with the Christian idea of God, Franklin was no atheist.

55. See the important studies gathered together in John Hedley Brooke and Ian McLean, eds., *Heterodoxy in Early Modern Science and Religion* (Oxford: Oxford University Press, 2006).

56. For example, see the recent revisionist account of Timothy Larsen, *Crisis of Doubt: Honest Faith in Nineteenth-Century England* (Oxford: Oxford University Press, 2006). For a nuanced version of the received view, see the variety of opinions in Richard J. Helmstadter and Bernard V. Lightman, eds., *Victorian Faith in Crisis: Essays on Continuity and Change in Nineteenth-Century Religious Belief* (Stanford, CA: Stanford University Press, 1990).

57. For example, see David N. Livingstone, *Darwin's Forgotten Defenders: The Encounter between Evangelical Theology and Evolutionary Thought* (Grand Rapids, MI: Eerdmans, 1987); Jon H. Roberts, *Darwinism and the Divine in America: Protestant Intellectuals and Organic Evolution, 1859–1900* (Madison: University of Wisconsin Press, 1988); Arthur R. Peacocke, *Evolution, the Disguised Friend of Faith? Selected Essays* (Philadelphia: Templeton Foundation Press, 2004).

58. See especially Richard Dawkins, "Universal Darwinism," in *Evolution from Molecules to Men*, ed. D. S. Bendall (Cambridge: Cambridge University Press, 1983), 403–25.

59. See Lance Workman and Will Reader, *Evolutionary Psychology: An Introduction* (New York: Cambridge University Press, 2004). For its application to archaeology, see Steven J. Mithen, *The Prehistory of the Mind: The Cognitive Origins of Art, Religion, and Science* (New York: Thames and Hudson, 1999). On evolution and archaeology, see Stephen Shennan, *Genes, Memes, and Human History: Darwinian Archaeology and Cultural Evolution* (London: Thames & Hudson, 2002).

60. For a good account, see Peter J. Bowler, *The Eclipse of Darwinism: Anti-Darwinian Evolution Theories in the Decades Around 1900* (Baltimore: Johns Hopkins University Press, 1983).

61. The landmark works here are Leslie A. White, *The Science of Culture: A Study of Man and Civilization* (New York: Farrar, Straus, 1949); Julian H. Steward, *Theory of Culture Change: The Methodology of Multilinear Evolution* (Urbana: University of Illinois Press, 1963).

62. See Alex Mesoudi, Andrew Whiten, and Kevin N. Laland, "Is Cultural Evolution Darwinian? Evidence Reviewed from the Perspective of *The Origin of Species*," *Evolution* (2004) 58: 1-11.

63. Parallels that continue to be explored; see R. S. Wells, "The Life and Growth of Language: Metaphors in Biology and Linguistics," in *Biological Metaphor and Cladistic Classification*, ed. H. H. Hoenigswald and L. F. Wiener (Philadelphia: University of Pennsylvania Press, 1987), 39-80.

64. Note especially the contribution of Sir Edward B. Tylor, in works such as *Researches into the Early History of Mankind and the Development of Civilization* (London: John Murray, 1865).

65. See especially M. I. Sereno, "Four Analogies between Biological and Cultural/Linguistic Evolution," *Journal of Theoretical Biology* (1991) 151: 467-507.

66. Mesoudi et al., *Evolution* (2004) 58: 1.

67. Donald T. Campbell, "Blind Variation and Selective Retention in Creative Thought as in Other Knowledge Processes," *Psychological Review* (1960) 67: 380-400.

68. Donald T. Campbell, "A General 'Selection Theory' as Implemented in Biological Evolution and in Social Belief-Transmission-with-Modification in Science," *Biology and Philosophy* (1988) 3: 413-63. The term was introduced by Campbell in 1974; this article sets out a later exposition of the notion. See also F. T. Cloak, "Is a Cultural Ethology Possible?" *Human Ecology* (1975) 3: 161-81.

69. L. L. Cavalli-Sforza and M. W. Feldman, *Cultural Transmission and Evolution: A Quantitative Approach* (Princeton, NJ: Princeton University Press, 1981).

70. Charles J. Lumsden and Edward O. Wilson, *Genes, Mind, and Culture: The Coevolutionary Process* (Cambridge, MA: Harvard University Press, 1981).

71. Dawkins, *The Selfish Gene*, 193.

72. Dawkins, *The Selfish Gene*, 192.

73. Cloak, *Human Ecol.* (1975) 3: 161-81.

74. Dawkins's subsequent correction of this error should be noted: Richard Dawkins, *The Extended Phenotype: The Gene as the Unit of Selection* (Oxford: Freeman, 1981), 109. Note also the comments of William Harms, "Cultural Evolution and the Variable Phenotype," *Biology and Philosophy* (1996) 11: 357-75.

75. This point was made forcefully by James W. Polichak, "Memes—What Are They Good For?" *Skeptic* (1998) 6: 45-54. For a sympathetic yet ultimately critical review, see Joseph Poulshock, "Universal Darwinism and the Potential of Memetics," *Quarterly Review of Biology* (2002) 77: 174-5. See further Bennett and Hacker, *Philosophical Foundations of Neuroscience*, 431-5. For a somewhat uneven collection of essays offering rival evaluations of the "meme," see Robert Aunger, ed., *Darwinizing Culture: The Status of Memetics as a Science* (Oxford: Oxford University Press, 2000).

76. See here Alister E. McGrath, "The Evolution of Doctrine? A Critical Examination of the Theological Validity of Biological Models of Doctrinal Devel-

opment," in *The Order of Things: Explorations in Scientific Theology* (Oxford: Blackwell Publishing, 2006), 117–68.

77. Dawkins, *God Delusion*, 196.

78. Dennett, *Darwin's Dangerous Idea*, 346.

79. Dennett, *Darwin's Dangerous Idea*, 366, summarizing Dawkins' approach. See also 370–1: "If human minds are nonmiraculous products of evolution, then they are, in the requisite sense, artifacts, and all their powers must have an ultimately 'mechanical' explanation. We are descended from macros and made of macros, and nothing we can do is anything beyond the power of huge assemblies of macros."

80. See the discussion of the "new replicators": Dennett, *Breaking the Spell*, 341–57, especially the taxonomy of memes on 344.

81. See especially Mark Lake, "Digging for Memes: The Role of Material Objects in Cultural Evolution," in *Cognition and Material Culture: The Archeology of Symbolic Storage*, ed. Colin Renfrew and Chris Scarre (Cambridge: McDonald Institute for Archaeological Research, 1998), 77–88.

82. Dan Sperber, *Explaining Culture: A Naturalistic Approach* (Oxford: Blackwell, 1996).

83. Alan Costall, "The 'Meme' Meme," *Cultural Dynamics* (1991) 4: 321–35.

84. Dorothy Nelkin, "Less Selfish Than Sacred? Genes and the Religious Impulse in Evolutionary Psychology," in *Alas Poor Darwin: Arguments against Evolutionary Psychology*, ed. Hilary Rose and Steven P. R. Rose (London: Jonathan Cape, 2000), 14–27.

85. For a detailed analysis of the difficulties, see Liane Gabora, "Ideas Are Not Replicators but Minds Are," *Biology and Philosophy* (2004) 19: 127–43.

86. For example, see Charles Kingsley, "The Natural Theology of the Future," in *Westminster Sermons* (London: Macmillan, 1874), v–xxxiii, especially the statement "We knew of old that God was so wise that He could make all things: but behold, He is so much wiser than even that, that He can make all things make themselves."

87. For the historical aspect of this question, see James R. Moore, *The Post-Darwinian Controversies: A Study of the Protestant Struggle to Come to Terms with Darwin in Great Britain and America, 1870–1900* (Cambridge: Cambridge University Press, 1979).

88. John Hick, *An Interpretation of Religion: Human Responses to the Transcendent* (London: Macmillan, 1989), 73: Nature "is capable from our present human vantage point of being thought of in both religious and naturalistic ways."

89. See Abigail Lustig, "Natural Atheology," in *Darwinian Heresies*, ed. Abigail Lustig, Robert J. Richards and Michael Ruse (Cambridge: Cambridge University Press, 2004), 69–83. The term "natural atheology" appears to have been coined by Alvin Plantinga: see Alvin Plantinga, *God and Other Minds: A Study of the Rational Justification of Belief in God* (Ithaca, NY: Cornell University Press, 1990), 115–85.

90. John C. Greene, *Science, Ideology, and World View: Essays in the History of Evolutionary Ideas* (Berkeley: University of California Press, 1981), 162.

91. See Dixon, *Studies Hist. Philos. Sci. C* (2002) 33: 337–59.

92. For a recent attempt to do this, see Alister E. McGrath, *The Open Secret: New Vision for Natural Theology* (Oxford: Blackwell, 2008), especially 198–209.

93. See the examples of such dogmatism provided by John Cornwell, *Darwin's Angel: An Angelic Riposte to "The God Delusion"* (London: Profile Books, 2007), 77–84, 99–110.

Contributors

DENIS R. ALEXANDER is director of the Faraday Institute for Science and Religion at St. Edmund's College, Cambridge, where he is a Fellow. He read biochemistry at Oxford University before researching for a Ph.D. in neurochemistry at the Institute of Psychiatry, University of London. Following this he spent fifteen years in academic positions in the Middle East, most recently (1981-86) as associate professor of biochemistry at the American University of Beirut, Lebanon. Upon his return to the UK he worked at the Imperial Cancer Research Fund (now Cancer Research UK) and from 1989 to 2008 at the Babraham Institute, Cambridge, where he was chair of the Molecular Immunology Programme and Head of the Laboratory of Lymphocyte Signalling and Development. Dr. Alexander has published numerous articles and reviews, particularly in the research field of lymphocyte signaling and development, most recently in the *New England Journal of Medicine* (2008). He is also editor of the journal *Science & Christian Belief* and contributes papers as part of the Cambridge Papers writing group. He is the author of *Rebuilding the Matrix: Science and Faith in the 21st Century* (Lion, 2001) and *Creation or Evolution: Do We Have to Choose?* (Monarch, 2008). He has also edited *Can We Be Sure about Anything? Science, Faith and Postmodernity* (Apollos, 2005) and co-authored (with Robert S. White) *Beyond Belief: Science, Faith and Ethical Challenges* (Lion, 2004).

PETER HARRISON is the Andreas Idreos Professor of Science and Religion at the University of Oxford. he has published extensively in the area of cultural and intellectual history, and has a particular interest in the philosophical, scientific, and religious thought of the sixteenth and seventeenth centuries. His books include *"Religion" and the Religions in the English Enlightenment* (Cambridge University Press, 1990), *The Bible, Protestantism, and the Rise of Natural Science* (Cambridge University Press, 1998), and *The Fall of Man and the Foundations of Science* (Cambridge University Press, 2007). A fellow of Harris Manchester College and director of the Ian Ramsey Centre, Oxford, his research presently focuses on disciplinary boundaries in early modern science, and on changing conceptions of science and of religion in the modern period.

NIKOLAI KREMENTSOV is associate professor in the Institute for the History and Philosophy of Science and Technology, Toronto University, Canada, and was previously in the Institute of the History of Science and Technology, the USSR Academy of Sciences. He has held a wide range of visiting fellowships, including those at the Massachusetts Institute of Technology, the Kennan Institute of Advanced Russian Studies (Washington, DC), the University of Edinburgh, the École des Hautes Études en Sciences Sociales (Paris), and Indiana University. Dr. Krementsov is the author of *Stalinist Science* (Princeton University Press, 1997), *The Cure: A Story of Cancer and Politics from the Annals of the Cold War* (University of Chicago Press, 2002), and *International Science between the World Wars: The Case of Genetics* (Routledge, 2005). His current research interests include the history of biomedical sciences in 1920s Russia and the history of Cold War science.

EDWARD J. LARSON is the University Professor in History and holds the Darling Chair in Law at Pepperdine University and is a senior fellow with the Institute of Higher Education at the University of Georgia. An author of seven books, co-author or co-editor of seven books, and author or co-author of over 100 published articles, Professor Larson writes mostly about issues of law, science, and medicine from a historical perspective. His first book, *Trial and Error: The American Controversy Over Creation and Evolution* (Oxford University Press, 1985, with expanded editions 1989 and 2002), chronicles the legal battles over teaching evolution in American public schools. His second book, *Sex, Race, and Science: Eugenics in the Deep South* (Johns Hopkins University Press, 1995), examines the legislative history of eugenics. For his 1997 book, *Summer for the Gods: The Scopes Trial and America's Continuing Debate Over Science and Religion* (Basic Books), Larson became the first sitting law professor to receive the Pulitzer Prize in History. In 2001, Larson published *Evolution's Workshop: God and Science on the Galapagos Islands* (Basic Books). His other recent publications include *Evolution: The Remarkable History of a Scientific Theory* (Random House, 2004) and *A Magnificent Catastrophe: The Tumultuous Election of 1800, America's First Presidential Campaign* (Free Press, 2007). His areas of expertise are history of biology and medicine, health, science, and technology law, and legal history.

ALISTER E. MCGRATH is professor of Theology, Ministry and Education at King's College, London, and has an academic background in both theology and biophysics. Professor McGrath is a prolific author and his best-known book is probably *Dawkins' God: Genes, Memes and the Meaning of Life* (Blackwell, 2004). Other recent books include *The Foundations of Dialogue in Science and Religion* (Blackwell, 1998), *Thomas F. Torrance: An Intellectual Biography* (T&T Clark, 1999), *A*

Scientific Theology, in three volumes (T&T Clark, 2001–3), *The Re-enchantment of Nature* (Hodder & Stoughton, 2002), *The Science of God: An Introduction to Scientific Theology* (T&T Clark, 2004), *The Twilight of Atheism* (Rider, 2004), *The Order of Things: Explorations in Scientific Theology* (Blackwell, 2006), *The Dawkins Delusion? Atheist Fundamentalism and the Denial of the Divine* (SPCK, 2007), and *The Open Secret: A New Vision for Natural Theology* (Blackwell, 2008). From 2006 to 2008, Professor McGrath was Senior Research Fellow at Harris Manchester College, Oxford, allowing him to concentrate on several major research projects, including the reformulation and renewal of natural theology, the 2009 Gifford Lectures at Aberdeen University, and the 2009–10 Hulsean Lectures at Cambridge University.

ERIKA LORRAINE MILAM is assistant professor in the Department of History, University of Maryland, and previously held a postdoctoral fellowship at the Max Planck Institute for the History of Science in Berlin. She is author of *Looking for a Few Good Males: Female Choice in Evolutionary Biology* (Johns Hopkins University Press, 2010). Her current research project, "Animal models of behavior: anthropomorphism, zoomorphism, and cultures of observation," addresses how and why zoological and primatological research on animal behavior came to compete with anthropological studies of human cultures as methods of providing reliable information about human nature.

RONALD L. NUMBERS is the Hilldale Professor of the History of Science and Medicine at the University of Wisconsin–Madison. He has written or edited more than two dozen books, including, most recently, *Science and Christianity in Pulpit and Pew* (Oxford University Press, 2007) and *Galileo Goes to Jail and Other Myths about Science and Religion* (Harvard University Press, 2009). For five years (1989–1993) he edited *Isis*. He is completing a history of science in America (for Basic Books), editing a series of monographs on the history of medicine, science, and religion for the Johns Hopkins University Press, and co-editing, with David Lindberg, the eight-volume *Cambridge History of Science*. A former Guggenheim Foundation Fellow, he is a fellow of the American Academy of Arts and Sciences and of the American Association for the Advancement of Science, and an effective member of the International Academy of the History of Science. He is a past president of both the American Society of Church History and the History of Science Society, which recently awarded him the Sarton Medal for lifetime scholarly achievement. In 2005 he was elected to a four-year term as president of the International Union of History and Philosophy of Science/Division of History of Science and Technology.

PETER HANNS REILL is professor and director of the Center for 17th and 18th Century Studies and William Andrews Clark Memorial Library at the University of California–Los Angeles. His books include *The German Enlightenment and the*

Rise of Historicism (University of California Press, 1975), *Visions of Empire: Voyages, Botany, and Representations of Nature* (co-editor, Cambridge University Press, 1996), *What's Left of the Enlightenment?: A Postmodern Question* (co-editor, Stanford University Press, 2001), *Vitalizing Nature in the Enlightenment* (University of California Press, 2005), and *Discourses of Tolerance and Intolerance in the Enlightenment* (co-editor, University of Toronto Press, 2008). Professor Reill's research interests are in modern European history, intellectual history, the history of science, Germany, and philosophy of history.

SHIRLEY A. ROE is professor of history and head of the History Department at the University of Connecticut. Her publications include "Radical Nature in the *Encyclopédie*," in *Science, History, and the Social Role of the Man of Knowledge*, edited by Garland Allen and Roy MacLeod (Kluwer Academic Publishers, 2002); *Science Against the Unbelievers: The Correspondence of Bonnet and Needham, 1760–1780*, with Renato G. Mazzolini (Studies on Voltaire and the Eighteenth Century, vol. 243; Oxford: The Voltaire Foundation, 1986); *Matter, Life, and Generation: Eighteenth-Century Embryology and the Haller–Wolff Debate* (Cambridge University Press, 1981). Professor Roe's current research interests include the social/political context of controversies over biological materialism in the late eighteenth century, the formation of racial views of humans, and science and social issues.

NICOLAAS RUPKE is Lower Saxony research professor of the History of Science at Göttingen University. His monographs include *Richard Owen: Biology without Darwin* (University of Chicago Press, 2009), and *Alexander von Humboldt: A Metabiography* (University of Chicago Press, 2008). Among his edited volumes is *Eminent Lives in Twentieth-Century Science and Religion* (Peter Lang Verlag, 2009). He is currently writing a monograph about the non-Darwinian tradition in nineteenth- and twentieth-century biology. Rupke is a fellow of Germany's National Academy of Sciences Leopoldina and of the Göttingen Academy of Sciences.

MICHAEL RUSE is the Lucyle T. Werkmeister Professor of Philosophy, Florida State University, was Gifford Lecturer, University of Edinburgh, Glasgow, Scotland, in 2001, and is a fellow of the Royal Society of Canada. Ruse has published widely on the evolution controversy in North America; his books include the following recent titles: *Debating Design: Darwin to DNA* (co-edited with William Dembski, Cambridge University Press, 2003), *Darwinian Heresies* (co-edited with A. Lustig and R. Richards, Cambridge University Press, 2003), *Darwin and Design: Does Evolution Have a Purpose?* (Harvard University Press, 2003), *The Evolution Wars: A Guide to the Debates* (Rutgers University Press, 2000), *Can a Darwinian be a Christian? The Relationship between Science and Religion* (Cambridge University Press, 2001), *The Darwinian Revolution: Science Red in Tooth and Claw* (2nd ed.,

University of Chicago Press, 1999), and *Taking Darwin Seriously: A Naturalistic Approach to Philosophy* (2nd ed., Blackwell, 1988). Ruse has continuing research interests in Darwinism and its reception, the philosophy and ethics of Darwinism, and the interactions between science and religion.

SUJIT SIVASUNDARAM is lecturer in International History at the London School of Economics and Political Science. He is completing a book about the impact of British colonization in Sri Lanka, with particular attention to science and medicine. He is the author of *Nature and the Godly Empire: Science and Evangelical Mission in the Pacific, 1795–1850* (Cambridge University Press, 2005). This was hailed as the first substantive account of the entanglement of science and Christianity outside the Western world in the nineteenth century. He has published articles in leading journals on a range of topics on the history of the British Empire and has held lectureships at the University of Cambridge and University College London.

JONATHAN R. TOPHAM is senior lecturer in History of Science at the University of Leeds. Among his co-publications are *Science in the Nineteenth-Century Periodical: Reading the Magazine of Nature* (Cambridge University Press, 2004), *Science in the Nineteenth-Century Periodical: An Electronic Index* (HRI Online, 2005), *Culture and Science in the Nineteenth-Century Media* (Ashgate, 2004), and two volumes of the *Correspondence of Charles Darwin* (Cambridge University Press, 1985–). His research relates mainly to the cultural history of science in early nineteenth-century Britain, and to the history of the life and earth sciences. He has published extensively on science and religion in early nineteenth-century Britain—more especially on natural theology and theologies of nature—and intends to complete a monograph on the Bridgewater Treatises shortly. He has also helped to pioneer a new approach to the subject of science and its publics, drawing particularly on the historiography of the book, and is currently preparing a book-length study of science and print culture in Britain, 1789–1832.

PAUL WEINDLING is the Wellcome Trust Research Professor in the History of Medicine in the Department of History, Oxford Brookes University. His publications include *Health, Race and German Politics between National Unification and Nazism* (Cambridge University Press, 1989), *Epidemics and Genocide in Eastern Europe 1890–1945* (Oxford University Press, 2000), and *Nazi Medicine and the Nuremberg Trials: From Medical War Crimes to Informed Consent 1945–55* (Palgrave Macmillan, 2004; paperback 2006). He recently completed *John W. Thompson: Psychiatrist in the Shadow of the Holocaust* (Rochester University Press, 2009). He edited a volume on *International Health Organisations and Movements 1918–1939* (Cambridge University Press, 1995), co-edited with Marius Turda *Blood and Homeland: Eugenics and Racial Nationalism in Central and Southeast Europe, 1900–1940* (CEU

Budapest, 2007), and edited the journal *Social History of Medicine* (1992–8). His research interests include international health organizations in the twentieth century, the medical emigration to Britain in the 1930s and '40s, and Nazi medical war crimes. He directs an Arts and Humanities Research Council project grant to reconstruct the life histories of victims of Nazi human experiments.

Index

Note: Italicized page numbers refer to figures.

Abderhalden, Emil, 205
Aberle, Sophie D., 409
abiogenesis, 146, 161–62
abolitionism, 120–23, 127
Aborigines Protection Society, 120, 124, 129
abortion, 193, 202
Academy of Natural Sciences of Philadelphia, 126
Adams, Mark B., 396, 398, 400
adaptation: functional, 88–89, 96–97, 100–103, 105–6, 112; perfect, 88, 102, 104, 112
Adventism, 314–15
Aesop, 20
Agassiz, Louis, 121–22, 143–44, 310
agnosticism, 111, 343, 350
Agol, Isaak, 230–31, 234, 236, 398
agrobiology, 239–41, 244
Aizupet, M. I., 237
Alexander, Cecil F., 404
Allen, Elizabeth, 412
Altick, Richard, 376
Altman, Jeanne, 298
Ambrose of Milan, 17
American Association for the Advancement of the Science (AAAS), 311–12, 314
American Breeders Association, 184
American Eugenics Society, 179, 186–88
American Genetics Association, 184
American Museum of Natural History, 312
American Neurological Association, 190
analogy, 84; between a watch and objects in nature, 51, 88–89, 95. *See also* body, as analogy

anarchism, 218
anatomy, 260, 263; comparative, 7, 97–98, 101–2, 106, 112, 122–23; Cuvierian, 135
Anderson, Wilda, 367
Andrews, James T., 395
animal film stars, 293
animal mechanism, 23–24
animal nature, and human culture, dichotomy, 298, 300
Anstey, Peter, 363
anthropogeography, 153
Anthropological Society of London, 124–25
anthropology, 122, 124, 135, 147, 152, 192, 200; biological, 204, 210
anti-Darwinism, 10, 310
Anti-Defamation League, 213
antimiscegenation statutes, 176
anti-Semitism, encyclical against, 200
Appel, Toby A., 377, 402
Aquinas, Thomas, 25
Arber, Werner, 420
archaeology, 124, 210, 345, 425
Ardrey, Robert, 411
Aristotle, 358, 401
Arnold, Matthew, 271
Aryanism, 131
Ashworth, William B., Jr., 359
astrology, 133
astronomy, 13, 26, 44, 84, 89, 95, 97, 108
atheism, 7, 26, 36–39, 43, 49, 51–52, 60, 95, 304, 308, 325, 330, 338, 342, 348, 350; natural, 350; scientific, 331

atheist apologetics, in the early twenty-first century, 329–31, 341, 346, 349–50
atheology, natural, 350
Atomic Energy Commission, 269
Augustine of Hippo, 17
Aunger, Robert, 426
Auschwitz, 195, 200, 205, 209–12
Austen, Ralph, 359
autogenesis, 8, 144–60, 162; as a German theory, 156
Aveling, Edward, 342
Ayala, Francis J., 424

Baartman, Sara, 119–20, 124
Baatz, Simon, 416
Babbage, Charles, 98, 110–11
baboons, 298, *299*
Bachman, John, 121
Bacon, Francis, 19, 29–34, *30*, 69
bacteriology, 200
Baer, Karl Ernst von, 147, 407
Ballantyne, Tony, 379
Banks, Joseph, 128
Banton, Michael, 377
baramins, 317
Barlow, Nora, 372
Barnes, Barry, 356
Baron, Samuel H., 395
Barthez, Paul, 64
Bartholomaeus Anglicus, 17
Bassin, Mark, 395
Bateman, A. J., 413
Bates, Henry Walter, 261, 263
Bateson, William, 169, 189, 218, 311–12, 314
Bathybius debacle, 162
Bauman, Zygmunt, 200
Baur, Erwin, 185, 208
Baur, Ferdinand Christian, 5
Bayle, Pierre, 36, 60
Beach, Frank Ambrose, 410
Beagle, 8, 135–36, *142*, 143, 258, 261
Beale, Lionel S., 310
beast-machine, Cartesian doctrine of, 21–24
Beckwith, Barbara, 412
Beckwith, Jon, 412
Beddoe, John, 124
Beer, Gillian, 379
Beeson, David, 365
behavior: animal, 225, 273, 277, 283, 287–88, 290–91, 293, 297; and evolutionary theory, 291, 332; goal-directed, 339; human, 260, 277, 287–90, 292–94, 297, 301; maternal, 298, 409; mating, 286 (*see also* mating displays; mating rituals); sexual, 289–91
Behe, Michael J., 326–27
Beketov, Andrei N., 217, 394
Belardelli, F., 394
Bell, Alexander Graham, 184
Bell, Charles, 97, 100–104, 106–7, 112
Bell, Daniel, 356
Bender, Bert, 408
Bennett, Max, 340, 426
Bentham, George, 260, 375
Bergman, Toberg Olaf, 64
Bergson, Henri, 338
Berlin, Isaiah, 69
Berlin Academy of Sciences, 48
Besterman, Theodore, 366
Bible, 91–92, 196, 319–20, 322
biblical passages: Genesis, 104, 116, 143, 308, 315, 317, 319, 325, 330, 349; Job, 358; Proverbs, 17, 21; Psalms, 24; Romans, 24
Bindman, David, 376
Binet-Simon tests. *See* intelligence tests
biochemistry, 326; under Nazism, 205
biodiversity, 278
biogenetic law, 263, 279
biogeography, 151–52
biography, 140
biologists, Jewish, 194
biology, 139, 147, 158, 195, 197; abuse of, 1, 3, 8, 61, 287, 333, 351; creationist, 142; evolutionary, 140, 143, 159, 161; German, 197, 200; globalization of, 116; history of, 6; molecular, 163; and natural theology, 88–113; and Nazi race theory, 193–94, 201, 203–5, 208–10, 212, 214; racial, 200–201; in the Soviet Union, 217, 244; as a surrogate for revealed religion, 198
biometry, 169
Birks, J. B., 358
Black, Joseph, 64, 70–71
Blacker, C. P., 213
Black Stork, The (movie), 186, *187*
Blackwell, Antoinette Brown, 281
Blair, Ann, 357–58
Blaisdell, Muriel, 374, 421

INDEX 437

Blake, C. Carter, 374
Bleier, Ruth, 412
Blest, David, 411
blood lines, 130, 171; and purity, 203, 206–9
Blood Protection Law, 203, 208
Bloor, David, 5
Blount, Thomas, 20
Blum, Deborah, 409
Blumenbach, Johann Friedrich, 64, 117–18, 146–48, 153–54, 370
Boas, Franz, 190, 204
Bock, Gisela, 392
body, as analogy, 84–85
body politick, 73–75, 79–80
Boehme, Jacob, 195, 359
Boerhaave, Hermann, 43
Bogdanov, Aleksandr, 220
Bole, S. James, 312
Bolsheviks, 220–24, 236–37
Bonar, James, 370
Bonhoeffer, Dietrich, 201
Bonnet, Charles, 37–38, 46–48, 50–51, 60, 64
bonobos, 298, 300
book of nature, 19, 27, 317
Boswell, 69
botany, 35, 142
Bougainville, Louis-Antoine de, 128
Bowler, Peter J., 364, 374, 384, 396, 425
Box, Joan F., 421
Boyle, Robert, 5, 20, 25, 27, 29, 363
Boyle Lectures, 25
brain research, 192
Brandeis, Louis D., 181
Brattain, Michelle, 406
Bridgewater Treatises, 25, 89, 92–93, 96–112, 261
Briggs, John Canning, 362
British Association for the Advancement of Science, 4
Brömer, Rainer, 380
Bronn, Heinrich Georg, 149, 155, 158, 381
Brooke, John Hedley, 90, 372–75, 424–25
Brougham, Henry, 97
Brown, Arthur I., 312–13
Brown, Frank Burch, 425
Brown, Vivienne, 369
Browne, Janet, 140, 379, 425
Bryan, William Jennings, 307–9, 316
Buch, Leopold von, 149
Buchanan, Francis, 134

Buchner, Ludwig, 221
Buckland, William, 97, 100–104, 106–7, 109, 111–12, 144
Buddhism, 114–15
Buell, John, 423
Buffon, Comte de (Georges-Louis Leclerc), 38, 43–47, 46, 49–53, 56, 59, 64, 67–68, 84, 117–18, 133, 144, 365
Buican, D., 394
Bukharin, Nikolai, 238–39
Bullough, Vern, 410
Burbank, Luther, 184
Burdach, Karl Friedrich, 147, 152–53
Burkhardt, Richard W., Jr., 409–10
Burleigh, Michael, 392
Burmeister, Hermann, 149, 151
Bury, J. B., 401
Bush, George W., 350
Butenandt, Adolf, 194
Butler, Larry G., 323
Butler, Pierce, 181
Bynum, William F., 376

Cacouacs, 57–59
Cain, A. J., 268
Cairns-Smith, Alexander Graham, 162
Calculating Engine, 98, 110–11
Calcutta Phrenological Society, 133
Campbell, Donald T., 345
Campbell, George, 131
Campbell, Roy Hutchenson, 83, 86, 369
Camper, Peter (Pieter), 64, 117–18
Candolle, Alphonse de, 152, 155
Canning, George, 254
Cannon, Susan Faye, 355
Cannon, W. F., 375, 402
Cantor, Geoffrey, 373
Carnegie Institution, Cold Spring Harbor genetics research station, 174, 177, 184, 191
Carroll, William E., 424
Carr-Saunders, A. M., 204
Carter, Julian B., 409
cartography, 166
Carus, Carl Gustav, 153–54
Casaubon, Meric, 28, 32, 363
Castle, Otis H., 388
Castle, William E., 184
castration, 180
catechism, 58

Cavalli-Sforza, Luigi Luca, 345
Center for Cognitive Studies, 334
Center for Interdisciplinary Studies, 420
Chain of Being. *See* Great Chain of Being
Chalmers, Thomas, 99–100, 107–8
Chambers, Robert, 111, 125, 154, 253, 257
Chapman, Matthew, 420
charity, Christian, as the goal of the study of nature, 20, 31, 362
charlatans, 2, 257, 340
Charron, Pierre, 19
chemistry, 64, 98; physiological, 146, 163
Chesterton, G. K., 188–89
Chetverikov, Sergei, 227–228, 239
Chorover, Steven, 412
Christian dogma, 5
Christian ethics, 200, 316
Christianity, 92, 94, 99–100, 130, 196–97, 304, 307, 343, 349–50; and science, 196–97
Christian tradition, 17
Churchill, Winston, 185
Cicero, 360
Circle of Materialist Physicians, 228, 230
Clark, Austin H., 312
Clark, Harold W., 316–17
Clark, John F. M., 356
Clark, J. W., 402
Clarke, Adele E., 410
classification systems, 35, 260
climate, influence on human variation, 118–20, 122, 133–34, 137
Cloak, F. T., 346–47, 426
Cockburn, Patrick, 361
Cold War, 213, 244
Colebrooke, Thomas Henry, 131
Coleman, William, 377, 402
collectivization, 236, 240
colonialism, 122, 127, 137
colony trade, 79
Combe, George, 115, 122, 133, 377
Commerçon, Philibert, 128
Communist Academy (Socialist Academy), 222, 228–30, 232, 234–38
Communist Party, 229; Central Committee, 222
comparative analysis, 67
Confessing Church, 200–201
Conklin, Edwin G., 389, 417
Conry, Yvette, 421

consciousness, 335
constructivism, 4–6
Cook, Captain (James), 128
Cook, Harold J., 359
Cook, R. C., 393
Cooke, A. B., 388
Coolidge, Calvin, 184
Cooter, Roger, 375
Copley, Stephen, 369
Corner, George W., 409
Cornwell, John, 428
Corsi, Pietro, 372
Cosans, Chris, 383
Costall, Alan, 427
Cottingham, John, 358, 360
Counter-Enlightenment, 69
courtship rituals. *See* mating rituals
Cowan, Ruth Schwartz, 384
Cravens, Hamilton, 389
creation, 42, 75–76, 104, 106, 148; Christian view, 329; divine (God's role or divine plan), 37, 91–92, 106–7, 110; versus evolution, 158, 160; history of, 112; progressive, 100, 104, 112; research, 322; science, 324; teaching in public schools, 142
creationism, 9–10, 142–44, 149, 310, 312, 315–16, 318–19, 321, 336; lunatic fringe, 323; scientific, 324–25; young-earth, 144, 149
Creation Museum, *324*
Creation Research Society (CRS), 321, 323
Creation Science Research Center, 323
Crimean War, 263
Crist, Eileen, 410, 412
Cronin, Helena, 407
Crook, D., 391
crystallization, 148
Cullen, William, 64, 70
cultural transmission, 295, 345–46, 348
Culver, David, 412
Curse of Ham, 121, 130
Curtis, W. C., 417
Cuvier, Georges, 101–3, 105–6, 120, 123–24, 126, 143, 149, 255, 261
Czolbe, Heinrich, 144

d'Alembert, Jean le Rond, 33, 55–56, 367
Damilaville, Étienne-Noël, 49, 53
Danilevskii, N. Ia., 394
Darré, Richard Walter, 208

INDEX 439

Darwin, Charles, 7–8, 88–89, 94, 98, 111–12, 125, 127, 135–37, 140–44, 157–64, 165, 167, 171–72, 195–99, 203, 213, 217–18, 220, 223–24, 227, 234, 237–39, 242, 248–49, 257–63, 266–68, 274, 302–4, 308–9, 326–27, 330–31, 335–36, 338, 341, 345, 349, 406; and the death of his daughter Annie, 342; his religious beliefs, 342–44; historiography, 160; theory of sexual selection, 277–80, 282, 296, 301 (*see also* sexual selection)
Darwin, Erasmus, 96, 248, 252, 254–55, 257–58, 262
Darwin, Francis, 372
Darwin, Leonard, 174, 189
Darwinism, 8–9, 115, 124–25, 135, 137, 139–43, 156–64, 168, 193, 196–97, 203, 213, 302, 329; alternative to, 143–46 (*see also* autogenesis); as an antitheistic ideology, 331, 333; Christian, 304; in the early twenty-first century, 330; equated with atheism, 303; failure of, 306, 314; German, 198, 307; and materialism, 305; rejected, 320, 328; social, 189, 199–200, 219; Soviet Creative, 215, 239, 243–44; in the Soviet Union, 216–24, 227–35, 237–39, 242–45; teaching of, 307–8; universal, 10, 344–46, 349; as a worldview, 334, 336–39, 350–51
Das, Kali Kumar, 133
Daubenton, Louis, 34
Davenport, Charles B., 177, 183–184, 188, 386, 417
David-Fox, Michael, 395
Davidson, Luke, 422
Davis, Joseph, 124
Dawkins, Richard, 10, 88, 294–97, 327, 329–39, *332*, 341–42, 344, 346–50, 357, 421
Dawson, John William, 310
Dawson, Virginia P., 365
death camps, 1, 171
Deborin, Abram, 232
De Brosses, Paul, 128
Declaration on Human Rights (United Nations), 213
degeneracy, 171–74, 177, 190, 204
Degler, Carl N., 389
Deichmann, Ute, 205, 390
deism, 37, 48–50, 52, 55–56, 92
Dembski, William, 326, 420

democracy, 20, 289, 307
Dennert, Eberhard, 306
Dennett, Daniel, 10, 329–30, 334–37, *335*, 341, 348–50, 421
Descartes, René, 7, 21–25, 22, 38–40, 43, 52, 84
design in nature, 88–92, 94, 100, 313, 339. *See also* divine design
Desmond, Adrian, 140, 357, 407
despotism, 54, 59
determinism, 156, 200; biological, 293, 295
Deutsch, Albert, 178, 387
Deutsche Forschungsgemeinschaft, 205
Deutscher, Penelope, 408
de Vries, Hugo, 169, 173, 218, 228
Dewar, Douglas, 320–21
Dewsbury, Donald A., 391
d'Holbach, Paul-Henri Thiry, 37, 45–46, 50–52, 54, 60
dialectical materialism, 222–23, 230–31, 234, 238
Diderot, Denis, 37–39, 45–46, 48–50, 52–60, 252
Digby, Kenelm, 24
Di Gregorio, Mario A., 380
Diogenes Laertius, 358
Discovery Institute, 325, 327–28
divination, 67
divine agency, 91, 101, 108, 327, 341
divine design, 24–25, 90, 98
Dixon, Thomas, 422, 428
DNA, 334, 342
Dobzhansky, Theodosius, 227, 243, 267–69, 318–20, 330, 397
Dodson, Michael, 379
double standards, sexual, 289–90
Dreger, Alice, 408
Drescher, Seymour, 376
Drickamer, Lee C., 413
Driesch, Hans, 423
Driver, Felix, 378
Du Chaillu, Paul Belloni, 125–26
Duchinskii, F., 397–99
Dugdale, Richard Louis, 171–75, 177
Duncan, Margaret, 412
Dupree, A. Hunter, 414
Durant, John, 421
d'Urville, Jules Dumont, 129

earth, physical history, 102, 104, 109
East, Edward M., 184

440 INDEX

East India Company, 130
Eddy, Matthew, 95
Edelman, Gerald M., 335, 423
Edinburgh Philosophical Society, 71
Edinburgh University Science Studies Unit, 4–5
Edwards, John, 20–21, 27, 361
Edwards, Karin, 360
Edwards, Mark, 358
Egerton, Francis Henry, 92
Ehrenberg, Christian Gottfried, 163
Eibl-Eibesfeldt, Irenäus, 293, 411
Einstein, Albert, 326, 336
elephants, 134–35
Eliot, Charles W., 2
Ellis, Havelock, 189
Ellis, William, 129
embryology, 194, 198, 260, 263
Empedocles, 252
empiricism, 119, 124–25, 127, 133; and race, 115; and science, 115
Encyclopédie, 39, 52, 54–57, 59
Engels, Friedrich, 215, 219, 221–23, 231–32
Enlightenment, 252; definition, 63
Enlightenment vitalism, 7, 64–71, 74, 76, 80, 83–85, 87
entelechy, 423
Epictetus, 360
epigenesis, 37, 43–46, 66, 146, 148, 370
Esdaile, James, 135
Estabrook, Arthur H., 385–86
Etheridge, Dr., of the British Museum (Robert Etheridge, Junr.), 310–11
ethnography, 134
Ethnological Society of London, 124–25
ethnology: Christian, 124; missionary, 129–30, 137
ethology, 194, 288, 290–93, 297, 332, 410–11; pop, 288
eugenic medicine, 201
eugenics (eugenic programs), 1, 8–9, 165–66, 169–72, 185, 195, 197, 200, 202, 205–6, 213, 245, 268, 282, 295, 301; British, 213; exterminatory, 206; laws (legislation), 185, 385 (*see also* sterilization: laws); modern, 185, 190–91; movements, 197; movements, American, 175–77, 180, 184; Nazi, 193, 203, 209–10; negative, 170, 174–75; popular support, 185–86; positive, 170, 173; racial, 208–9; and religion, 186–89; socialist, 230, 239, 242
Eugenics Education Society, 174
Eugenics Record Office, 174, 182–83, 191
euthanasia, 180, 185, 192, 197, 206; Nazi, 198, 209–10
evangelicalism, 91–92
Evans, Richard J., 392
evolution, 9, 36, 144, 157–58, 160, 162, 192; and abortion, 323; behavioral, 411; and Calvinism, 304; and Christianity, 307; and Christian theology, 304, 316; cultural, 292, 344–46; Darwinian, 141, 143, 201; Darwin's theory of, 112, 329, 334–35; definition, 247; and democracy, 307; and the doctrine of biblical inerrancy, 304; and feminism, 324; genetic, 346; and humans, 171, 266, 287, 305; human versus animal, 283; memetic, 346; not a science, 309–10; as a pseudo-science, 274; and race, 197; social, 220; in the Soviet Union, 216–19, 221–24, 232–35, 242, 245; in the Soviet Union, of plants, 225; teaching, 142, 309; theistic, 326
evolutionary biology, 8, 10, 194, 197, 213, 301, 329–30, 333, 337, 339, 349–51
evolutionary hierarchy, 279, 281
evolutionism, 140
evolutionary psychology, 11, 329, 344–45
evolutionary synthesis, 141, 244
evolutionary theory, 8–9, 112, 125, 169, 193, 213, 247, 249, 274, 306, 320, 322–23, 327–28, 345; and the origin of sex differences, 276–77, 279, 294, 297
Evolution Protest Movement, 321
exhibitions, of peoples, in London, 119
experimentation, on humans, 1, 192, 211
extermination, 196, 198, 209, 212
extinction, 102, 146, 148, 151–52, 155
eye: and atheism, 416; evolutionary origin of, 260, 302, 308–9, 416

facial angle. *See* skulls, study of
Fäger, Benedikt, 391
faith, religious, Victorian crisis, 343
Famintsyn, Andrei, 218
Fancher, Raymond E., 384
Fara, Patricia, 406
Farley, John, 381

INDEX

Fausto-Sterling, Anne, 406, 412
Fedigan, Linda Marie, 413
Feingold, Mordechai, 361
Feldman, Marcus, 345
Fellow, Otis, 367
feminism: and evolutionary biology, 281, 296–98, 300; and primatology, 413
Ferguson, Adam, 69
Ferri, Enrico, 221
Filipchenko, Iurii, 224–27, 234, 240
fingerprinting, pioneered by Galton, 166
Fisch, H., 361
Fischer, Eugen, 204, 208, 210
Fisher, Ronald Aylmer, 173–74, 191, 214, 251, 267–69, 283–86, 301, 330
Fitzroy, Robert, 135–37
Fleischmann, Albert, 310–12
Flood, Noah's, 116, 123, 131, 144, 248, 305, 314–15, 317, 322, 325
Fontenelle, Bernard de, 43, 56
Forbes, Edward, 152, 155, 162
Ford, Clellan S., 410
Ford, E. B., 267–68, 270
Forrest, Barbara, 420
Forster, Georg, 128–29, 131
Forster, Johann Reinhold, 128–29
Fosdick, Harry Emerson, 188
fossils (fossil record), 102–3, 254, 260–61, 263, 266, 320–21
Franck, R., 361
Franklin, Benjamin, 252, 256, 425
Franks, Robert, 308
Fray, J.-B., 382
Fredman, J., 394
freedom, 81, 83
Fresnel, Augustin-Jean, 254
Freyhofer, Horst H., 423
Friedan, Betty, 296
Frisch, Karl von, 194, 391
Fry, Iris, 381
Furbank, P. N., 367
Fyfe, Aileen, 372–73

Gaissinovitch, Abba E., 394, 396, 399
Galileo, 39
Gall, Johann Franz, 115
Galton, Francis, 8, 165–74, 188, 190–91, 218, 407
Gamble, Clarence, 181, 191

Gamble, Elisa Burt, 281
gamete physiology, 280, 296
Gascoigne, John, 34
Gasking, Elizabeth, 364
Gassendi, Pierre, 28, 358
Gassert, Phillip, 391
Gatewood, Willard B., Jr., 417
Gathorne-Hardy, Jonathan, 410
Gaukroger, Stephen, 362
Geddes, Patrick, 280–81, 296
gemmules, 165, 167
gender: as a biological variation, 278, 287, 293; and sex, 9, 406
generation (reproduction), 36–39, 41, 57, 146; autochthonous, 156; autogenous, 151–52, 161; metaphors, 76–77; and political controversy, 38, 54–60; spontaneous, 47, 52–54, 56, 145–47; theories, 43–47
generic divinity, 425
genes, 8, 174, 267, 294–96, 332–33, 346–47
genetic engineering, 245
genetic recombination, 296
genetics, 8–9, 165–66, 173–74, 193, 195, 200, 205–6, 214, 251; agricultural, 240; of animal breeding, 195; ecological, 267; human, 192, 205–6, 209, 214; Mendelian, 139, 200, 204, 209, 274, 314, 334; modern, 168; population, 267, 274; in the Soviet Union, 215–16, 218, 224, 226–31, 233–34, 236, 239–44
genetic screening, 191
genocide, 171, 192, 195, 203–4, 213, 287, 292
Genocide Declaration (United Nations), 213
Geoffroy Saint–Hilaire, Étienne, 106, 123–24, 154, 254–55
geography, 116, 146, 151, 166; Humboldtian physical, 151
Geological Society, 258
geology, 35, 52, 89, 97–98, 100, 102, 104, 109, 116, 122, 148, 158, 160, 314; Flood, 316, 321–22, 324
Gérard, Frédéric, 382
German Research Fund (DFG), 212
German Society for Racial Hygiene, 209
germs (germ plasm), 167–69, 199–200, 218; preexistence of, 37–45, 48
Gesner, Conrad, 18, 20

Gibson, Daniel G., 384
Gieryn, Thomas F., 356
Gilchrist, George W., 420
Gilkey, Charles W., 188
Gillespie, Neal C., 360, 372, 383
Gillies, Mary Ann, 423
Gillispie, Charles C., 140–41, 143, 421
Gilman, Charlotte Perkins, 281
Giry de Saint-Cyr, Odet Joseph de Vaux de, 57–59
Glanvill, Joseph, 32, 361–63
Glass, Bentley, 159, 380
Gliboff, Sander, 397
Glick, Thomas F., 421
Gliddon, George, 121
God: belief in, 37, 51, 58, 329, 331, 333–36, 338, 341–43, 346, 348–49; existence of, 24, 37, 52, 58, 98, 106
Goddard, Henry H., 176, 178–79, 183–84, 385, 388
Goebbels, Joseph, 194
Goethe, Johann Wolfgang von, 68, 154
Goldsmith, Oliver, 96
Golinski, Jan, 356
Goodall, Jane, 411
Goodrick-Clarke, Nicholas, 391
gorilla debates, 125, *126*
Gosney, E. S., 184, 385
Gould, Stephen J., 273, 377, 407, 412, 414
government: forms of, 20 (*see also* monarchy); French, 54, 59
Gowaty, Patricia, 298–99
Graebner, Theodore, 417
Graf, G. Ia., 395
Grafen, Alan, 422
Graham, Billy, 248
Graham, Loren R., 355, 393, 399
Grant, Madison, 184
Grant, Peter and Rosemary, 249
Gray, Asa, 302–4, 309
Great Break of 1928–30, 236, 239–40
Great Chain of Being, 23, 76–77, 117–18, 127, 253, 370
Greek culture, 73
Greene, John C., 156, 164, 423, 428
Griffin, Roger, 393
Grisebach, August, 152
Grob, Gerald N., 177
Gross, Paul R., 420
Grotjahn, Alfred, 201–2

Grumett, David, 423
Guerrini, Anita, 360
Gul'be, D., 397
Gunson, Neil, 379
Günther, Hans F. K., 195, 200, 204
Guyot, Arnold, 310
Gypsies, 202, 208, 211–12

Haas, J. W., Jr., 326–27
Hacker, Peter, 340, 426
Haddon, A. C., 204
Haeckel, Ernst, 162, 171, 198–200, *199*, 263, *264*, 279, 407
Haiselden, Harry, 186
Haldane, J. B. S., 173–74, 190, 251–52, 267, 269, 320–21, 412
Hall, G. Stanley, 279–80
Haller, Albrecht von, 37–38, 45–47, 50–51, 60
Haller, Mark H., 386–87, 389
Hamilton, William, 191, 332, 412–13
Hanson, Norwood Russell, 3
Haraway, Donna, 406, 409, 412–13
Hardy–Weinberg law, 272
Harlow, Harry, 409
harmonic mediation, 68, 80–81, 83
harmony, 37, 48, 51, 80–81, 84
Harms, William, 426
Harriman, E. H., 184
Harrington, James, 362
Harris, Sam, 421
Harrison, Mark, 379
Harrison, Peter, 6, 11–35, 359–62
Harting, Pieter, 158
Harwood, John, 361
Harwood, Jonathan, 393
Hassencahl, Frances, 389
Hastings, H. L., 305, 415
Hatch, F. W., 181, 387
Haycraft, John Berry, 198–99
Hearn, Virginia, 423
Hearst, William Randolph, 186
Helmstadter, Richard J., 425
Helvétius, Claude Adrien, 55, 58
Henderson, L. J., 268
Henslow, John, 261
Herder, Johann Gottfried, 61–62, *65*, 65, 68–70, 72–73, 75–77, 80–81, 83–87, 131
heredity, 172, 209, 213, 260; hard, 167–70, 173; soft, 167, 173–74; in the Soviet

Union, 216, 218–19, 224–25, 227, 229, 232–33, 235–36, 240, 242
hereditary degeneracy, 171–74, 177, 190
hereditary health, 208; indices (*Erbkarteien*), 202–3
hereditary pathology, 212
Herero (Bantu tribe), 204
heresy, 5
hermaphrodites, 281
Herschel, John F. W., 258
Hertwig, Oskar, 391
Hertz, Friedrich, 200
Hesse, M., 401
Hessen, Boris, 2
heterogenesis, 218
Hick, John, 427
Hill, Christopher, 356
Himmler, Heinrich, 195, 208–10
Hinde, Robert, 410
Hirt, August, 210
history, 13–14; philosophy of, 73 (*see also* natural history)
Hitchcock, Edward, *150*, 305
Hitchens, Christopher, 421
Hitler, Adolph, 192, 194–97, 201, 206, 208, 212–13, 295; library, 195
Hobbes, Thomas, 5, 358
Hodge, Charles, 303–4
Hodge, M. J. S., 374
Hodgkin, Thomas, 121, 129
Hoff, K. E. A. von, 148
Hogben, Lancelot, 190
Hollinger, David A., 2
Holmes, Christopher M., 413
Holmes, Oliver Wendell, 181
Holocaust, 192, 195–97, 203–4, 206, 209–10, 214
homology, 263
homosexuals, 192, 203, 206
Hont, Istvan, 369
Hooke, Robert, 16, 26
Hooker, Joseph Dalton, 155, 160, 162
Hossfeld, Uwe, 380, 393
Hottentot Venus, 119
Howell, Kenneth J., 359
Hrdy, Sarah Blaffer, 297–98, 412
Hubbard, Ruth, 412
Hudson, Nicholas, 376
Hughes, T. M., 402
human genome mapping, 191

humanism, in the Renaissance, 11–12
humanitarianism: British, 121, 124–25; Christian, 137
humanity: history of, 62, 72; science of, 83, 87
human nature, 11, 37, 54, 60, 82, 86, 331
human origin, 118, 245, 329, 331
Humboldt, Alexander von, 64, 68
Hume, David, 62, 69, 81, 90, 118, 341
Huneman, Philippe, 424
Hunt, James, 124
Hunter, George William, 389
Hunter, John and William, 64, 70
Hunter, Michael, 363
Hutcheson, Francis, 69
Hutchinson, John, 99
Hutton, James, 70–71
Huxley, Aldous, 214
Huxley, Julian Sorrell, 190, 204, 214, 243, 270–71, 283–87, 291, 403, 411
Huxley, Thomas Henry (T. H.), 125–26, 155–56, 159–60, 162, 198, 200, 262–65, 270, 283, 338, 344
hybridization, 149

Iakovlev, Iakov, 242
ideology, 61, 70, 139; origins of the term, 2–3, 6
Ignatieff, Michael, 369
imagination, 67–68, 72, 84–85
Immerwahr, John, 376
immigration, limitations, 175, 183–84, 201
Immigration Act of 1924, 190
Indian Mutiny of 1857, 131
Inge, William Ralph, 188
Innere Mission (Inner mission), 200
Inouye, Hiroshi, 412
insects, study of, 20–21, 26
instinct, 98, 197, 260, 287; definition, 196–97
Institute for Hereditary Biology and Racial Hygiene, 211
Institute of Experimental Biology, 227
Institute of Medical Genetics, 240
Institute of Natural Sciences, 236
intelligence tests, 178, 189
intelligent design, 9–10, 325–28, 336; legal test, 327
International Council for Science, 420
intuition, 67
in vitro fertilization, 191
isolationism, 224, 227, 231

Jaccheaus, Gilbert, 13–14
Jackson, Stevi, 406
James, Jesse, 308
Jansen, Sara, 393
Johnson, Alexander, 178
Johnson, James, 133
Johnson, Phillip E., 325–26
Johnson, Roswell Hill, 176–77, 179, 181, 184, 387
Johnson, Virginia, 296
Johnston, John, 12–13, 27
Johnstone, E. R., 179, 388
Joly de Fleury, 55–57
Jones, James H., 410
Jones, John E., III, 327
Jones, Matthew, 358
Jones, William, 130–31
Joravsky, David, 393, 397, 399
Jordan, David Starr, 184
Judson, Horace F., 423
Jukes family, 171–75, *172*, 177, 179
Junker, Thomas, 393
jurisprudence, 369

Kaiser Wilhelm Gesellschaft (KWG), 205
Kaiser Wilhelm Institute for Anthropology (KWIA), 200, 204–6, 209–10, 212
Kames, Lord (Henry Home), 119
Kamin, Leon J., 357
Kammerer, Paul, 227–28, 232
Kansas State Home for the Feeble-Minded, 180
Kant, Immanuel, 75, 90, 129
Kapila, Shruti, 379
Kappitipola, 114–15, 122
Kater, Michael H., 392
Kautsky, Karl, 221
Kazin, Michael, 415
Keill, James, 96
Keller, Boris, 238
Kellogg, Vernon L., 306–7, 311
Kelly, Alfred, 421
Kepler, Johannes, 19, 27, 84, 361
Kessler, Karl, 218
Kettlewell, H. B. D., 268
Kevles, Daniel J., 385–90
Keynes, Randal, 424
Kholodnyi, Nikolai, 238
Kidd, Benjamin, 199, 307
Kidd, Colin, 375–77

Kidd, John, 99
Kingsley, Charles, 427
kin selection, 294, 332, 413
Kinsey, Alfred, 289, 301
Kirby, William, 91, 98–99, 106–7
Kline, Wendy, 388, 408
Klinghoffer, David, 420
Knott, Sarah, 407
knowledge, human, 63
Knox, Robert, 107, 124
Koch, Hannsjoachim, 391
Koerner, Lisbet, 407
Koertge, Noretta, 422
Koestler, Arthur, 398
Kohn, David, 424
Kol'tsov, Nikolai, 224–27, 234, 241
Komarov, Vladimir, 227
Koonz, Claudia, 392
Kors, Alan Charles, 367
Korzhinskii, Sergei, 218
Kotek, D., 394
Kotek, J., 394
Kovalevskii, Aleksandr, 217
Kovalevskii, Maksim, 220, 394
Krayer, Otto, 194
Krementsov, Nikolai, 8, 215–46, 396, 400
Kropotkin, Petr, 218
Kuhn, Richard, 194
Kuhn, Thomas, 3–4, 262
Kuzin, Boris, 227

labor, 75, 78, 85; physiological division of, 78, 252, 279, 281
Ladd, Everett C., 367
Ladous, R., 394
Lagrange, 50
Laipson, Peter, 409
Lake, Mark, 427
Laland, Kevin N., 426
Lamarck, Jean-Baptiste, 9, 106, 154–55, 165, 173–74, 200, 227, 247, 252–53, 255, 257, 262
Lamarckism, 198–200, 209, 315, 333–34, 347; in the Soviet Union, 218, 226–31, 234–36, 242
Lamétherie, Jean-Claude de, 148, 382
Lammerts, Walter E., 321–22
Laplace, Pierre-Simon, 108
Larsen, Timothy, 425
Larson, Edward J., 8, 165–91, 385, 388–90

INDEX 445

Laughlin, H. H., 183–84, 388
Lavoisier, Antoine, 254, 256, 326
Leclerc, Georges-Louis. *See* Buffon, Comte de (Georges-Louis Leclerc)
Lecourt, D., 394
Leeds, Anthony, 412
Lehmann, Julius, 208
Leibniz, Gottfried Wilhelm, 43, 84
Lemkin, Raphael, 213
Lemov, Rebecca, 409–10
Lenin, Vladimir Illych, 221–22, 231–32
Lenin All-Union Academy of Agricultural Sciences (VASKhNIL), 240–42
Leninism in Medicine, 228
Lenz, Fritz, 201, 208–10, 21
Lenz, Widukind, 201
Leopold, Nathan, 308
Leuba, James H., 308
Leuckart, Rudolf, 381
Levens, Joshua, 410
Levin, Maks, 229
Levine, Lawrence W., 415–16
Levins, R., 393
Levinus Lemnius, 18
Levit, Solomon, 230, 234, 238, 240
Lewin, Roger, 413
Lewontin, Richard C., 357, 393, 412, 422
liberty, perfect, 79
Lichtenberger, Henri, 392
Liebau, Siegfried, 210, 212
Liebenfels, Lanz von, 195
Liebersohn, Harry, 378
life, origin of, 144, 149, 154, 157–58, 161–63
life sciences, 63–64, 71, 96–97, 112, 124–25
Lightman, Bernard V., 407, 425
linguistics, 345
Linnaean system, 35
Linnaeus, Carl, 117, 142–43
Linnean Society, 112
Lippmann, Edmund O. von, 381
Livingstone, David N., 140, 375, 414, 422, 425
Llana, J., 364
Lloyd, Elizabeth, 413–14
Locke, John, 341
Loeb, Richard, 308
logical positivism, 269
Long, Edward, 120
Lorenz, Konrad, 194, 214, 288, 292, 332, 410–11

Lough, John, 367
Louis XV, 38, 55, 59
louse, *16*, 28
Lovejoy, A. O., 376
Lowe, M., 412
Lucretius, 50
Ludmerer, Kenneth M., 387, 389
Lumsden, Charles, 345
Lunacharskii, Anatolii, 232–33
Lunar Society, 252
Lunn, Arnold, 419
Lustig, Abigail, 427
Lyell, Charles, 109, 152, 155, 160, 162, 258
lynching, 180
Lysenko, Trofim D., 8–9, 139, 215–16, 240–44
Lysenkoism, 139, 216

Mackenzie, Colin, 134
MacLeay, William Sharp, 106–7
MacLeod, Roy, 378
macroevolution, 318–19
Madansky, Chuck, 412
Magendie, François, 256–57
magnifying instruments. *See* microscopes
Magnussen, Karin, 212
Makea, 130
Malebranche, Nicolas, 26, 40–42, 52, 358, 360
Malesherbes, Chrétien-Guillaume de Lamoignon de, 57
Malpighi, Marcello, 41, *42*, 52
Mandeville, Bernard, 83
Mandler, Peter, 378
marriage, 201–3, 282–86, 290
marriage laws, 175–77, 208
Marsh, Frank Lewis, 317–20
Marshall, Alan John, 291
Marshall, Henry, 114, 375
Martin, Brian, 423
Martin, Thomas Theodore (T. T.), 309–10
Marx, Karl, 2–3, 215, 219–23, 231–32, 342
Marx–Engels Institute, 232
Marxism, 9, 157, 215–17, 219–23, 228–29, 231–37, 242, 244–45, 334, 336
Marxism-Leninism, 236
Marxist-Darwinism, 216–17, 235, 237–38, 240–45
Mason, John Hope, 367
Masters, William, 296
mate choice (female choice), 278–80, 282–84, 286–87, 291–92, 294, 297

materialism: biological, 37–39, 45–46, 49–50, 52–54, 56–57; Marxist, 342; militant, 222; revolutionary, 91
mathematics, 70
mating displays, 286, 291; in great crested grebes, 285, 291
mating rituals, 291–93
matter: active, 44–50; concept of, 64, 66; definition of, 62; living versus dead, 53, 57–59; and motion, in reproduction, 38–39, 44
Mattern, D. W., 391
Maupeou, René Nicolas de, 59
Maupertuis, Pierre-Louis Moreau de, 43, 46–48, 50–51, 58–59, 64
Max Planck Society, 192
Maxwell, Anne, 376
Mayr, Ernst, 141, 243, 267–70, 318, 339
Mazzolini, Renato G., 364–66, 368
McConnell, R. J., 188
McCook, Stuart, 378
McGrath, Alister, 10, 329–51, 422, 426, 428
McLean, Ian, 425
McMullin, Ernan, 401
Mead, Margaret, 190
mechanism, 62, 68, 84; Cartesian, 38–39, 42; human, 95; Newtonian, 43–44, 47–48, 258
Mechau, Otto, 212
Mechnikov, Il'ia, 218
Medawar, Peter, 271, 340, 344
medicine, 64
Megatherium, 103
Meijer, M., 376
meme, 346–49
memetics, 347
Mendel, Gregor, 209, 218, 241
Mendelism, 173–74, 251
Mendel's laws, 173, 224, 240
Mengele, Josef, 192, 200–201, 209–12
mental health hospitals, in the United States, 177–78
Mentzel, Rudolf, 205
Menzbir, Mikhail, 234
mercantilism, 74
Merchant, Carolyn, 406–7
Merezhkovskii, Konstantin, 218
Merton, Robert K., 4
mesmerism, 135, 256
Mesoudi, Alex, 426

Mestergazi, M. M., 398
metagenesis, 149
metaphysics, 223, 231, 331, 340
Metcalf, Maynard M., 306
Metcalf, Thomas R., 379
meteorology, 108, 166, 257
Meyen, Franz, 152, 155
Meyer, Konrad, 212
Meyer, Stephen C., 327
Meyerowitz, Joanne, 410
Michurin, Ivan, 239, 241
microevolution, 317, 319
microscopes, 14–17, 19, 26
microscopic research, 41, 45, 54
microscopic world, 26, 38, 45
Midgley, Mary, 422
Mikskevich, M., 238
Mikulak, M. W., 394
Milam, Erika Lorraine, 9, 276–301
Miller, Larry, 412
Miller, Stanley, 163
Milner, Benjamin, 362
Milton, John, 20–21
mineralogy, 52, 97
miracles, 37, 49, 104, 149, 160, 322
Mithen, Steven J., 425
Mitin, Mark, 243
Mitman, Gregg, 415
Mizuta, Hiroshi, 370
mnemone, 345
Moffett, Thomas, 20
monarchy, 54, 59
money, 82, 85
Monod, Jacques, 339
monogenesis (monogenism), 116–19, 121–22, 125, 153
Montagu, Ashley, 411
Montaigne, Michel de, 19
Moore, A. J., 413
Moore, Aubrey L., 304
Moore, James R., 140, 161, 164, 357, 379, 407, 424, 427
Moore, P. J., 413
morality, 37–38, 47, 57, 60, 61; materialist view, 54; naturalistic, 54; as the ultimate goal of the sciences, 24
More, Henry, 25
Moreau, Jacob Nicolas, 57, 59
Morgan, David T., 425
Morgan, Thomas Hunt, 169, 224, 226, 230

morphology, 89, 104–7, 112, 124, 291
Morrell, Jack, 4
Morris, Desmond, 411
Morris, Henry M., 321–22, 324–25
Morton, A., 393
motion, laws of, 39–41
Moyer, Albert E., 355
Muckermann, Hermann, 200
mulattos, 121
Muller, Herman J., 190, 230
Müller, Johannes, 147
Müller, Max, 131
Müller-Hill, Benno, 204–5
museums, of natural history, 2, 184, 266–67, 272
mutation, 168–69, 173, 251, 322, 339
mutationism, 224–25, 227–28, 230–31
Myerson, Abraham, 190

Nachtsheim, Hans, 205
Nagel, Ernst, 424
Narkompros, 220–21, 223–24, 232
Narkomzdrav, 224, 240
Narkomzem, 224
nationalism, Romantic, 69
National Research Council-Committee for Research in Problems of Sex (NRC-CRPS), 288
National Science Foundation, 288
National Socialism, in Germany, 8, 192, 209
natural history, 6, 13–15, 25, 33–34, 64, 70, 115–17, 122; and biblical exegesis, 18–19, 21; British, 101; medical uses of, 18; and moral edification, 15–21, 28; philosophical approach, 89; in physician training, 18; practical utility of, 29–33; religious aims, 26; in the Renaissance, 12, 18; study in India, 134; and theology, 27
naturalism, scientific, 98, 325–26, 328
natural rights, 369
natural science, 62, 330, 334, 339; limits, 340
natural selection, 8, 88–89, 91, 98, 111, 139, 141, 161–62, 171, 174, 195–99, 201–2, 212, 249, 258–65, 267–68, 277–78, 280, 282–86, 290–91, 302, 306, 309, 314–16, 319, 334, 345–46, 349; British origin, 156; in the Soviet Union, 224, 227, 234
natural theology, 7, 15, 19, 25, 350; defining, 89; literature of, 89, 96, 106, 112; role in biology, 88–113

Natural Theology (Paley), 25, 88–89, 92–94, 96–97, 99–101
nature, 9, 42, 48, 61; divine economy of, 117; as a dynamic system, 7; facts of, 141; historicization of, 63, 102; its goals, 71; language of, 68, 70; laws of, 67, 81, 104, 106–8; and morality, 70 (*see also* morality); as religion, 196; religious ambiguity of, 350; separated from God, 37; study of, 14, 31; women's, 277–78
nature, mechanical philosophy of. *See* mechanism
Nazi eugenics practices, 191
Nazi ideology, 8
Nazi propaganda, 207
Nazi science and pseudo-science, 193–94, 196, 204, 213–14
Nazism, 192, 200–202, 205, 208, 210; and Darwinism, 197
nebular hypothesis, 108–9
Needham, John Turberville, 37–38, 44–51, 46, 53–54, 60, 64
Nelkin, Dorothy, 427
Nelson, G. Blair, 377
neo-Darwinism, 267
neo-Lamarckism, 224–25
neuroscience, 340
Newcomb, Simon, 2
Newman, John Henry, 256
Newton, Isaac, 48, 84, 90, 141, 258, 326, 341
Newtonianism. *See* mechanism: Newtonian
Niemoeller, Martin, 200
Nietzsche, Friedrich, 197
Nieuwentijt, Bernard, 26
9/11, 329
Nitsche, Paul, 209
Noah's Ark, 117, 316, 318
Noll, Mark A., 414
Nott, Josiah, 121–22
Novikov, P. A., 396
Nuffield Foundation, 270
Numbers, Ronald L., 9, 140, 302–28, 414–20
Nuremberg Code, 214
Nuremberg laws (1935), 193, 203
Nuremberg Trials, 213–14
nutrition, 146
Nyiszli, Myklos, 393

O'Brien, Michael, 377
Obukh, Petr, 397

Ochsner, A. J., 180
Oken, Lorenz, 155
Olden, Marian S., 385
Oliver, Mary Laack, 386
ontogeny, recapitulating phylogeny, 263, 279, 290
Ord, George, 126
orientalism, 130–31
original sin, 292
Origin of Species (Darwin), 7–8, 88–89, 92–93, 111, 127, 141, 143–44, 146, 152, 155–61, 166–67, 196, 278, 302, 309, 336, 338; in Russian, 217–18, 223, 238, 243, 249–51, *250*, 254, 258–66, 274
orthogenesis, 224–25, 228, 231, 270
Osborn, Henry Fairfield, 184, 311–12
Ospovat, Dov, 104, 375
Outram, Dorinda, 356, 377
Owen, Richard, 106, 112, 126, 149, 155, 161–62
Oyster Club, 71

Packham, Catherine, 369, 371
Paine, Thomas, 92
paleontological chart, *150*
paleontology, 97, 102, 142, 260
Paley, William, 25, 88–105, 107–8, 112, 260, 326, 416
Palmer, Craig, 300
palmistry, 133
Paracelsus, 18
Paris Academy of Sciences, 44
Passelecq, Georges, 392
passions, 81–82
Pasteur, Louis, 145, 326
patriotism, 80
Patterson, Alexander, 308, 311, 416–17
Patterson, George Murray, 133
Paul, Diane, 191, 408
Pavlov, Ivan, 219, 225–27, 229–30
Peacocke, Arthur R., 425
Pearson, Karl, 169, 174, 189
Peckham, M., 403
Peiresc, Nicolas-Claude Fabri de, 15, 17, 28
People's Commissariat of Agriculture. *See* Narkomzem
People's Commissariat of Enlightenment. *See* Narkompros
People's Commissariat of Public Health. *See* Narkomzdrav

Pérez-Ramos, Antonio, 362
Pernick, Martin S., 389
Perrault, Claude, 42–43, 363
Petrarch, Francesco, 11, 13
Peyrére, Isaac La, 116–17
phenotype, 346–47
philology, comparative, 131
philosophes, 7, 36, 38, 48, 52, 54–59, 253
philosophy, 14, 16, 72; deductive, 62; mechanical, 38, 62, 90 (*see also* mechanism); of the mind, 335; moral, 13, 369; natural, 14–15, 21, 25, 32–34, 62, 108, 125; natural, during the Enlightenment, 63–64, 66, 69–70; pantheistic nature, 195. *See also* Stoicism
photography, of India, 131, *132*, 137
Phrenological Society of Edinburgh, 114
phrenology, 114–15, 122–23, 132–33, 257; Spurzheimian, 135
physico-theology, 360
physics, 14, 34, 41, 51, 70, 97–98, 108; Newtonian, 141
physiology: animal and vegetable, 98, 105, 112; biomedical, 146–47; human, 39, *40*, 263
Piccolomini, Alessandro, 13
Pieris, P. E., 375
Pinto-Correia, Carla, 364
Pittendrigh, C. S., 339
Pitts, Jennifer, 369
Plantinga, Alvin, 427
Plato, 24
Plekhanov, Georgii, 219, 228
Ploetz, Alfred, 198–99, 201
Plot, Robert, 363
Pluche, Noël Antoine, 19, 26
poetry, 13–14
Poker Club, 71
Polichak, James W., 426
political economy, 65, 74, 99
political mechanics, 71–72
politics, 61
Pollock, Ethan, 394
polygenism (polygeny), 119–20, 122, 129, 153–54
Pomeau, René, 366
Poniatskii, N. S., 399
Popenoe, Paul, 176–77, 179, 181, 184, 387
Pope Pius XI and Pope Pius XII, 200
Popkin, Richard, 368

Popper, Karl, 262, 269, 340
population: control, 191, 193; evolutionary change, 227, 283
Porter, Roy, 3, 160
Post, George E., 311
Poteat, William Louis, 310
Poulshock, Joseph, 426
Power, Henry, 27, 361
Prater Vivarium (Institute for Experimental Biology), 194
preformation (preformationism, preformationists), 36–39, 41–42, 45–48, 50–51, 60, 226
preformation–mechanism synthesis, 43–44
Prezent, Isaak, 236–38, 240, 242, 400
Price, George McCready, 314–17, 321
Prichard, James Cowles, 120–21, 124, 129–30, 137, 153
primatology, 298, 301
process, 76, 78
progress (Progress), 76, 80–82, 87; definition, 247; ideology of, 247
projectors, 31–32
Protestantism, 200
Prout, William, 373
Providence, 253, 265; definition, 247
Provine, W. B., 405
Przibram, Hans, 194
pseudo-science, 5, 256–57, 274. *See also* Nazi science and pseudo-science
psychological types, 8, 203
psychology, 166, 257, 331; moral, 24
public health, biologization of, 202
Purdon, John E., 387
Pyeritz, Reed, 412

Quesnay, François, 74
Qureshi, Sadiah, 376

race, 7, 123, 127–28, 138, 194, 245; biology of, 116–17; classification (typology), 115, 134, 204; Darwinian idea, 127; and gender, 118–19, 123, 125; in India, 132, 134; origin of, 8, 122, 195, 203–4; science of, 115–16, 119–20, 127–28, 132, 134–35, 138, 171, 214; as a stable biological variation, 278, 289, 406; theory, 171, 200, 202–4; tropical, 134
Race and Settlement Office of the SS, 205, 211
Races et Racisme, 204
racial hygiene, 6, 198–201, 208–9
Racial Political Office of the Nazi Party, 205, 210
racial superiority, 1
racism: Nazi (Nazi race theory), 197, 202–5, 210–11, 242; scientific (biologized), 125, 127, 131, 137–38
Rangarajan, Mahesh, 379
rape, 300
Rascher, Sigmund, 210
Rathke, Martin Heinrich, 147
rationalism, 69, 94, 100; mechanistic, 69
Raven, C. E., 360
Ravich-Cherkasskii, Moisei, 221, 223
Ray, John, 13–14, 25, 27, 32, 34, 361, 363
Ray Society, 155
Reader, Will, 425
reasoning: abstract, 62–63, 84; analogical, 67–68
Réaumur, René Antoine Ferchault de, 20, 43, 47
Re:Design (Menagerie Theatre Company), 303
Reed, John, 398
Rees, Amanda, 406, 413
Reformation, 19
regeneration, 146
Rehbock, Philip F., 374, 378, 384
Reich Citizenship Law, 207
Reich Health Office, 211
Reill, Peter Hanns, 7, 61–87
Reilly, Philip R., 388–89
religion, 37, 47, 51, 56–57, 59–60, 125, 265, 335, 340, 351; natural, 19, 37, 94; naturalist account, 348; revealed, 37, 51, 94; role in cultural development, 338
reproduction: rational, 283; sexual, 145, 193 (*see also* generation)
republicanism, 369
Resis, Albert, 394
Reumann, Miriam, 410
Reustow, Edward, 364
Reuter, Christoph, 421
revolution, 77, 90; American Revolution, 252, 254; Bolshevik Revolution of 1917, 219–20, 223, 230, 233; French Revolution, 92, 123, 254–55; Russian revolution of 1905, 219
Riazanov, Dmitrii, 228

Rice, E. F., 358
Richards, Eveleen, 407
Richardson, Angelique, 408
Riddle, John, 359
Ridley, Mark, 422
Riley, William Bell, 309, 415–16
Rimmer, Harry, 313–14
Ritter, Robert, 204, 208, 211
ritualization, 290–91
Rivet, Paul, 204
Roberts, Jon H., 425
Robinet, Jean Baptiste, 68
Robitscher, Jonas, 388
Rockefeller Foundation, 184, 191
Roe, Shirley A., 7, 36–60, 364–66, 368
Roger, Jacques, 42, 364–65, 376
Rogers, J. A., 394
Roget, Peter Mark, 98, 105–7, 112
Roma, 202–3, 211–12
Roman Catholic Church, opposed to eugenics, 189
Romanticism, 69
Roosevelt, Theodore, 184
Rose, Steven, 357
Rosen, Christine, 389
Rosenthal, Miriam, 412
Ross, Edward A., 184
Roth, Karl Heinz, 206
Roughgarden, Joan, 406
Rousseau, Jean Jacques, 58, 83
Royal Asiatic Society of Bengal, 130
Royal Geographical Society, 125
Royal Society, 31–33, 44, 92, 98, 120
Royden, Maude, 403
Rüdin, Ernst, 204, 206, 209–10, 213
Rudolphi, Karl Asmund, 153
Rudwick, Martin J. S., 377, 383
Rupke, Nicolaas, 8, 139–64, 378, 380–84
Ruse, Michael, 9, 142, 247–75, 401–4, 407
Russell, Colin A., 421
Russett, Cynthia Eagle, 406–8
Russian Eugenics Society, 239
Russian Social Democratic Labor Party. *See* Bolsheviks
Rutherford, Ernest, 14

Saint-Evremond, Charles de, 358
Saint Paul, 24
salpingectomy, 180
Samson, Jane, 378

samudrika (*samudrikvidya*), 133
Sanger, Margaret, 184, 385
Sanskrit, 130–31
Sapp, Jan, 394
Sarab'ianov, Vladimir, 228
Sarton, George, 1, 3
Saussure, Horace-Bénédict de, 64
Savage-Rumbaugh, Sue, 300
Savioz, Raymond, 365
Sayers, J., 412
Schaffer, Simon, 5
Schallmayer, Wilhelm, 201
Schiebinger, Londa, 376, 406–7
Schmarda, Ludwig, 152
Schneider, Carl, 210
Schopenhauer, Arthur, 154, 195, 197
Schreier, Herb, 412
Schreiner, Olive, 281
Schrödinger, Erwin, 163–64, 326
science, 253, 273; Aristotelian, 12–13; history of, 1–3, 5, 140, 337; human (of human beings), 61–63, 65, 68, 70, 125 (*see also* humanity: science of); ideological abuse of, 1; and ideology, as a tautology, 380; and ideology, linked, 2; moral, 24; and natural theology, 90; as the norm of truth, 1, 64, 85–86, 160; popular, 262, 266, 274; professional, 254, 261–66, 269, 274–75; and race, 115, 122, 125, 127, 133 (*see also* race: science of); and religion, 330, 349; as a value-neutral enterprise, 3, 6
sciences, of life. *See* life sciences
scientific knowledge, as a cultural resource, 216–17, 230, 235, 237, 242–43, 245–46
scientific method, 2, 340
scientific theory, 337–39
Scottish Royal Society, 71
Scripture, 17–19, 31, 104, 119, 121 (*see also* Bible; biblical passages); Buddhist, 114; Hebrew, 99
Sechenov, Il'ia, 217
Secondat, Charles de, Baron de Montesquieu, 76
Second International Congress of the History of Science and Technology, 2
Secord, James, 378
Sedgwick, Adam, 144, 255–56, 260
Sedley, David, 401
Segerstråle, Ullica, 357

INDEX 451

Segraves, Nell, 323
segregation (sexual, compulsory), 8, 170, 175, 177–80, 183; racial, 193
selection, 262–65 (*see also* natural selection); relativism of, 251
selectionism, Galton's, 384
Select Society, 71
Sellars, John, 358
semiotics, 66, 84
Semon, Richard, 200
Sepkoski, Jack, 251
Serebrovskii, Aleksandr, 227–30, 234, 240–41, 397
Sereno, M. I., 426
serology, 206
Severtsov, Aleksei, 234
sex, 276–78, 295–97; energetics of, 280; and gender, 9; research, 288; as a stable biological variation, 276, 278, 281
sexual difference, scientific theories, 281–82
sexuality: in animals, 286; in humans, 286, 300–301
sexually transmitted disease, 197, 206, 208. *See also* venereal disease
sexual orientation, 289
sexual response, female, 296
sexual revolution, 296
sexual selection, 201–2, 277–80, 282–87, 294–97, 301; runaway, 284
sexual stereotypes, 293, 296–97, 300
Shanahan, Timothy, 338
Shanker, Stuart, 413
Shapin, Steven, 5, 362
Sharp, H. C., 387
Shennan, Stephen, 425
Sheppard, P. M., 268
Shmidt, O. Iu., 399
Shteir, Ann, 407
Sidney, Philip, 13–14
Sierra Leone, 124
Simonyi, Charles, 333
Simpson, George Gaylord, 243, 267–69, 272
Sinti, 202, 212
Sivasundaram, Sujit, 7, 114–38, 378–79
skepticism, in the late eighteenth century, 62, 64
Skinner, Andrew S., 83, 86, 369
skulls, study of, 117–18, 124, 133
skyhooks, 341
slavery, 120–22, 127

Sleigh, Charlotte, 411
Slepkov, Vasilii, 230
Smirnov, Evgenii, 227, 229, 397
Smith, Adam, 64, 65, 68–87; library, 71, 370
Smith, Bernard, 378
Smith, Crosbie, 374
Smith, Henry B., 416
Smith, Stevenson, 385–86
Smuts, Barbara, 298
social hygiene, 201
Socialist Academy. *See* Communist Academy (Socialist Academy)
social workers, 178
Society of Marxist Biologists, 229, 236, 238
sociobiology, 277, 295, 297–98, 300, 357
sociology, 166
solar system, formation, 108
South, Robert, 362
South Pacific, as a laboratory, 128
Soviet science system, 235, 237
Spallanzani, Lazzaro, 38, 49
special creation, doctrine of, 8, 143–45, 313, 318–20, 326
speciation, 225, 227
species: evolution of, 168, 173; extinction, 102; fixity of, 316; origin of, 8, 104, 109, 111, 143, 147–48, 158, 227, 278, 306, 312–17; origin of, four theories, 144
Spencer, Herbert, 198–200, 266, 295
Sperber, Dan, 427
Sprat, Thomas, 27, 29–32, 362
Spurzheim, Johann, 115, 134
Sri Lanka, 114–15
SS, and the academy, 210
SS Ahnenerbe (Ancestral Heritage Society), 205, 210
Stalin, Joseph, 216, 236, 244
statistics, 166, 168
Stebbins, G. Ledyard, 267
Steichen, Edward, 409
Stenger, Victor J., 421
Stepan, Nancy, 376–77, 407
Stephens, Lester D., 377, 414
sterilization, 179–80, 190, 192, 206–7, 210, 213; enforced (involuntary, compulsory, eugenic), 1, 8, 175, 181–83, 191, 200–201, 203–4; laws, 181–85, 189, 206; Roman Catholic opposition, 188; voluntary, 190
Steward, Julian H., 344
Stewart, Dugald, 369

Stocking, George, 377–78
Stoicism, 17, 19, 24
Stopes, Marie, 286
stratigraphy, 102
Strebel, Bernhard, 390
Strick, James, 381
struggle for existence, 156, 159, 174, 195–98, 202, 218–20, 244, 259
Strugnell, Anthony, 367
Strum, Shirley Carol, 413
Stubbe, Henry, 362–63
Studentsov, Nikolai, 225–26
Suchecky, Bernard, 392
survival of the fittest, 156, 193, 195, 198, 319, 324
Sutherland, Kathryn, 369
Sverdlov Communist University, 228
Swammerdam, Jan, 41
Swift, Jonathan, 28
Swiney, Frances, 281
symbiogenesis, 218
Symons, Donald, 422

Taft, William Howard, 181
Taguieff, Pierre-André, 392
Talbot, Margaret, 420
Taschenberg, Otto, 381
Taschwer, Klaus, 391
taxonomy, of humans, 117–18, 120
Taylor, Barbara, 407
Taylor, Talbot, 413
Teilhard de Chardin, Pierre, 271, 338
Teilhard de Chardin Society, 268
teleology, 98, 100, 102, 107, 112, 127, 255, 338–39
teleonomy, 339
telogeny, 193
Temple, Frederick, 304
Terrall, Mary, 365
Thackray, Arnold, 4
Theophrastus of Eresos, 18
thermodynamics, first law, 266, 279–80
Thomas, Nicholas, 378
Thompson, J. Arthur, 280–81, 296
Thornhill, Randy, 300–301
Thurman, Joseph, 124
Thurs, Daniel Patrick, 355
Tiedemann, Friedrich, 120
Tierra del Fuego, 135–37, *136*
Tille, Alexander, 198

Timiriazev, Kliment, 217–19, 223, 232, 239, 241
Timiriazev Biological Institute (Timiriazev Scientific Research Institute for the Study and Propoganda of the Scientific Foundations of Dialectical Materialism), 228–31, *229, 233*, 236
Tinbergen, Nikolaas, 288, 290–92, 332, 410
Tobey, Ronald C., 4
Todes, Daniel, 394, 396
Tokin, Boris, 236–38
Topazio, Virgil W., 367
Topham, Jonathan R., 7, 88–113, 372, 374
Topsell, Edward, 18, 20, 27, 357, 359, 363
Townsend, Luther T., 306, 310–11
training schools, for the mentally retarded, 177–78
transmutation, of species, 96, 106, 110–11, 155, 162, 197, 278
Trautmann, Thomas, 379
tree of life, 150–51, 247, *264*
Trembley, Abraham, 38, 44
Treviranus, Gottfried Reinhold, 147
Trivers, Robert, 297
Trotsky, Leon, 221
Turner, Frank Miller, 421
twins: comparative studies, 170, 206, 211; experimentation on, 192, 212 (*see also* Mengele, Josef)
Tylor, Edward B., 426

UNESCO, 214, 271
uniformitarianism, 160
USSR Academy of Sciences, 238, 240, 243–44
utopia, 79, 129; racial, 203, 205

Vagner, V. A., 399
Valeskaln, P. I., 400
van der Zande, Johan, 368
Vardiman, Larry, 420
Vartanian, Aram, 44, 367
vasectomy, 180
Vavilov, Nikolai, 224–25, 234, 240–43
venereal disease, 175–76, 201
Ventner, J. Craig, 163
Vermel', Iulii, 227
Verschuer, Otmar von, 200–201, 206, 209–14
Vickers, Brian, 363
Vico, Giovanni Battista, 69

Vicziany, Maria, 379
Vineland Training School for Feeble-Minded Children, 178–79
Virchow, Rudolf, 144, 147, 204, 310
virology, 195
vitalism, biological, 65. *See also* Enlightenment vitalism
Vögler, Albert, 205
Vogt, Carl, 123, 149, 155, 160
Volland, Sophie, 53
Voltaire, François-Marie Arouet de, 36, 38–39, 45–46, 48–53, 58, 60
Vucinich, Alexander, 394

Waal, Frans de, 300
Waddington, Conrad Hal, 294–95
Wagar, W. Warren, 405
Wagner, Gerhard, 206
Wagner, Jens-Christian, 390
Wagoner, Louisa C., 385–86
Wainright, I., 394
Wallace, Alfred Russel, 141, 161, 198, 280, 283
Wallace, William A., 12–13
Wallin, J. E. Wallace, 386
Warfield, Benjamin B., 304
Watanabe, Masao, 362
Watson, James, 191, 342
Webster, Charles, 360
Webster, John, 31
Wedgwood, Josiah, 189, 257, 261
Weeks, David F., 388
Weikart, Richard, 391
Weindling, Paul, 8, 192–214, 390–93
Weinreich, Max, 389
Weismann, August, 168–69, 173, 199–200, 218
Weitz, Wilhelm, 209
Weldon, W. F. R., 169
welfare, social, biologization of, 202
welfare institutions, 167, 183, 196–97, 200–203
Wellesley, Lord, 134
Wells, H. G., 265
Wells, R. S., 426
Wesley, John, 91
Westmacott, William, 359

Wheeler, Roxanne, 376
Whewell, William, 97–100, 108–11, 258, 261
Whitcomb, John C., 321
White, Ellen G., 314–15, 317
White, Leslie A., 344
Whiten, Andrew, 426
Whytt, Robert, 70, 369
Wilberforce, William, 91, 120
Wilkins, John, 362
Wilkinson, Madge W., 385–86
Williams, George C., 412
Williams, John, 129
Williams, Wayne C., 416
Wilson, Arthur, 367–68
Wilson, David Sloan, 423
Wilson, Edward O., 273–74, 294–97, 345, 357, 421
Wilson, Woodrow, 184
Winch, Donald, 369
Winter, Alison, 379
Wipperman, Wolfgang, 392
Wolfe, B. D., 393
Woltmann, Ludwig, 221
women, in science, 300
women's movement, 281–82. *See also* feminism
Woodrow, James, 304
Woodward, W. H., 357
Workman, Lance, 425
World's Christian Fundamentals Association, 309
Wright, George Frederick, 302
Wright, Sewall, 173–74, 252, 267–69

Yeo, Richard R., 355, 363
Young, G. L., 415
Young, Robert M. (Bob), 2–3, 93, 139–40, 156, 164, 394

Zambon, Francesco, 358
Zavadovskii, Boris, 228, 234–35, 238
Zavadovskii, Mikhail, 234–35
Zirkle, Conway, 393
Zoological Society of London, 320
zoology, 35, 142
Zoonomia, 112